ANNALS OF
THE NEW YORK ACADEMY
OF SCIENCES

Volume 867

EDITORIAL STAFF

Executive Editor
BILL M. BOLAND

Managing Editor
JUSTINE CULLINAN

Associate Editor
STEPHANIE J. BLUDAU

The New York Academy of Sciences
2 East 63rd Street
New York, New York 10021

THE NEW YORK ACADEMY OF SCIENCES
(Founded in 1817)

BOARD OF GOVERNORS, October 1998 – September 1999

ELEANOR BAUM, *Chairman of the Board*
BILL GREEN, *Vice Chairman of the Board*
RODNEY W. NICHOLS, *President and CEO* [ex officio]

Honorary Life Governors
WILLIAM T. GOLDEN   JOSHUA LEDERBERG

JOHN T. MORGAN, *Treasurer*

Governors

D. ALLAN BROMLEY	LAWRENCE B. BUTTENWIESER	PRAVEEN CHAUDHARI
JOHN H. GIBBONS	RONALD L. GRAHAM	HENRY M. GREENBERG
ROBERT G. LAHITA	MARTIN L. LEIBOWITZ	JACQUELINE LEO
WILLIAM J. McDONOUGH	KATHLEEN P. MULLINIX	SANDRA PANEM
CHARLES RAMOND	SARA LEE SCHUPF	JAMES H. SIMONS
	TORSTEN WIESEL	

RICHARD A. RIFKIND, *Past Chairman of the Board*
HELENE L. KAPLAN, *Counsel* [ex officio]   CRAIG PURINTON, *Secretary* [ex officio]

NONLINEAR DYNAMICS AND CHAOS IN ASTROPHYSICS: A FESTSCHRIFT IN HONOR OF GEORGE CONTOPOULOS

ANNALS OF THE NEW YORK ACADEMY OF SCIENCES
Volume 867

NONLINEAR DYNAMICS AND CHAOS IN ASTROPHYSICS: A FESTSCHRIFT IN HONOR OF GEORGE CONTOPOULOS

Edited by J.R. Buchler, S.T. Gottesman, and H.E. Kandrup

The New York Academy of Sciences
New York, New York
1998

Copyright © 1998 by the New York Academy of Sciences. All rights reserved. Under the provisions of the United States Copyright Act of 1976, individual readers of the Annals *are permitted to make fair use of the material in them for teaching or research. Permission is granted to quote from the* Annals *provided that the customary acknowledgment is made of the source. Material in the* Annals *may be republished only by permission of the Academy. Address inquiries to the Executive Editor at the New York Academy of Sciences.*

Copying fees: *For each copy of an article made beyond the free copying permitted under Section 107 or 108 of the 1976 Copyright Act, a fee should be paid through the Copyright Clearance Center, Inc., 222 Rosewood Drive, Danvers, MA 01923. The fee for copying an article is $3.00 for nonacademic use; for use in the classroom it is $0.07 per page.*

⊖*The paper used in this publication meets the minimum requirements of the American National Standard for Information Sciences—Permanence of Paper for Printed Library Materials, ANSI Z39.48-1984.*

Library of Congress Cataloging-in-Publication Data

Nonlinear dynamics and chaos in astrophysics : a festschrift in honor of George Contopoulos / edited by J. R. Buchler, S.T. Gottesman, and H. E. Kandrup.
    p. cm. — (Annals of the New York Academy of Sciences, 0077-8923 ; v. 867)
  Includes bibliographical references and index.
  ISBN 1-57331-162-6 (cloth : alk. paper)
  ISBN 1-57331-163-4 (pbk. : alk. paper)
  1. Chaotic behavior in systems—Congresses. 2. Astrophysics—Congresses. 3. Dynamics—Congresses. 4. Nonlinear theories—Congresses. 5. Kontopoulos, Georgios Ioannou, 1928– I. Kontopoulos, Georgios Ioannou, 1928– II. Buchler, J. R. (J. Robert) III. Gottesman, S. T. (Stephen T.) IV. Kandrup, Henry E. V. Series.
  Q11 .N5 vol. 867 QB466.C45
  523.01—dc21
                                                               98-51749
                                                                      CIP

GYAT / PCP
Printed in the United States of America
ISBN 1-57331-162-6 (cloth)
ISBN 1-57331-163-4 (paper)
ISSN 0077-8923

ANNALS OF THE NEW YORK ACADEMY OF SCIENCES
Volume 867
December 30, 1998

NONLINEAR DYNAMICS AND CHAOS IN ASTROPHYSICS: A FESTSCHRIFT IN HONOR OF GEORGE CONTOPOULOS[a]

Editors
J.R. BUCHLER, S.T. GOTTESMAN, AND H.E. KANDRUP

Conference Organizers
J.R. BUCHLER, S.T. GOTTESMAN, J.H. HUNTER, JR., AND H.E. KANDRUP

Advisory Board
E. ATHANASSOULA, A. DRAGT, C. HUNTER, D. KAZANOS, D. LYNDEN-BELL, AND J. MEISS

CONTENTS

Preface. *By* J.R. BUCHLER, S.T. GOTTESMAN, AND H.E. KANDRUP	vii
Laudation. *By* HEINRICH EICHHORN	1
From Quasars to Extraordinary N-body Problems. *By* D. LYNDEN-BELL	3
Dynamical Spectra and the Onset of Chaos. *By* G. CONTOPOULOS	14
Orbital Complexity, Short-Time Lyapunov Exponents, and Phase Space Transport in Time-Independent Hamiltonian Systems. *By* CHRISTOS SIOPIS, BARBARA L. ECKSTEIN, AND HENRY E. KANDRUP	41
Bifurcations of Periodic Orbits in Axisymmetric Scalefree Potentials. *By* C. HUNTER, BALŠA TERZIĆ, AMY M. BURNS, DONALD PORCHIA, AND CHRIS ZINK	61
Irregular Period-Tripling Bifurcations in Axisymmetric Scalefree Potentials. *By* BALŠA TERZIĆ	85
Negative Energy Modes and Gravitational Instability of Interpenetrating Fluids. *By* A.R.R. CASTI, P.J. MORRISON, AND E.A. SPIEGEL	93
Invariants and Labels in Lie–Poisson Systems. *By* JEAN-LUC THIFFEAULT AND P.J. MORRISON	109

[a]This volume is the result of the thirteenth conference of a series entitled *Florida Workshops in Nonlinear Astronomy* held on February 12–14, 1998 in Gainesville, Florida.

From Jupiter's Great Red Spot to the Structure of Galaxies: Statistical
  Mechanics of Two-Dimensional Vortices and Stellar Systems. *By*
  PIERRE-HENRI CHAVANIS... 120

N-Body Simulations of Galaxies and Groups of Galaxies with the Marseille
  GRAPE Systems. *By* E. ATHANASSOULA............................ 141

On Nonlinear Dynamics of Three-Dimensional Astrophysical Disks. *By*
  A.M. FRIDMAN AND O.V. KHORUZHII................................ 156

Satellites as Probes of the Masses of Spiral Galaxies. *By* LANCE K. ERICKSON,
  S.T. GOTTESMAN, AND JAMES H. HUNTER, JR........................ 173

Chaos in the Centers of Galaxies. *By* ISAAC SHLOSMAN, CLAYTON HELLER,
  AND INGO BERENTZEN... 200

Counterrotating Galaxies and Accretion Disks. *By* R.V.E. LOVELACE........ 217

Global Spiral Patterns in Galaxies: Complexity and Simplicity. *By* C.C. LIN.. 229

Candidates for Abundance Gradients at Intermediate Red-Shift Clusters. *By*
  RENATO A. DUPKE... 253

Scaling Regimes in the Distribution of Galaxies. *By* G. MURANTE,
  A. PROVENZALE, E. A. SPIEGEL, AND R. THIEBERGER................ 258

Recent Progress in the Study of One-Dimensional Gravitating Systems. *By*
  BRUCE N. MILLER, KENNETH YAWN, AND PAIGE YOUNGKINS............ 268

Modeling the Time Variability of Black Hole Candidates. *By* DEMOSTHENES
  KAZANAS AND XIN-MIN HUA....................................... 283

Stellar Oscillons. *By* O.M. UMURHAN, L. TAO, AND E.A. SPIEGEL........... 298

Chaos in Cosmological Hamiltonians. *By* HENRY E. KANDRUP AND
  JOHN DRURY.. 306

Phase Space Transport in Noisy Hamiltonian Systems. *By* HENRY E. KANDRUP. 320-1

Papers by George Contopoulos... 321

Index of Contributors.. 337

Financial assistance was received from:
- COLLEGE OF LIBERAL ARTS AND SCIENCES AT THE UNIVERSITY OF FLORIDA
- DEPARTMENTS OF PHYSICS AND ASTRONOMY AT THE UNIVERSITY OF FLORIDA
- NATIONAL SCIENCE FOUNDATION
- OFFICE OF RESEARCH, TECHNOLOGY, AND GRADUATE EDUCATION AT THE UNIVERSITY OF FLORIDA

The New York Academy of Sciences believes it has a responsibility to provide an open forum for discussion of scientific questions. The positions taken by the participants in the reported conferences are their own and not necessarily those of the Academy. The Academy has no intent to influence legislation by providing such forums.

Preface

This Florida Workshop was the thirteenth in a series that has been devoted to problems in nonlinear astronomy and physics. This year's workshop, entitled "Nonlinear Dynamics and Chaos in Astrophysics," brought together a collection of eighteen senior scientists from England, France, Greece, Holland, Italy, Russia, and the United States, as well as a number of postdoctorate and graduate students, most of whom have contributed to the proceedings. The workshop was held in honor of George Contopoulos, for ten years a part-time graduate research professor of astronomy at the University of Florida, who, together with Heinrich Eichhorn, then Chair of the Department of Astronomy, was instrumental in beginning this workshop series back in 1985.

Although the workshop's title might suggest a narrow focus on astrophysics, the talks ranged over a wide range of topics, extending from plasma physics, accelerator dynamics, and formal nonlinear dynamics to more conventional astronomical problems involving cosmology, accretion phenomena, and the structure and evolution of galaxies. The workshop was organized by J. Robert Buchler, Stephen Gottesman, Henry Kandrup, and James H. Hunter as a collaborative effort involving the Department of Physics and the Department of Astronomy at the University of Florida.

Support for the speakers and workshop infrastructure was provided by the Office of Research, Technology, and Graduate Education, the College of Liberal Arts and Sciences, the Department of Astronomy, and the Department of Physics at the University of Florida. The National Science Foundation provided additional support for graduate students and scientists from outside the United States. Special thanks are due to the Editorial Department of the New York Academy of Sciences for their usual excellent job; and especially to Stephanie Bludau for her skill in seeing this volume through the press.

The proceedings of the Second through Twelfth Florida Workshops have also appeared in the *Annals of the New York Academy of Sciences*:

BUCHLER, J.R. & H. EICHHORN, Eds. 1987. Chaotic Phenomena in Astrophysics. Ann. N.Y. Acad. Sci. **497**.

BUCHLER, J.R., J.R. IPSER & C.A. WILLIAMS, Eds. 1988. Integrability in Dynamical Systems. Ann. N.Y. Acad. Sci. **536**.

BUCHLER, J.R., S.T. GOTTESMAN & J.H. HUNTER, Eds. 1989. Galactic Models. Ann. N.Y. Acad. Sci. **596**.

BUCHLER, J.R. & S.T. GOTTESMAN, Eds. 1990. Nonlinear Astrophysical Fluid Dynamics. Ann. N.Y. Acad. Sci. **617**.

BUCHLER, J.R., S.L. DETWEILER & J.R. IPSER, Eds. 1991. Nonlinear Problems in Relativity and Cosmology. Ann. N.Y. Acad. Sci. **631**.

DERMOTT, S.F., J.H. HUNTER, & R.E. WILSON, Eds. 1992. Astrophysical Disks. Ann. N.Y. Acad. Sci. **675**.

BUCHLER, J.R. & H.E. KANDRUP, Eds. 1993. Stochastic Processes in Astrophysics. Ann. N.Y. Acad. Sci. **706**.

KANDRUP, H.E., S.T. GOTTESMAN & J.R. IPSER, Eds. 1995. Three-Dimensional Systems. Ann. N.Y. Acad. Sci. **751**.

HUNTER, J.H. & R.E. WILSON, Eds. 1995. Waves in Astrophysics. Ann. N.Y. Acad. Sci. **773**.

BUCHLER, J.R. & H.E. KANDRUP, Eds. 1997. Nonlinear Signal and Image Analysis. Ann. N.Y. Acad. Sci. **808**.

BUCHLER, J.R., J.W. DUFTY & H.E. KANDRUP, Eds. 1998. Long-Range Correlations in Astrophysical Systems. Ann. N.Y. Acad. Sci. **848**.

— J. ROBERT BUCHLER
— STEPHEN T. GOTTESMAN
— HENRY E. KANDRUP

NONLINEAR DYNAMICS AND CHAOS IN ASTROPHYSICS: A FESTSCHRIFT IN HONOR OF GEORGE CONTOPOULOS

George Contopoulos

Laudation

When I look back on the long time I have known George Contopoulos, I feel like Franz Grillparzer, the Austrian poet, whose Emperor Rudolf II says in the fourth act of "Ein Bruderzwist in Habsburg": "Mein Geist verirrt sich in die Jugendzeit." (My mind wanders back to the time of my youth). George is indeed, like Odysseus, an ἀνὴρ πολύτροπος ὅς μάλα πολλὰ πλάγχθη, who πολλῶν ἀνθρώπων ἴδεν ἄστεα καὶ νόον ἔγνω.

I first met George Contopoulos around 1960 when he was a guest at the Yale University Astronomy Department. I was living in Middletown, about 20 miles north of New Haven, and regularly attended the colloquia at Yale. George was the speaker on one of these occasions; the subject was stellar dynamics. He had been much encouraged in the pursuit of this area by Bertil Lindblad during his soujourn in Stockholm. We spoke briefly after the end of his talk, and he suggested that I might do some work in the field, but alas, this never happened.

Still, I followed his output with fascination. His discovery of a "third integral" in dynamical systems had founded his fame, and he was appointed (full) Professor at the University of Thessalonike at an unheard of young age. One of his older colleagues there remarked then that this would be the end of his career as an active scientist, if he was true to pattern. But George Contopoulos was not true to pattern, his research output grew and grew, and he gathered a group of collaborators and students around him. The equations that govern self-consistent stellar systems — as I surely need not remind this of all groups — are so intractable, that a purely analytical approach, such as exemplified by Chandrasekhar's "Stellar Dynamics," will not get very far, but George Contopoulos had been one of the few students of Poincaré's "Les méthodes nouvelles de la mécanique celèste" and realized that the then fairly newly available computers provided the means — nonexistent during Poincaré's lifetime — to harvest the planted seed and explore the topological properties of dynamical systems, and he took ample advantage of this opportunity.

This might not be seen as a big deal by some contemporary audiences who grew up with computers, but I remember a colloquium talk at the Georgetown University Astronomy Department in 1957, right after the launch of the first sputnik, given by Gerald Clemence, the distinguished celestial mechanician and Director of the U.S. Naval Observatory. During the discussion, Clemence was asked how many IBM 650ies (remember these?) the American astronomical community would need. The answer: 100 would be ten times as many as could properly be utilized. Indeed, it took the community a while before they realized that computers made it possible to address problems that had up to then been completely intractable, in particular, that computers could do more than integrate nasty differential equations or invert large matrices.

George's merits and the ingenuity with which he provided insight into dynamical problems were rewarded with one of the earliest Brouwer Prizes of the American Astronomical Society's Division on Dynamical Astronomy. Nothing could stop or brake the predictably steady output of his always important and significant contributions to dynamical systems, not even his term as Secretary General of the IAU, as its concurrent president, Leo Goldberg, fully realized and acknowledged.

Anyone who has been privileged to collaborate with George has learned to admire his uncompromising integrity (in matters scientific as well as civilian) and the impeccably high and demanding standards he imposes on his own and the work of his collaborators, of whom there are many. As in the many other places he visited, so he was also during his stay here in Gainesville a veritable sparkplug, generous with his ideas and suggestions that proved seminal to a significant part of the output from this department, and his influence and the inspiration he provided still linger and will continue to do so.

But what is science, what is mathematics without a place in the general frame of human experience? While many brilliant scientists pay no attention to epistemology and are thus comparable to virtuoso violinists who know nothing about the theory of sound, George Contopoulos has been deeply interested in, is familiar with, and has contributed profound insights into the philosophy of science. He leaves an indelible impression of strength of character and profundity of conviction on those who have an opportunity to discuss philosophical and ethical questions with him.

He has had his share of honors. Foremost and most important among these, his name is a household word among astronomers and others concerned with dynamical systems. He recently retired from the chair of astronomy at Athens University and the Directorship of the Greek National Observatory in Athens. Without question, our world famous colleague is the foremost among the living Greek astronomers and it is very fitting that he was elected to membership in the Greek Academy of Sciences, after having already been elected to the European and other National Academies, and after the University of Chicago had conferred upon him a doctorate *honoris causa*, a most rarely bestowed signal honor.

Even though he is approaching the "big seven-oh," I cannot imagine a retired George Contopoulos. I have had the honor to be his friend for many years, and he has not aged one day during this period. We predict with confidence that his work will continue full speed and look forward eagerly to all the inspiration and insights he will still provide us with, *ad multos annos*.

— HEINRICH EICHHORN

From Quasars to Extraordinary N-body Problems

D. LYNDEN-BELL[a]

Institute of Astronomy, Cambridge CB3 0HA, United Kingdom

ABSTRACT: We outline reasoning that led to the current theory of quasars and look at George Contopoulos's place in the long history of the N-body problem. Following Newton we find new exactly soluble N-body problems with multibody forces and give a strange eternally pulsating system that in its other degrees of freedom reaches statistical equilibrium.

INTRODUCTION

The conference was held to honor George Contopoulos for his great contributions to dynamical systems theory and the N-body problem. I shall pay my tribute to him in three parts:

(1) Placing his contributions in the proud history of those who have made major contributions to the N-body problem.
(2) Since nothing but the best is good enough to honor George, I present to him a copy of my best paper [24] and include here a résumé of its arguments that led to the current theory of quasars. There are questions my paper raised 29 years ago which are still unexplored.
(3) With Prof. Ruth Lynden-Bell (my wife) I present our new extraordinary N-body problems which we solve for all initial conditions. These problems can also be solved in quantum mechanics when the hyper-keplerian potential energy is

$$V = -N\tilde{Z}e^2/r, \tag{1}$$

where

$$r^2 = \sum_{i=1}^{N} \frac{m_i}{M}(\mathbf{x}_i - \bar{\mathbf{x}})^2, \tag{2}$$

$$M = -\sum m_i, \tag{3}$$

i.e., r is the mass-weighted-root-mean-square radius of the N-body system. The energy and degeneracy of the nth quantum state are quoted here, but derivations will be published elsewhere.

---

[a]Visiting Professorial Fellow, The Queen's University, Belfast, Ireland.

$$E_n = -\frac{2M\hbar^{-2}(N\tilde{Z}e^2)^2}{[2n+3(N-2)]^2},\qquad(4)$$

the degeneracy of this N-particle state is

$$g(n,N) = \frac{[n+3(N-2)]!}{(n-1)!(3N-4)!}[2n+3(N-2)],\qquad(5)$$

for $N=2$, g reduces to n^2 and the energy reduces that of the hydrogen atom for which $M = m_p + m_e$. To see this, $\tilde{Z}$ is replaced by $\frac{1}{2}(m_p m_e)^{1/2}/M$ when r is replaced by the separation of the electron from the proton. We have written the coefficient of the potential energy in the clumsy form $N\tilde{Z}e^2$ so that the analogy to hydrogenic atoms can be readily seen by physicists.

1. CONTRIBUTIONS TO THE N-BODY PROBLEM (EXCLUDING AGATHA CHRISTIE'S)

The N-body problem probably started with Newton, although Hooke would undoubtedly dispute it as he seems to have conceived the idea independently, but had not the mathematical ability to work out its consequences. As Chandrasekhar [8] has shown, Newton's Principia [35], [7] has much to teach us even today (see reference [30]). Recent studies of the Portsmouth papers have shown that Newton developed most of the perturbation theory that was hitherto attributed to the mathematical astronomers of the 18th and 19th centuries. Newton's method was to store up the momentum generated by perturbations and then deliver it as an impulse that changed the motion from one ellipse to another. This of course gives him the equations for the variations of the orbital elements which are the meat of perturbation theory. My brief résumé of the N-body problem's history is:

Newton 1687	Orbit theory and the general solution of the first extraordinary N-body problem
Laplace 1795	Perturbation theory for near circular orbits
Poincaré 1892–1899	Topological methods [36]
Whittaker 1913 (1959)	Adelphic integrals as series [43]
Contopoulos 1956–	Third integrals and chaos
Kolmagorov-Arnold-Moser	Invariant tori and Arnold diffusion

To these theoretical studies we must add the numerical computation of the N-body problem. Here Aarseth's name stands out as a persistent pioneer exploring this problem [1], although many others have contributed, especially Heggie [18] through his work on triple interactions. Both in globular cluster theory and in dynamical systems Henon's work stands out for its beauty [19]–[21], while Antonov [2] was responsible for a fundamental advance in the understanding of gravitational thermodynamics later popularized and extended to negative specific heats and gravitational phase transitions by the author [33], [28] and by Thirring [42]. Betteweiser and Sugimoto [4] were responsible for giving the gravothermal instability a delightful new twist in their discovery of the inverse gravothermal catastrophe that leads to giant thermal oscillations. But let me return to what George Contopoulos taught me in our many discussions since 1961.

To set the scene, I had written my thesis in 1960 which contained a new derivation of what potentials had local first integrals of the motion besides the energy and the angular momentum about the axis. The main part of the work was the derivation of these different classes of potential, while other parts of the thesis contained the time dependent evolution of accretion disks (a name given them only later) and a first attempt to apply Jeans's [22] gravitational instability to make a theory of the spiral structure of galaxies. The beliefs of those times are well-illustrated by the first edition of Landau and Lifshitz's book on classical mechanics; either a dynamical system was separable and integrable or it was ergodic (by which was meant that almost all orbits visited all volumes of the phase space accessible under the energy constraint). Having classified the special forms of potential that had local integrals, I expected that most other potentials would show ergodic behavior. From the inequality of the z and R dispersions of the stars in the Galaxy it was clear that there must be another integral other than E and h for the Milky Way, so I began trying to fit Eddington [16] (now called Stakle) potentials to galaxies. It was quite shattering when at the 1961 IAU general assembly in Berkeley, George Contopoulos [10] showed that orbits in most smooth potentials behaved as though there were third integrals. Suddenly, the special interest of the special potentials fell away — they were not the only systems with third integrals, merely those for which we knew the exact analytical form of those integrals. They now seemed to be mathematical curiosities rather than systems fundamental to the dynamics of real galaxies.

Three years later George organized a very instructive IAU symposium (No. 25) at Thessaloniki on the theory of orbits in the solar system and stellar systems. Here he brought into contact the celestial mechanics fraternity, with their long history of analytically calculating orbits in the solar system by perturbation theory, with us new boys who were attempting to understand the statistics of the orbits in the more complicated potentials of galaxies; George Contopoulos [11], [12] here taught us that many of the problems were common to both fields and showed how fertile it was to bring different communities who knew different things to the same conference — his wide interests have made him especially good at that throughout his life and this 1998 conference is no exception.

For brevity, I shall skip contacts at Besançon on the N-body problem where George presented Poisson Bracket series for third integrals and we were introduced to Lie series.

In 1973 at Saas Fée, George gave lectures in which he introduced me to the wonders of modern dynamical theory — topological methods, incomplete chaos, and the KAM theorem. It opened my eyes to so much that was new to me that I retreated back to more directly astronomical topics, preferring the contact with astronomy to the unchartered seas revealed by this new alliance between computers and topology.

Two years ago at Salsjobaden in a conference on the dynamics of barred spirals, George again broke open a new field [15]. His invariant dynamical spectra (also described in his contribution here) taught us how to measure and classify chaos, even complete chaos!

I have picked out a tiny fraction of George Contopoulos's work [14] and mentioned things I learned from our direct contacts. He will no doubt deduce that I am not a very attentive pupil, but it would be mean not to mention a lovely paper on the light distributions of elliptical galaxies [9] because it is a beautiful work to which I constantly have to refer my astronomical colleagues!

The essence of this paper can be deduced by the following argument. Consider a spherical galaxy with any radial light profile. Now flatten its density distribution by linear contraction along any axis. This contraction can be resolved into one along and one perpendicular to the line of sight. The one along makes no difference while the one perpendicular flattens the circular distribution of observed light into one stratified on similar ellipses. If a further contraction is made along another axis we can apply the same argument again since ellipses contracted along any direction remain ellipses. So we arrive at George's beautiful theorem that if the density distribution of an elliptical galaxy is stratified on similar concentric ellipsoids, then the light seen will be stratified on similar concentric ellipses whatever the orientation of the galaxy to the line of sight.

2. BACKGROUND TO THE ACCRETION DISC THEORY OF QUASARS

My own best work is "Galactic Nuclei as Collapsed Old Quasars" written in 1969. Then the discovery of quasars by Schmidt using Hazard's accurate position for one of Ryle's radio sources was still recent and quasars themselves were enigmatic objects more especially so because even the brightest 3C273 and 3C48 too did not seem to be associated with clusters of galaxies.

No one then knew that Michell [34] in his wonderfully percipient paper of 1784 had predicted both giant black holes and how they would be discovered! Even the name black hole only came into general use in 1970! In my 1969 paper I refer to "Schwarzschild throats." Laplace's translation [23] of Michell's work into French (without attribution!) was not common reading among astronomers either.

Among the modern works on quasars as accretion discs, priority goes to Salpeter's fine 1964 letter to the *Astrophysical Journal*. Turning against the then common view that quasars were not associated with clusters of galaxies, he worked out the consequences of a large black hole moving through a galaxy and accreting according to the Hoyle and Lyttleton formula. He derived the power emitted per unit accretion rate by considering the binding energy of the last stable circular orbit and deduced a number of consequences of such black holes accreting as they wandered through the interstellar gas of a galactic disc. Five years elapsed before I wrote my paper. Originally unaware of Salpeter's [39] note, I luckily learned of it before the proofs came and so was able to add a sentence and a reference to his work. My aim was to show that the very small nuclei already known in the centers of galaxies were likely to be stars gathered around the giant-black-hole remnants of quasars. At the time, 1969, we already knew that the optical violently variable (OVV) quasars could change by a magnitude from one night to the next. Geoffrey Burbridge [6] had been insistent that the giant radio sources needed 10^{61} ergs in fast electrons and magnetic field, while Ryle [38] had emphasized that quasars would not be distinguished from such sources by radio measurements.

Now 10^{61} ergs weigh $\frac{1}{2} 10^7 M_\odot$. If one entertained the idea that these ergs came from nuclear energy, then the 1% mass conversion efficiency of nuclear burning means that $10^9 M_\odot$ are needed. However, putting $10^9 M_\odot$ within the light-variation-time-length-scale of 10 light hours gives a gravitational binding energy of 10^{62} ergs — on such a hypothesis 10^{62} ergs of gravitational energy would have been lost, all in order to burn $10^9 M_\odot$ of hydrogen into helium and thereby get the mere 10^{61} ergs needed. This shows that in assuming nuclear power we nevertheless conclude that most of the energy comes from gravity. So the nuclear idea is not sensible and we should assume a preponderant gravity power and a somewhat smaller mass $\sim 10^8 M_\odot$. If conversion of mass into radiation is not 100% efficient, quasars must leave behind massive remnants of $> 10^7 - 10^8 M_\odot$ and, because they have radiated their binding energy, they have insufficient energy to reexpand. Since the masses are far beyond the Chandrasekhar limit, there are no other final resting places other than giant black holes. Turning to the numbers of quasars derived by Sandage and estimating possible lifetimes, I deduced

Number of clusters of galaxies < Number of dead quasars < Number of galaxies.

Thus the nearest dead quasar must be nearer than M87 and there may be as many dead quasars as there are massive galaxies. How could we hide dead quasars of $10^8 M_\odot$ when they still gravitate? They would naturally be centers of attraction for

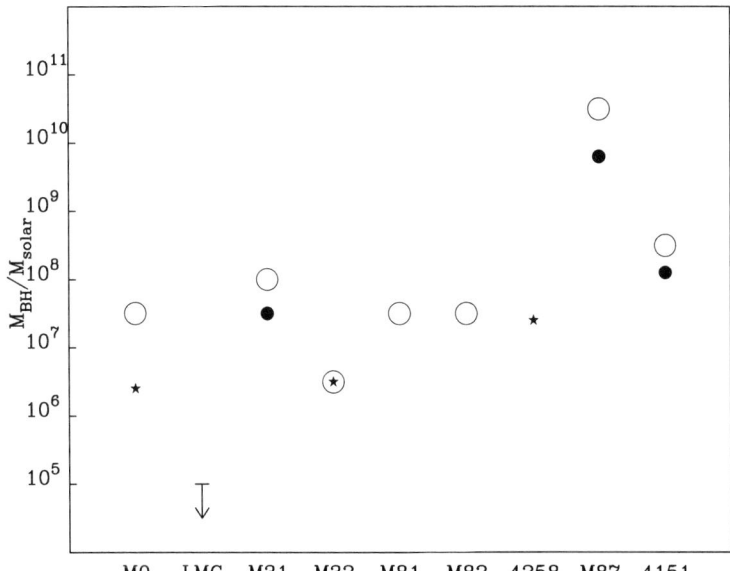

FIGURE 1. Comparison of my 1969 estimates of black hole masses ○ with modern determinations ★ and less precise modern estimates ●. Evidently, the early estimates were somewhat over enthusiastic!

stars so it is natural to find such a body at the center of an exceptionally dense region. Galactic nuclei then became the obvious candidates so I looked at the Galaxy, the Magellanic clouds, M31, M32, M81, M82, NGC4151, M87, etc., and estimated possible black hole masses from the 1969 data on their nuclei, many of which were due to pioneering work by Merle Walker (see FIGURE 1). I also drew on the accretion discs of my thesis and, finding the gaseous viscosity too low, I estimated a magnetic viscosity. This was based on the shearing of the disc causing magnetic reconnection and continual flaring above the disc. Indeed, I found that the protons got most of the energy as they more readily achieved "runaway." Particle energies up to 10^{13} eV were readily generated and hard emission would result when this hit the disc material. While the energy was primarily dissipated into such fast cosmic rays they would collide with the disc and heat it to temperatures $T \propto r^{-3/4}$ for $r \gg 2GM_0/c^2$. Adding together such black body rings of emission, I got the disc spectrum $S_\nu \sim \nu^{1/3} \exp - (h\nu/kT_{max})$ where T_{max}, the maximum temperature in Kelvin, was $6.6 \times 10^4 \, F_{-3}^{1/4} \, M_7^{-1/3}$; here F_{-3} is the mass flux in units of $10^{-3} M_\odot/\text{yr}$ with M_7 the black hole's mass in units of $10^7 M_\odot$. (I did not estimate how much hard emission would come from the initial collisions of the cosmic rays with the disk, but a 10^{13} eV cosmic ray is certainly capable of emitting hard γ rays at its first few collisions.) Even today, 29 years later, I think this model deserves more attention as a serious rival to the currently popular advection models.

The following year Jim Bardeen [3] wrote a particularly fine paper which showed how accretion would spin up a Schwarzschild hole and, after a finite mass was accreted, leave it growing as a near-limiting Kerr hole of significantly greater efficiency. I gave a paper on these models at the 1970 Vatican Symposium on the nuclei of galaxies and in 1971 reestimated the luminosity functions of quasars and mini quasars by developing the C^- method. That year Ekers was stimulated to look with higher radio resolution at the galactic center [17] and I reviewed the then known data on it with Rees [32]. A year or two later, attention turned to lower mass black holes with the discovery of many X-ray binaries by the UHURU satellite. Papers by Pringle and Rees [37] and Shakura and Sunyaev [40], [41] applied such ideas on a smaller scale and with Pringle I applied them (1974) to star formation both with and without magnetic fields. In 1978 I introduced the thick Kerr and Schwarzschild vortices in the hope of getting a more natural collimation mechanism than that of Blandford and Rees [5], but the very narrow jets are still inadequately understood.

3. GENERAL EXACT SOLUTION TO AN EXTRAORDINARY N-BODY PROBLEM

I now return to the N-body problem and the little known fact that in *Principia* Newton [35] solved an N-body problem in which every body attracts every other one and he solved it for all initial conditions!

His was the first of the class of extraordinary N-body problems which Ruth Lynden-Bell and I have been studying. Newton took the force between two bodies,

i and j, to be $F = km_im_j(\mathbf{x}_j - \mathbf{x}_i)$. To get the total force in particle i he summed over j and since the $j = i$ term is zero, we may sum over all j to obtain

$$\mathbf{F}_i = \sum_j \mathbf{F}_{ij} = km_iM(\bar{\mathbf{x}} - \mathbf{x}_i), \tag{6}$$

where $\bar{\mathbf{x}}$ is the position vector of the center of mass, which of course moves uniformly in a straight line, and M is the total mass of the system. Thus with this linear mass-weighted law, that Newton would never have ascribed to Hooke, the total force on the ith body is directed to the center of mass and proportional to the distance from it. Therefore, Newton found that each body describes a centered ellipse about the center of mass, which itself moves uniformly. This completes Newton's solution. In his case, the potential energy is

$$V = -\tfrac{1}{2} K \sum_{i<j} \sum m_i m_j (\mathbf{x}_i - \mathbf{x}_j)^2 = -\tfrac{1}{2} kM \sum_i m_i(\mathbf{x}_i - \bar{\mathbf{x}})^2 = -\tfrac{1}{2} kM^2 r^2. \tag{7}$$

Generalizing some work on statistical mechanics by Ruth Lynden-Bell we were led to consider the dynamics of N-body systems with the more general potential energy $V = V(r)$, where r is given above (cf. Eq. (2)). We define a mass weighted radius $\mathbf{r}$ in $3N$ dimensions by

$$\mathbf{r} = \left(\sqrt{\tfrac{m_1}{M}}(\mathbf{x}_1 - \bar{\mathbf{x}}), \sqrt{\tfrac{m_2}{M}}(\mathbf{x}_2 - \bar{\mathbf{x}}), \sqrt{\tfrac{m_N}{M}}(\mathbf{x}_N - \bar{\mathbf{x}})\right), \tag{8}$$

so the first 3 of the N coordinates tell us where particle 1 is, the next 3 where particle 2 is, etc. Notice that $|\mathbf{r}|$ is the r we defined previously. Equations of motion of the particles in center-of-mass coordinates then lead directly to the equation

$$M\ddot{\mathbf{r}} = -V'(r)\hat{\mathbf{r}}, \tag{9}$$

where $\hat{\mathbf{r}} = \mathbf{r}/r$ is the unit radial vector in $3N$ space.

One readily sees that

$$r_\alpha \ddot{r}_\beta - r_\beta \ddot{r}_\alpha = 0, \tag{10}$$

so

$$r_\alpha \dot{r}_\beta - r_\beta \dot{r}_\alpha = L_{\alpha\beta} = -L_{\beta\alpha} = \text{const}. \tag{11}$$

Furthermore,

$$L^2 = \tfrac{1}{2}L_{\alpha\beta}L_{\alpha\beta} = \tfrac{1}{2}(r_\alpha \dot{r}_\beta - r_\beta \dot{r}_\alpha)(r_\alpha \dot{r}_\beta - r_\beta \dot{r}_\alpha) = [r^2(\dot{\mathbf{r}})^2 - (\mathbf{r} \cdot \dot{\mathbf{r}})^2], \tag{12}$$

where L^2 is the constant defined by the first equality.

The energy in center-of-mass coordinates is therefore given by

$$\tfrac{1}{2}M\dot{\mathbf{r}}^2 + V(r) = E = \tfrac{1}{2}M(\dot{r}^2 + L^2 r^{-3}) + V(r). \tag{13}$$

This determines $r(t)$ as a periodic function if $E < 0$, so there is no violent relaxation in these systems and they vibrate eternally.

Differentiating (13) we find

$$M(\ddot{r} - L^2 r^{-3}) = -V'(r). \tag{14}$$

This is the same equation of motion as that for the central distance to an object in planar motion with angular momentum L about a center of force with potential $V(r)$. It is natural to imagine such a planar orbit and to invent an angle ϕ such that $\phi = 0$ at some pericenter and

$$r^2 \dot{\phi} = L. \tag{15}$$

We may then imagine an orbit in two-dimensional polar coordinates r, ϕ and following Newton, we shall cling to the geometry by eliminating the time in favor of ϕ. Now

$$\ddot{\mathbf{r}} = \frac{d^2}{dt^2}(r\hat{\mathbf{r}}) = \frac{d}{dt}\left(\dot{r}\hat{\mathbf{r}} + \frac{L}{r}\frac{d\hat{\mathbf{r}}}{d\phi}\right) = \ddot{r}\hat{\mathbf{r}} + \frac{L^2}{r^3}\frac{d^2\hat{\mathbf{r}}}{d\phi^2}, \tag{16}$$

where two terms in $\dot{r}Lr^{-2}d\hat{\mathbf{r}}/d\phi$ cancel at the last step. Inserting this result into our equation of motion (9) and using (14), we deduce the wonderfully simple equation

$$d^2\hat{\mathbf{r}}/d\phi^2 + \hat{\mathbf{r}} = 0, \tag{17}$$

whose solution is

$$\hat{\mathbf{r}} = \mathbf{A}\cos\phi + \mathbf{B}\sin\phi, \tag{18}$$

where $\mathbf{A}$ and $\mathbf{B}$ are constant $3N$-vectors which obey $A^2 = B^2 = 1$ and $\mathbf{A} \cdot \mathbf{B} = 0$ in order that $\hat{\mathbf{r}}$ should be a unit vector for all ϕ. Three further constraints on $\mathbf{A}$ and $\mathbf{B}$ follow from the fixed center of mass. They are detailed in our paper but need not concern us here.

We now have the general solution, the center of mass moves uniformly in a line, and the particles pursue orbits about it of the form

$$\mathbf{r} = r(\phi)(\mathbf{A}\cos\phi + \mathbf{B}\sin\phi), \tag{19}$$

where $r(\phi)$ is the form of the two-dimensional orbit governed by Eqs. (13) and (15). These can be integrated explicitly for the isochrone potential $V \propto k/(b+s)$, $s^2 = r^2 + b^2$ and for the Kepler and harmonic oscillator potentials. For the Kepler case $r(\phi) = \ell/(1 + e\cos\phi)$ so the solution is of the pleasing form

$$\mathbf{r} = \ell(1 + e\cos\phi)^{-1}(\mathbf{A}\cos\phi + \mathbf{B}\sin\phi).$$

If we concentrate on the particle i, we find its orbit lies in the plane perpendicular to $\mathbf{A}_i \times \mathbf{B}_i$ where i denotes the three components corresponding to particle i. Taking x, y coordinates in that plane and eliminating ϕ, we find that the orbit is quadratic. If e

were zero it would be a central ellipse, while if $|\mathbf{A}_i|$ and $|\mathbf{B}_i|$ are equal and orthogonal it gives a Keplerian eccentric ellipse. In the general bound case the ellipse has neither its center nor its focus at the center of mass $r = 0$. These systems obey the equilibrium virial theorem in the form $2\mathcal{T} - rV' = 0$, so for the hyper-Keplerian case $V \propto r^{-1}$ it takes the more familiar form $2\mathcal{T} + V = 0$.

One may work out the microcanonical statistical mechanics and find that $E = -\frac{3}{2}(N-2)kT$ so that the heat capacity $C = -\frac{3}{2}(N-2)k$, which is clearly negative as for other gravitating systems [33] and black holes. If V takes the form

$$V = \begin{cases} \infty & r < b \\ -kM^2/r & b < r < R \\ \infty & r > R \end{cases}$$

corresponding to a gravitating system which cannot get too small or too big, then a canonical ensemble is possible and the negative specific heat region of the microcanonical ensemble is replaced by a giant first-order phase transition as in our earlier model [28].

3B. GENERALIZATION

We may extend these extraordinary N-body problems by taking V to be of the more general form

$$V = V_0(r) + r^{-2} V_2(\hat{\mathbf{r}}),$$

the only restriction on the second term being that it scales under expansion as r^{-2}. Those familiar with separable systems in 3 dimensions will know that for such potentials $\frac{1}{2}mh^2 + V_2$ is constant along an orbit where for that case $V_2 = V_2(\theta, \phi)$ and $\mathbf{h} = \mathbf{r} \times \mathbf{v}$. The generalization to $3N$ dimensions is the first integral $\frac{1}{2}M\mathcal{L}^2 + V_2(\hat{\mathbf{r}}) = \frac{1}{2}M\mathcal{L}^2$, for example (note that due to the V_2 term, $\mathcal{L}^2$ does not have to be positive).

The energy equation now reads

$$E = \tfrac{1}{2} M \dot{\mathbf{r}}^2 + V_0 + r^{-2} V_2 = \tfrac{1}{2} M (\dot{r}^2 + L^2 r^{-2}) + V_0 + r^{-2} V_2 = \tfrac{1}{2} M (\dot{r}^2 + \mathcal{L}^2 r^{-2}) + V_0$$

so the r motion pulsates for ever as before. These systems show no violent relaxation in their breathing mode which pulsates (or evolves $E > 0$) independently of the complication of the $\hat{\mathbf{r}}$ motion. Since $V_2(\hat{\mathbf{r}})$ is still free to choose, that motion can be as complicated as we like to make it. Defining a new time τ by $d/d\tau = r^2 d/dt$ the equations of motion for $\hat{\mathbf{r}}$ as a function of τ are totally independent of the r motion, having a reduced Lagrangian system of their own in τ-time. An interesting case to consider is the statistical mechanics of a "hard cone" gas in which V_2 is large and repulsive only in very small regions where two particles are nearly in the same direction as seen from the mass center. This corresponds to the small hard sphere gas so beloved of textbooks. Carrying out the statistical mechanics, which is totally independent of any r motion that may be going on, we obtain a new system at equilibrium in its $\hat{\mathbf{r}}$ coordinates but pulsating or evolving in r.

We have shown [29] this equilibrium to be best described in terms of the peculiar velocity v_i relative to a "Hubble flow" $H(x_i - \bar{x})$ where $H = \dot{r}/r$, that is,

$$v_i = \dot{x}_i - \dot{\bar{x}} - H(x_i - \bar{x})$$

$$f(v_i, x - \bar{x}) \propto \exp\left[-(\tilde{\beta} r^2 \tfrac{1}{2} m_i v_i^2) - \frac{\tilde{\beta} r_i^2}{2r^2}\right].$$

Thus the distribution is Maxwell-Boltzmann relative to be mean Hubble flow with a temperature proportional to $r^{-2}(t)$ and the profile is gaussian with a dispersion proportional to $r(t)$. It is notable that the "equilibrium" of the $\hat{r}$ coordinates is maintained throughout the pulsation just as the Planck distribution of cosmic black-body radiation in the Universe is maintained *without interaction* during the expansion of the Universe. Thus whether the relaxation to equilibrium of the angular coordinates is longer than or shorter than the pulsation time of r is not relevant because "equilibrium" once attained is maintained throughout the pulsation, it does not have to be recreated as each radius r is attained.

REFERENCES

1. AARSETH, S.J. 1974. Dynamical evolution of simulated star clusters. Astron. & Astrophys. **35**: 237.
2. ANTONOV, V.A. 1962. Most probable phase distribution in spherical star systems and conditions for its existence. Originally Vest Leningrad Univ. **7**: 135. Translation, 1995. IAU Symposium **113**: 525.
3. BARDEEN, J.M. 1970. Kerr metric black holes. Nature **226**: 74.
4. BETTEWEISER, E. & D. SUGIMOTO. 1984. Post collapse evolution and gravothermal oscillation of globular clusters. Mon. Not. R. Astr. Soc. **208**: 493.
5. BLANDFORD, R.D. & M.J. REES. 1974. A twin-exhaust model for double radio sources. Mon. Not. R. Astr. Soc. **169**: 395.
6. BURBRIDGE, G.R. 1958. Possible sources of radio emission in clusters of galaxies. Ap.J. **128**: 1.
7. CAJORI, F. 1939. Newton's Principia. Motte's Translation Revised. University of California Press.
8. CHANDRASEKHAR, S. 1995. Newton's Principia for the Common Reader. Oxford University Press.
9. CONTOPOULOS, G. 1956. On the isophotes of ellipsoidal nebulae. Z.f. Astrophysik **39**: 126.
10. CONTOPOULOS, G. 1960. A third integral of motion in a galaxy. Z.f. Astrophysik **49**: 273.
11. CONTOPOULOS, G. 1965. Periodic and tube orbits. Astron. J. **70**: 526.
12. CONTOPOULOS, G. 1966. Recent developments in stellar dynamics. *In* IAU Symposium No. 25, The Theory of Orbits in the Solar System and Stellar Systems. G. Contopoulos, Ed.
13. CONTOPOULOS, G. 1973. Topological methods in stellar dynamics. *In* Dynamics of Stellar Systems, Saas Fée. Martinet & Mayor, Eds.:52–87.
14. CONTOPOULOS, G. 1975. Integrals of motion. *In* IAU Symposium No. 64, Dynamics of Stellar Systems. Reidel.
15. CONTOPOULOS, G. 1997. Chaos in barred spiral galaxies. *In* Nobel Symposium No. 98, Barred Galaxies and Grain Nuclear Activity. Springer. New York.

16. EDDINGTON, A.S. 1915. The dynamics of a stellar system III. Mon. Not. R. Astr. Soc. **76:** 37.
17. EKERS, R.D. & D. LYNDEN-BELL. 1971. High resolution observations of the galactic center at 5 GHz. Ap. J. Lett. **9:** 189.
18. HEGGIE, D. 1975. Binary evolution in stellar dynamics. Mon. Not. R. Astr. Soc. **173:** 729.
19. HENON, M. 1961. Sur l'evolution dynamique des amas globulaires. Annales d'Astrophysique **24:** 369.
20. HENON, M. 1969. Numerical study of quadratic area preserving mappings. Q. J. App. Math. **27:** 291.
21. HENON, M. 1974. Integrals of the Toda Lattice. Phys. Rev. B. **9:** 1921.
22. JEANS, J.H. 1928. Astronomy & Cosmology 337. Cambridge Univ. Press. Cambridge.
23. LAPLACE, P.S. 1795. Le Systéme du Monde. Paris.
24. LYNDEN-BELL, D. 1969. Galactic nuclei as collapsed old quasars. Nature **223:** 690.
25. LYNDEN-BELL, D. 1970. Formation and evolution of bright black holes. *In* Vatical Symposium on Activity in the Nuclei of Galaxies. D.J. O'Conell, Ed.
26. LYNDEN-BELL, D. 1971. The C^- method and N-galaxies as mini-quasars. Mon. Not. R. Astr. Soc. **155:** 95 & 119.
27. LYNDEN-BELL, D. 1978. Gravity power. Physica Scripta **17:** 185.
28. LYNDEN-BELL, D. & R.M. LYNDEN-BELL. 1977. On the negative specific heat paradox. Mon. Not. R. Astr. Soc. **181:** 405.
29. LYNDEN-BELL, D. & R.M. LYNDEN-BELL. 1998. Exact general solutions to extraordinary N-body problems. Proc. R. Soc. (London) A **474**.
30. LYNDEN-BELL, D. & M. NOURI-ZONOZ. 1998. Classical monopoles: Newton, NUT-space, gravomagnetic lensing and atomic spectra. Revs. Mod. Phys. **70:** 427.
31. LYNDEN-BELL, D. & J.E. PRINGLE. 1974. The evolution of viscous discs and the origin of the nebular variables. Mon. Not. R. Astr. Soc. **168:** 603.
32. LYNDEN-BELL, D. & M.J. REES. 1971. On quasars dust and the galactic centre. Mon. Not. R. Astr. Soc. **152:** 461.
33. LYNDEN-BELL, D. & R. WOOD. 1968. The gravo-thermal catastrophe in isothermal spheres and the onset of red-giant structures in stellar systems. Mon. Not. R. Astr. Soc. **138:** 495.
34. MICHELL, J. 1784. On the means of discovering the distance magnitude of the fixed stars etc. Phil. Trans. R. Soc. (London) **74:** 35–57.
35. NEWTON, I. 1687. Principia. Royal Society, London.
36. POINCARÉ, H. 1892–99. Les Methodes Nouvelles de la Méchanique Celeste, Paris. Translated History of Modern Physics & Astronomy, 1993. Vol. 13. D.L. Goroff, Ed. American Institute of Physics.
37. PRINGLE, J.E. & M.J. REES. 1972. Accretion disc models for compact X-ray sources. Astron. & Astrophys. **21:** 1.
38. RYLE, M. 1968. Radio astronomy and quasars I. *In* IAU Highlights of Astronomy. L. Perek, Ed.
39. SALPETER, E.E. 1964. Notes: Accretion of interstellar matter by massive objects. Ap. J. **140:** 796.
40. SHAKURA, N.I. & R.A. SUNYAEV. 1973. Black holes in binary systems. Astron. & Astrophys. **24:** 337.
41. SHAKURA, N.I. & R.A. SUNYAEV. 1976. A theory of instability of disc accretion onto black holes, galactic nuclei, quasars. Mon. Not. R. Astr. Soc. **175:** 613.
42. THIRRING, W. 1972. Negative specific heat. Essays in Physics **4:** 125.
43. WHITTAKER, E.T. 1959. Analytical Dynamics, 4th Edit. Chap. 16. Cambridge University Press. Cambridge.

Dynamical Spectra and the Onset of Chaos

G. CONTOPOULOS

Research Center for Astronomy, Academy of Athens, Anagnostopoulou 14, 10673 Athens, Greece

INVARIANT SPECTRA

In order to decide whether a given orbit is chaotic or ordered, one usually calculates the (maximal) Lyapunov characteristic number (LCN)

$$LCN = \lim_{t \to \infty} \frac{\ln\left|\frac{\xi}{\xi_0}\right|}{t}, \quad (1)$$

where ξ_0 is the initial infinitesimal deviation from an orbit and ξ is the deviation at time t. The orbit is chaotic if this limit is positive and ordered if it is zero.

If the deviation $|\xi|$ is exponential

$$|\xi| = |\xi_0| = e^{at} \quad (2)$$

with constant $a > 0$, then

$$LCN = a. \quad (3)$$

In general, however, a varies along an orbit, and in order to find a convergent value for the Lyapunov characteristic number one has to calculate the equations of motion, and the variational equations that give ξ, for a very long time. In galactic dynamics this time is usually thousands of times longer than the age of the Universe.

Thus in recent years people have used "short-time Lyapunov characteristic numbers" for a fixed short time t. The first who used such "short-time LCN" were Nicolis et al. [1] and Fujisaka [2]. Several astronomers [3]–[7] have since used such "short-time LCNs."

We studied in detail the case when t is equal to one iteration of a map given analytically (or a map defined on a Poincaré surface of section [5]), and we called the quantity

$$a_i = \ln\left|\frac{\xi_{i+1}}{\xi_i}\right| \quad (4)$$

a "stretching number." The "spectrum of the stretching numbers" gives then the number dN of values of a_i within an interval $(a, a + da)$, divided by da and by the total number of iterations N

$$S(a) = \frac{dN(a, a + da)}{N da}. \quad (5)$$

In that paper [5] we have shown that $S(a)$ is invariant with respect to the initial conditions along an orbit. In the case of chaotic orbits, the spectrum is invariant also with respect to initial conditions in the same connected chaotic domain. But in the case of ordered orbits the spectrum is invariant only with respect to initial conditions along the same invariant curve.

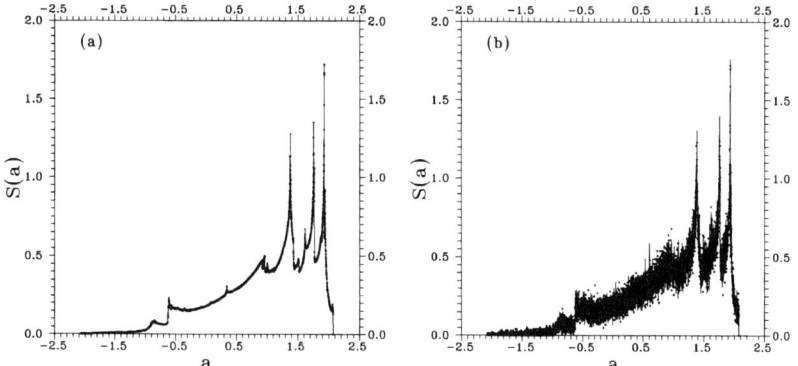

FIGURE 1. The spectra of two chaotic orbits in the standard map with $K = 5$ (solid line and dots). The orbits are calculated for 10^8 periods in (**a**), and for 10^5 periods in (**b**).

An example is given in FIGURE 1, where we see the spectra of two chaotic orbits in the standard map

$$x_{i+1} = x_i + y_{i+1}$$
$$y_{i+1} = y_i + \frac{K}{2\pi}\sin 2\pi x_i \quad \text{(mod1)} \tag{6}$$

for $K = 5$.

The two spectra are superimposed after N iterations. If N is large ($N = 10^8$, FIG. 1a), the deviations of the two spectra are insignificant, while for smaller N ($N = 10^5$, FIG.1b) there are some discrepancies.

Another way to construct spectra is by taking the distribution of the angles ϕ_i formed by the infinitesimal vectors ξ_i with a given axis (say, the x-axis). We call the angles ϕ_i "helicity angles." Such spectra are also invariant as above.

It is easy to show that in 2D maps the direction of the vector becomes practically tangent to an invariant curve after a few iterations. In the case of an ordered orbit (FIG. 2a) ξ becomes tangent to a closed invariant curve (i.e., the angle of ξ with the tangent of the invariant curve becomes very small after a few iterations). In the case of a chaotic orbit ξ becomes tangent to an unstable asymptotic curve (FIG. 2b), which is an open invariant curve. Thus in 2D systems the spectra are also invariant with respect to the initial direction of ξ. We will see below that this is also true in more dimensions for chaotic orbits, but not for ordered orbits.

In FIGURE 3a we see the distribution of 10^4 iterates of the standard map for $K = 10$, and we compare it with the corresponding distribution of 10^4 iterates in the Hénon map

$$x_{i+1} = 1 - K'x_i^2 - y_i$$
$$y_{i+1} = x_i \quad \text{(mod1)} \tag{7}$$

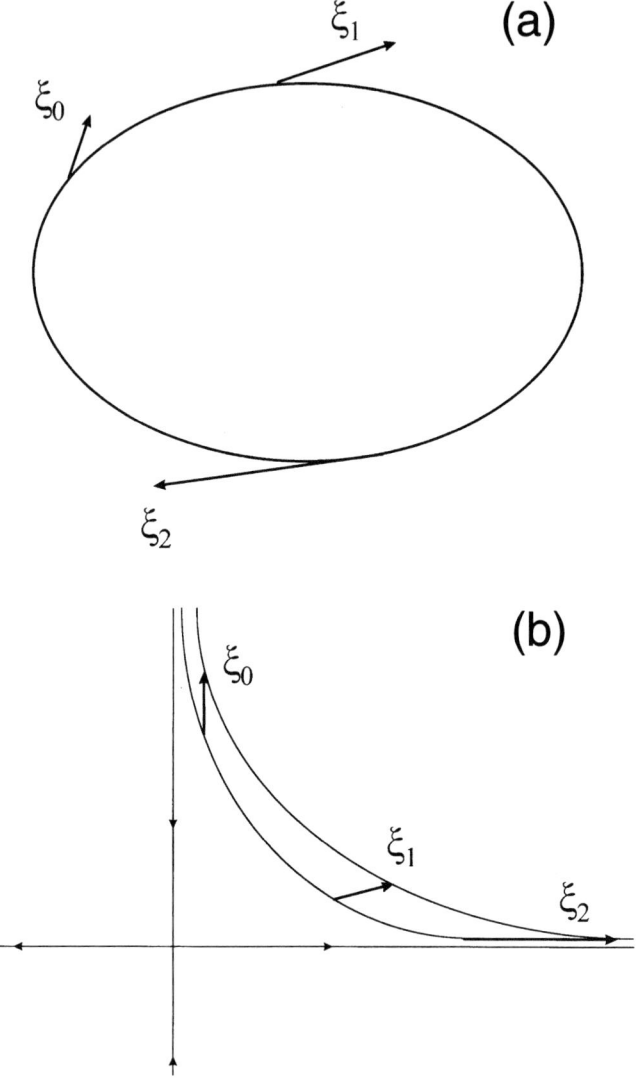

FIGURE 2. The deviation ξ becomes tangent to an invariant curve after some iterations, both in the ordered case (**a, top**) and in the chaotic case (**b, bottom**). In the latter case the invariant curve is an unstable asymptotic curve from an unstable periodic orbit.

for $K' = 7.407$, and the same initial conditions (FIG. 3d). The two distributions look identical. They appear to be uniform (random) over the square $(0, 1) \times (0, 1)$. Furthermore they have the same Lyapunov characteristic number $LCN = 1.62$ (this was secured by choosing appropriately K', for a given K).

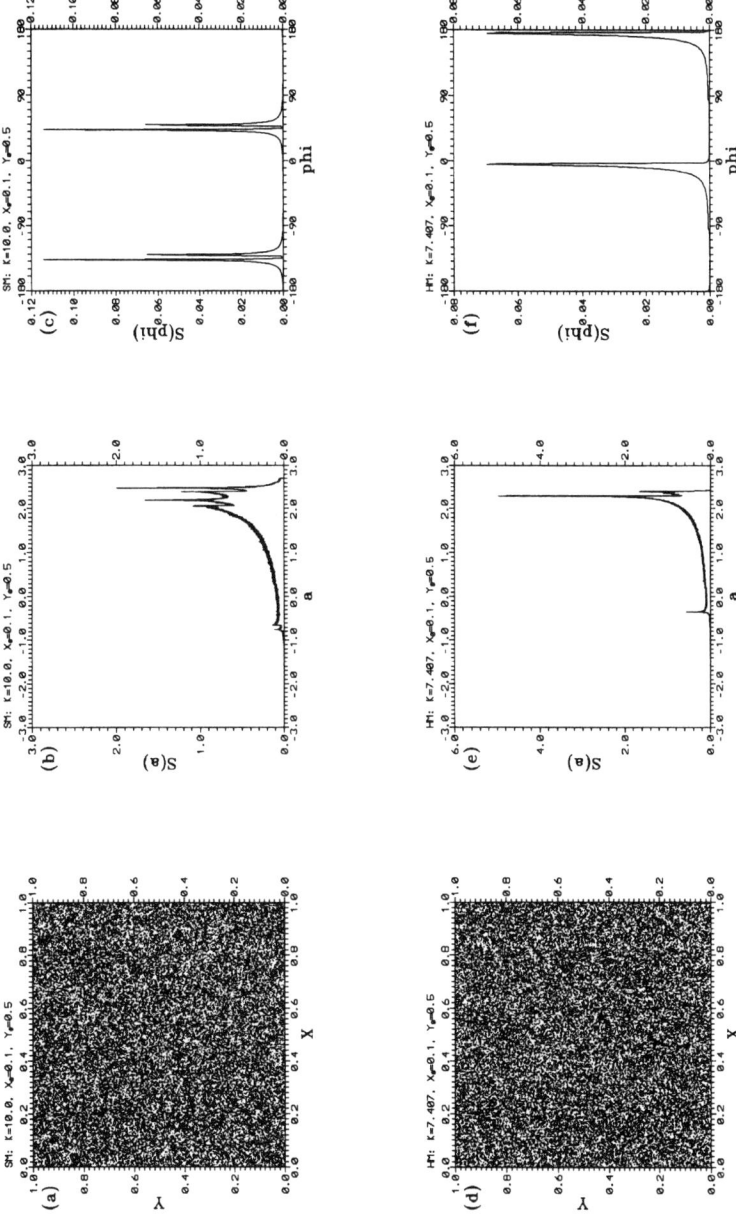

FIGURE 3. The distribution of 10^4 consequents in the standard map (**a**) and in the Hénon map (**d**) and the corresponding spectra of stretching numbers (**b**), (**e**) and helicity angles (**c**), (**f**). In the standard map $K = 10$ and in the Hénon map $K'' = 7.407$, and the initial conditions are the same.

However, the spectra of stretching numbers (FIGS. 3c, 3e) and of helicity angles (FIGS. 3c, 3f) are quite different.

The Lyapunov characteristic numbers are the mean values of a in FIGURES 3c and 3e. They are equal, but the forms of the spectra around the mean values are different.

Although the distributions of the points in FIGURES 3a and 3d are very similar, and the initial point is the same, the successive points are different. In fact the whole structure of chaos in FIGURES 3a and 3d is quite different. The main difference refers to the periodic orbits of the two maps, and their asymptotic manifolds.

Most of the periodic orbits in both maps are unstable, but their positions are different. The simplest unstable periodic orbit of the standard map is ($x = y = 0$). This is of period 1. Its unstable asymptotic manifold is shown in FIGURE 4. Because of the modulo 1 this curve continues at $x = 0$ when it reaches $x = 1$, and at $y = 0$ when it reaches $y = 1$. According to a well-known theorem this curve never intersects itself. It fills the whole square $(0, 1) \times (0, 1)$, but its main directions are near $\phi = 42°$ and $\phi = -138°$ (i.e., $42°-180°$), and to a lesser degree near $\phi = 55°$ and $\phi = -135°$ (i.e., $55°-180°$). These angles correspond to the maxima of the spectrum of FIGURE 3c. All other values of ϕ are represented in FIGURE 4, and in the corresponding spectrum (FIG. 3c), but only rarely, thus the amplitude of the spectrum for such angles is low.

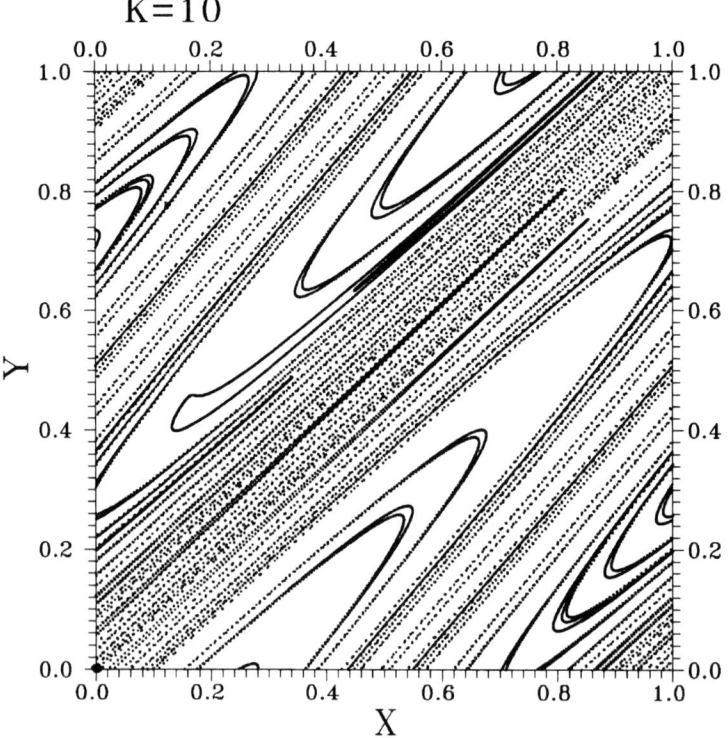

FIGURE 4. The unstable asymptotic curve of the simple periodic orbit ($x_0 = y_0 = 0$) in the standard map for $K = 10$.

The 180° periodicity of the spectrum is due to a property of the asymptotic curves, namely that very close to each segment of such a curve there are segments almost parallel to it in the same and in the opposite direction. It is of interest to notice, in FIGURE 4, some (almost) angular points, at which the curve returns along almost exactly the opposite direction.

What is even more important is the fact that the unstable asymptotic curves of other periodic orbits in FIGURE 4 cannot intersect the unstable asymptotic curve of the periodic orbit (0, 0) shown in this figure. Therefore, these higher-order asymptotic curves lie in the gaps between the various segments of the given curve, and thus they are parallel to them.

If we start an orbit at the original asymptotic curve near (0, 0) with a direction ϕ along the same curve we find the successive points of the orbit along the asymptotic curve and the directions ϕ_i are tangent to this curve. If we start with initial conditions at another unstable asymptotic curve we find practically the same directions, hence the same invariant spectrum of helicity angles. Even orbits starting with very different initial positions (x_0, y_0) and directions ϕ_0, soon become almost tangent to the asymptotic directions of FIGURE 4 and thus produce the same spectrum.

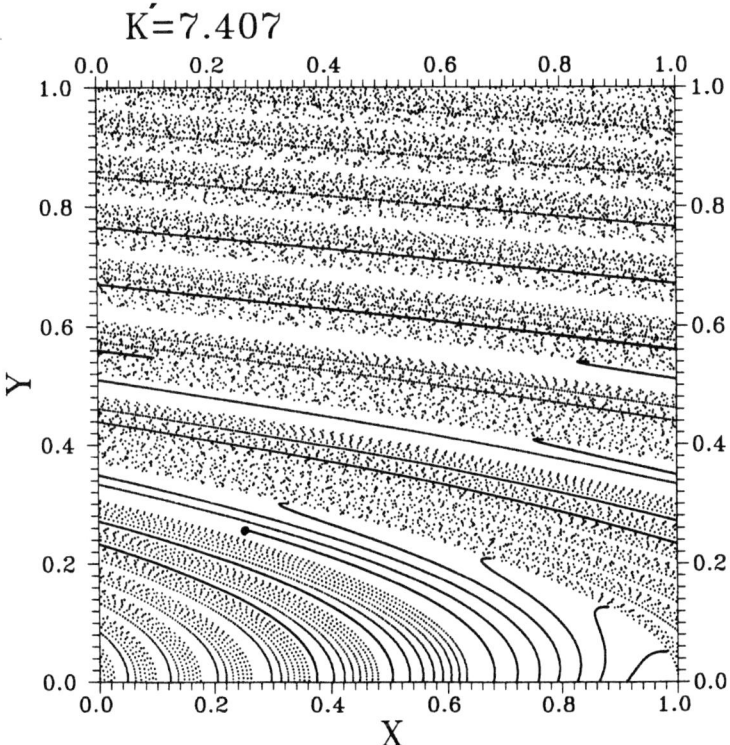

FIGURE 5. The unstable asymptotic curve of the simple periodic orbit $x_0 = y_0 = 0.256444$) in the Hénon map for $K_R = 7.407$.

But if we calculate the periodic orbits and the asymptotic curves of the Hénon map (FIG. 5) we find a very different pattern. The period 1 periodic orbit is now at

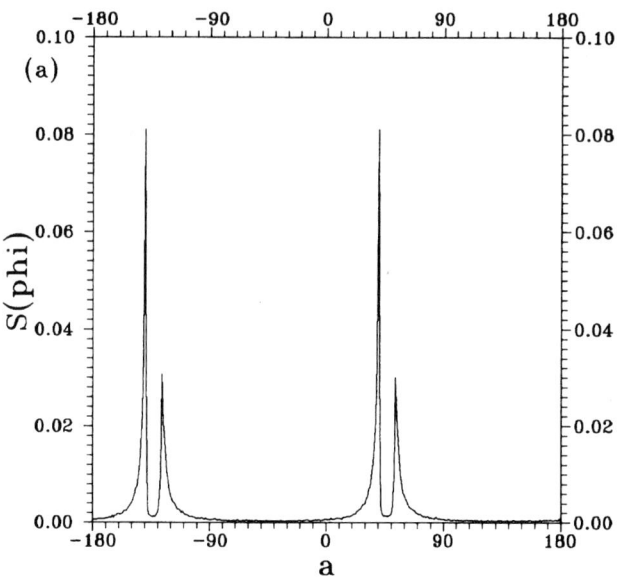

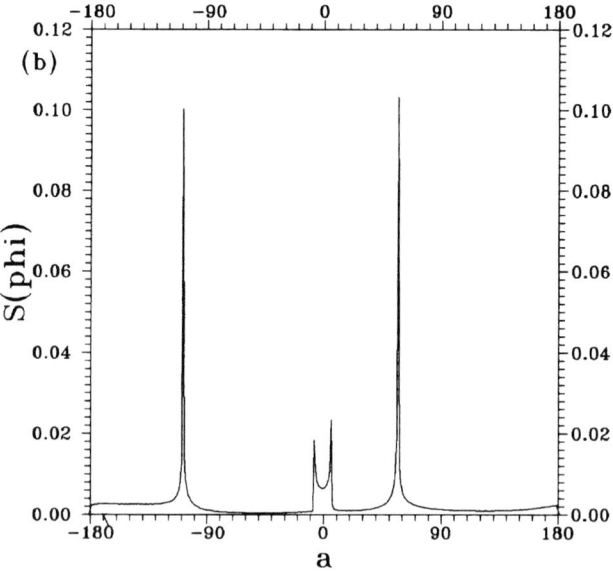

FIGURE 6. The spectra of helicity angles of a chaotic orbit (**a**) and of an ordered orbit (**b**) in the standard map with $K = 5$.

$x_0 = y_0 = 0.256444$, and the asymptotic curves have very different directions. The spectrum of the Hénon map is again invariant, but very different from the spectrum of the standard map.

We conclude that although the standard map and the Hénon map give very similar overall distributions of points and have the same Lyapunov characteristic numbers, nevertheless they have a very different underlying order.

DISTINCTION BETWEEN ORDERED AND CHAOTIC DOMAINS

The spectra of chaotic and ordered orbits are quite different in general. For example, the average value of the stretching numbers $\langle a \rangle$, which is the Lyapunov characteristic number, is positive for chaotic orbits, and constant in the same chaotic domain, while it is zero for ordered orbits. Similarly the spectra of helicity angles of chaotic orbits (in the same chaotic domain) are the same, while the spectra of ordered orbits are different. This is seen in FIGURE 6, where we compare the helicity spectra of a chaotic orbit (FIG. 6a) and of a regular orbit (FIG.6b).

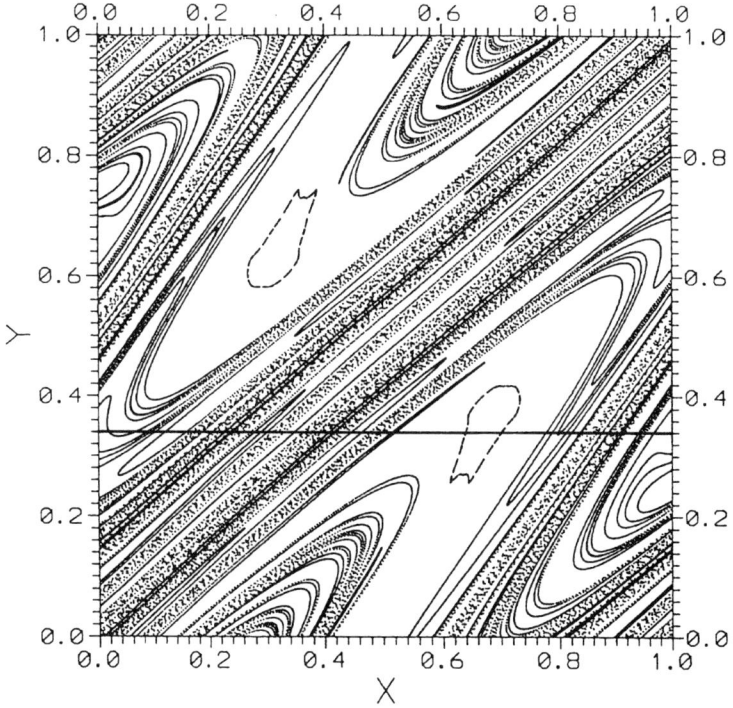

FIGURE 7. The unstable asymptotic curve from the unstable orbit (0, 0) in the standard map for $K = 5$ is very similar with the corresponding curve for $K = 10$ (FIG. 5) but does not enter the regions of the islands. The line $y = 0.34$ passes through ordered and chaotic domains.

Both orbits have been calculated in the standard map with $K = 5$. The first orbit fills most of the square $(0, 1) \times (0, 1)$ but not all of it. It leaves two blank islands where lie regular orbits (FIG. 7). An orbit starting in one blank region forms two closed invariant curves. However, if we take only every second point of the map, we have only one closed invariant curve.

The chaotic spectrum has a periodicity of 180°, as we explained in the previous section (FIG. 6a). But the spectrum of the second orbit (FIG. 6b) does not have such a periodicity. The reason is that, while the invariant curve of the ordered orbit of FIGURE 7 has for each direction ϕ an opposite direction $\phi - 180°$, the values of ϕ and $\phi - 180°$ do not appear with equal frequency, because of the asymmetry of the invariant curve with respect to its center.

Only very close to the center of the island the invariant curve is close to an ellipse and its helicity spectrum is almost periodic with period 180°. Also the spectrum of the invariant curve, in the upper left blank region of FIGURE 7 is exactly the same as the spectrum of FIGURE 6b, transposed by 180°. Thus if we had taken both spectra together, we would have a spectrum periodic by 180°. But by taking only every second invariant point we find a spectrum without this periodicity.

As a conclusion we can in general clearly distinguish an ordered from a chaotic orbit by its spectrum.

But this method is rather lengthy, requiring enough points to construct a spectrum (at least 10^3 or 10^4 points).

A much faster method to distinguish between ordered and chaotic domains is based on the fact that the spectra are invariant with respect to initial conditions in a connected chaotic domain, while they change gradually from one invariant curve to the next in an ordered domain. For example, the average value of the helicity angle is constant in the chaotic domain, while it varies smoothly in the ordered domain.

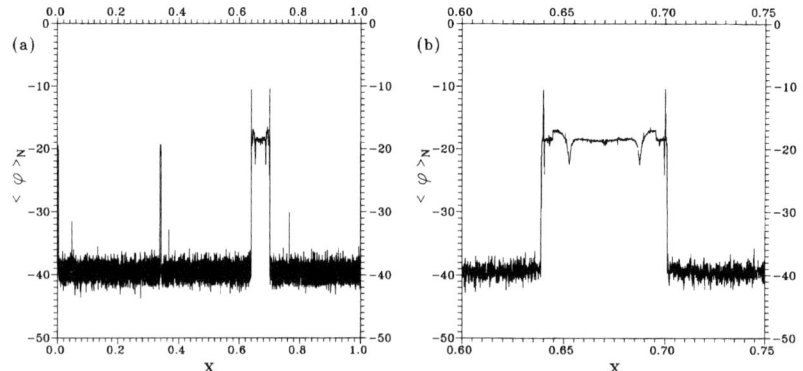

FIGURE 8. (a) The average value of the helicity angle $\langle \phi \rangle_N$, calculated for $N = 10^4$ periods, for 10^4 orbits at equal intervals along the line $y = 0.34$ for $0 < x < 1$. There is some scatter around the average $\langle \phi \rangle$. The average $\langle \phi \rangle$ is constant in the chaotic domain, but it varies smoothly in the ordered domain. (b) A magnification of the region of the main island of FIGURE 8a.

As an example we take 10^4 orbits along the horizontal line $y = 0.34$ in the standard map for $K = 5$ (FIG. 7). This line passes through both chaotic and ordered domains. For each orbit we calculate the value of $\langle\phi\rangle_N$ after $N = 10^4$ iterations. These values deviate from the limiting constant $\langle\phi\rangle$, which is derived after $N \to \infty$ iterations, but the deviations are small (FIG. 8a). On the other hand, the values of $\langle\phi\rangle$ vary smoothly inside an island of stability. More important is the fact that the change of $\langle\phi\rangle_N$ at the borders of the island is abrupt. Thus, even a rather small island can be identified (see the almost vertical line at about $x = 0.34$ in FIGURE 8a).

The structure of the island can be seen in magnification in FIGURE 8b. We see, first, that the deviations from a smooth line are extremely small in general. However we see some larger deviations near the borders of the island corresponding to secondary chaotic zones, or higher-order islands.

The clear separation of the chaotic from the ordered domain allows us to do the same calculation for only 10 iterations after the first 10 transient points (FIG. 9). The dispersion of the values of $\langle\phi\rangle_N$ is now much larger, but again we can clearly distinguish between the chaotic and ordered domains. Even the small island near $x = 0.34$ can be distinguished.

Instead of the average values of $\langle\phi\rangle_N$ one may use the average stretching numbers $\langle a\rangle_N$, which are positive on the average for chaotic orbits and around zero for ordered orbits [8]. The number of $N = 20$–10 iterations is again sufficient.

However, with only 10 or 20 iterations one cannot say with certainty whether one particular orbit is chaotic or not. In fact individual chaotic orbits may have very dis-

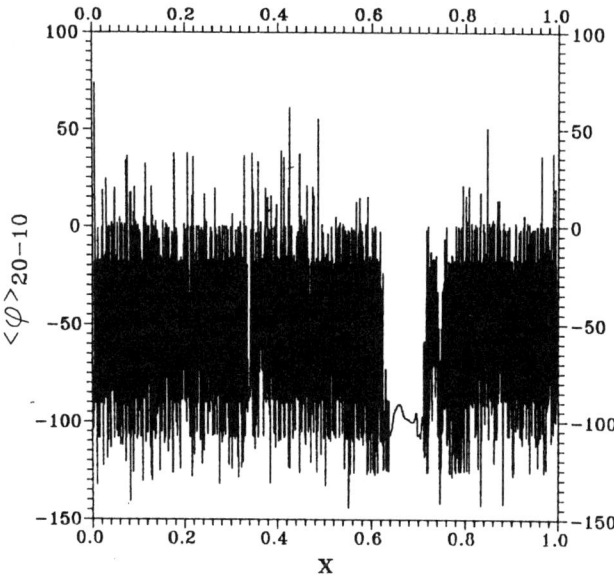

FIGURE 9. The average value of $\langle\phi\rangle_{20-10}$ (i.e., for 10 iterations after the first 10 transients). The scatter in the chaotic domain is very large, but the ordered domains are clearly separated.

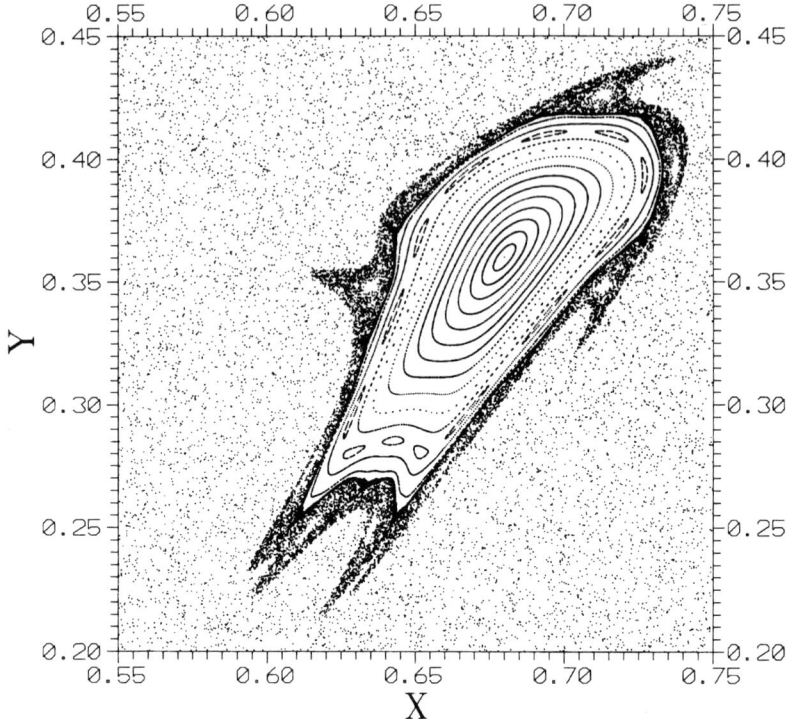

FIGURE 10. The right island of FIGURE 7 (standard map for $K = 5$) in detail. We see several invariant curves and two sticky zones, a large dark zone, and an inner thin very dark zone. The island and the sticky regions are surrounded by the large chaotic sea.

cordant values of $\langle\phi\rangle_N$, or of $\langle a\rangle_N$, for $N = 20$–10, that reach the range of values of $\langle\phi\rangle_N$, or of $\langle a\rangle_N$, of ordered orbits, making their distinction difficult. Thus for individual orbits, much larger values of N would be needed.

But as regards domains of chaotic and ordered orbits our method with $N = 20$–10 iterations is very effective. It is faster by a factor about 50 than Laskar's frequency analysis method [9]. It is even faster than a systematic calculation of Poincaré surfaces of section.

Furthermore, the calculation of 20 points from each orbit allows a rough delineation of the invariant curves, so that it is not necessary to use a dense grid of parallel lines, each with constant y.

The method of fast separation of chaotic and ordered orbits has been used effectively in realistic systems, like galaxies [10]. In such cases the usefulness of our method is even more evident, because the potentials used are rather complicated, and the time required for long-time calculations of orbits becomes prohibitive.

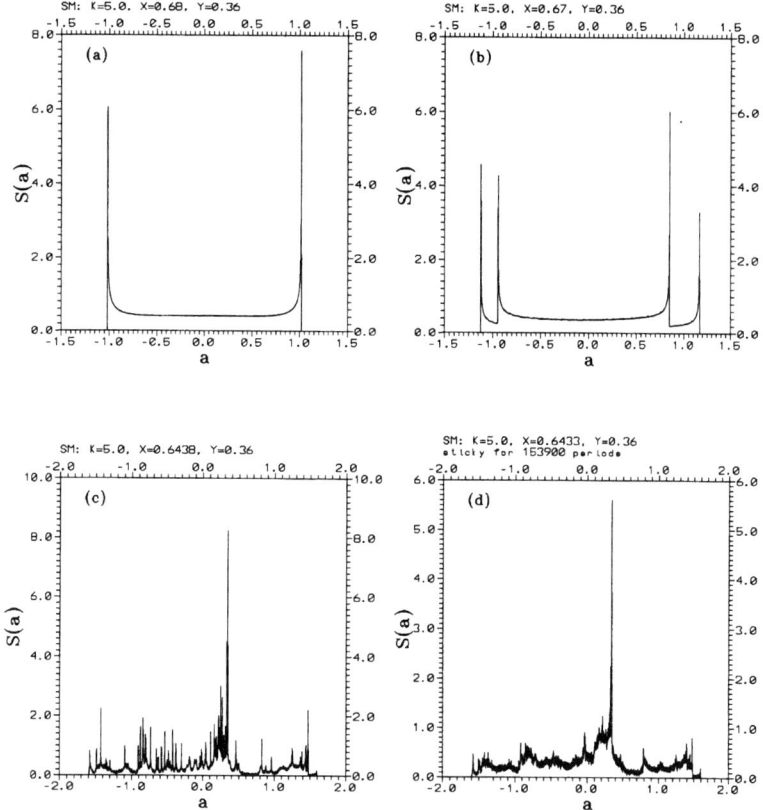

FIGURE 11. Spectra of orbits around an island in the standard map with $K = 5$. The spectra (a), (b), (c), (d) correspond to orbits at increasing distances from the center. (d) is a temporary sticky spectum.

TRANSITION FROM ORDER TO CHAOS: STICKINESS

We find the transition from order to chaos by calculating orbits from the center of an island outward until we reach the large chaotic sea (FIG. 10). If one tries to find the outermost KAM curve around an island, one reaches the region of sticky orbits, that stay for a long time close to the island but then escape to the large chaotic sea. This phenomenon was observed already in 1971 [11], and was later called stickiness [12].

The difference of the density of points in the sticky region is only temporary. After a long time the density in the large chaotic sea and in the sticky zone are equalized. The distribution tends to a microcanonical invariant measure in the whole chaotic domain that surrounds the islands of stability [13].

The pnenomenon of stickiness is due to the existence of cantori, which produce a temporary barrier to the orbits. Orbits inside the sticky region have difficulty to communicate with orbits outside it. Similarly orbits starting outside the cantorus have difficulty to enter the sticky region. However, the barriers are not absolute as in the case of invariant tori. The orbits can cross a cantorus, albeit after a long time.

If we calculate the spectra of stretching numbers of orbits from the center on an island outward we have a sequence like in FIGURE 11. Very close to the center the spectrum has only two maxima. Further away new maxima appear. These new maxima can be explained theoretically by the form of the curve $a(\theta)$ [14] (stretching number as a function of the azimuth). In particular, whenever $a(\theta)$ has a maximum or minimum the spectrum $S(a)$ has an infinite peak.

When we approach the outermost KAM curve the spectrum has a large number of peaks (FIG. 11c).

A sticky orbit is just outside the last KAM curve, but for a long time it remains close to the last KAM curve. Thus its transient spectrum (during the stickiness period, FIG. 11d) is similar to that of an orbit just inside the last KAM curve. However, after a long time this orbit goes out of the sticky zone into the large chaotic sea, and its spectrum tends to the invariant spectrum of the large chaotic sea (FIG. 1a).

The time of stickiness increases as we approach an island of stability, but not smoothly. In FIGURE 12 we plot the stickiness time (escape time) as a function of x along a line of constant $y = 0.36$ for $K = 5$.

Near the right end of FIGURE 12 we see an island, where the stickiness time goes to infinity. This island is on the right of $x_1 = 0.643378$. It is not the main island of FIGURE 10, but a satellite of type 14/31 (FIG. 16).

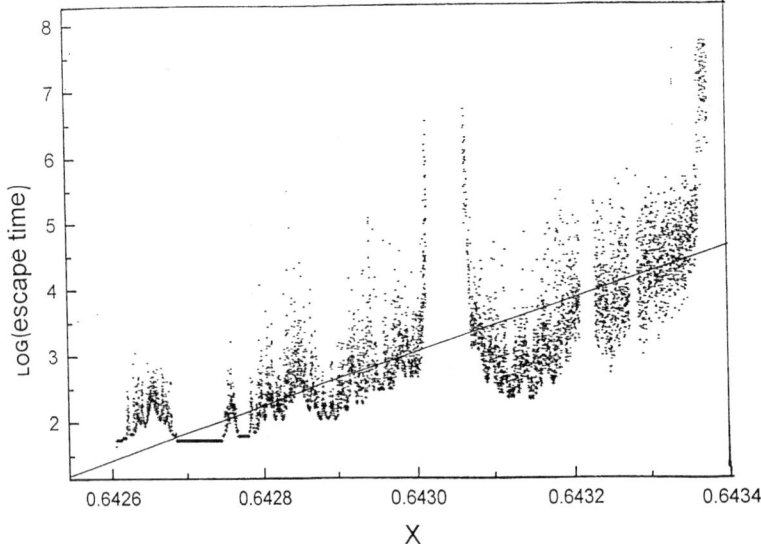

FIGURE 12. The stickiness time as a function of the distance from an island. The island under consideration is to the right of $x = 0.643378$. The straight line gives an average stickiness time that increases exponentially as we approach the island.

As we approach the island the stickiness time increases on the average, almost exponentially, according to the empirical formula

$$T = 10^{4036(x-x_1)+4.67}. \tag{8}$$

This is represented as a straight line in the logarithmic scale of FIGURE 12.

However there are various exceptions to this exponential law:
(1) Very close to the island the stickiness time increases even faster than exponentially. Such a superexponential stickiness time appears near the last KAM curve of every island. This phenomenon seems to be connected with the superexponential regime introduced theoretically by Morbidelli and Giorgilli [15] for the neighborhood of every KAM torus, by extending the theory of Nehoroshev.
(2) Inside the sticky zone there are higher-order islands, where the orbits are trapped for ever (FIG. 10). In their neighborhood there is again a superexponential increase of the stickiness time.
(3) There are some localized regions of low stickiness time in the left part of FIGURE 12. These are explained by the forms of the asymptotic curves of the periodic orbits near the cantori surrounding the sticky zone, that form some regions of fast escape [16].
(4) Beyond the left end of FIGURE 12 there are some further stickiness regions, where the law (8) does not apply, because these regions are quite far from the island on the right of FIGURE 12. They form fractals of relatively long stickiness time, separated by intervals of low stickiness [16].

THE ONSET OF CHAOS

The onset of chaos is connected with the destruction of tori, i.e., the change of tori into cantori, as the perturbation increases.

The last KAM torus around an island is normally a "noble" torus [17], that separates an outer and an inner chaotic domain. Noble tori are those with a "noble rotation number," i.e., a number that is represented as a continuous fraction

$$a = \cfrac{1}{a_1 + \cfrac{1}{a_2 + \cfrac{1}{a_3 + \ldots}}}, \tag{9}$$

where all a_i above a certain order $i = N$ are equal to 1. This is written in the form $a = [a_1, a_2, a_3, \ldots]$. A particular case of a noble number is the golden mean $[1, 1, 1, \ldots]$ which is equal to $\frac{1}{2}(\sqrt{5} - 1)$. According to Greene's conjecture [17] the noble tori are the last KAM tori (or last KAM curves) to be destroyed, as a perturbation parameter K increases beyond a critical value K_c. A destroyed KAM curve is a cantorus. For small positive values of $(K - K_c)$ the gaps of the cantorus are small and the cantorus provides a partial barrier for the communication of the chaotic domains on both its sides. However, for larger $(K - K_c)$ the gaps become large and no effective barrier is provided to the diffusion from one chaotic domain to another.

As K increases, the size of a torus with a given rotation number increases. If this is the last KAM torus, this means that the overall size of the island increases. However inside and close to the last KAM torus there are some higher-order periodic orbits, both stable and unstable. The unstable orbits are followed by a chaotic domain, which is completely separated from the outer chaotic sea. For example, in FIGURE 13a we see five secondary islands inside the last KAM curve and five unstable points between them. As the perturbation increases beyond a critical value $K = K_c \approx 4.80$ the last KAM curve is destroyed and the chaotic sea enters into the chaotic regions around the five unstable points. Then the five islands do not belong to the main island any more, which shrinks discontinuously, and is limited then by a new last KAM curve inside the five islands (FIG. 13b). For larger values of K the 5 islands decrease in size and they finally disappear after a cascade of infinite period doubling bifurcations.

As K increases beyond K_c the size of the new last KAM curve increases. But inside it new islands are formed. These islands are generated in general at the center of the main island and move outward, at the same time growing in size. For example, for $K = 4.92$ one sees 7 main islands inside the new last KAM curve. Again, chaos is generated near the 7 unstable points between the islands, and for another critical value $K_c' \approx 4.93$ the new last KAM curve is destroyed and the outer chaos reaches the chaotic domain near the 7 unstable orbits. Then again the size of the main island decreases abruptly.

This process is repeated again and again until the periodic orbit at the center of the island becomes unstable, and its stability is transferred to a pair of stable periodic orbits of equal period. The whole set (unstable periodic orbit plus two stable periodic orbits surrounded by their own islands) is still surrounded by a "final" last KAM curve. When this final curve is also destroyed at some critical value K_c^∞ the outer chaotic sea reaches the center of the main island.

For still larger K the two stable periodic orbits, on each side of the previous "center of the main island" (which is now unstable), also become unstable and produce 4 islands of stability by period doubling.

This process is repeated again and again, and after an infinite number of period-doubling bifurcations all the KAM curves and all the islands generated from the original main island are destroyed. Only some cantori remain, that in some cases may act as temporary barriers. (This does not exclude the appearance of stable "irregular" periodic orbits [18] that are not produced by bifurcation from stable periodic orbits that existed for smaller values of K).

R. Dvorak (Vienna) has produced a movie, showing the variations of the forms of the islands and the surrounding chaos in the standard map, as the value of K increases. One sees the successive expansions and abrupt contractions of the main island, and the subsequent disappearance of the outer multiple islands.

We have studied in detail the growth of the chaotic domains both outside and inside the last KAM curve in a typical case. This case is the standard map for K a little above 4.79.

The last KAM curve is a noble torus with rotation number [2, 1, 1, 3, 1, ...]. For $K = 4.79$ this torus is surrounded by other noble tori. The important characteristic of

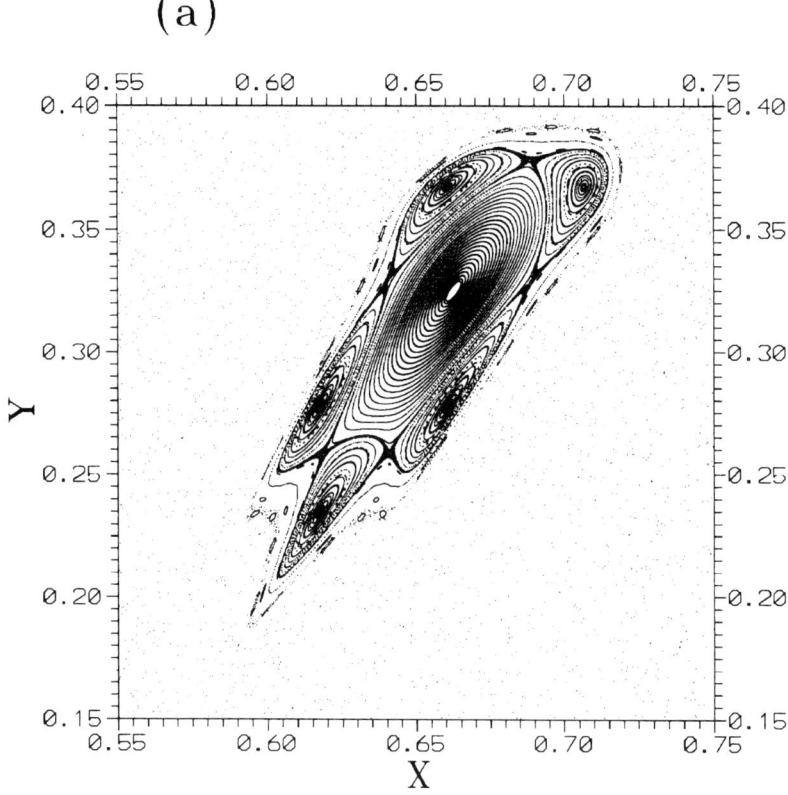

FIGURE 13. A set of five secondary islands around the main island of the standard map. Between these islands there are 5 points representing an unstable periodic orbit, which are followed by smaller chaotic domains. These are seperated from the outer chaotic sea for $K = 4.79$ **(a)** by a closed last KAM curve, surrounding also the five islands. But for $K = 4.80$ **(b)** this KAM curve is destroyed and the chaotic domains near the 5 unstable periodic points communicate with the outer chaotic sea.

this case is the fact that inside this torus there is an appreciable chaotic domain, corresponding to the unstable periodic orbit 2/5 (FIG. 14), that we considered already in FIGURE 13. As the perturbation increases, chaos increases both from outside and from inside. For $K = 4.791$ the outer noble torus [2, 1, 1, ...] has been destroyed, and for $K = 4.793$ the outer noble torus [2, 1, 1, 2, 1, ...] inside it, is also destroyed. At the same time the inner noble tori [2, 1, 1, 5, 1, ...] and [2, 1, 1, 4, 1, ...] are successively destroyed from inside outward for $K = 4.791$ and $K = 4.793$.

When $K = 4.793$ the noble torus [2, 1, 1, 3, 1, ...] still separates the inner and the outer chaos. But for $K = 4.794$ even this torus has been destroyed and the outer chaotic sea has invaded the chaotic domain around the 2/5 periodic orbit.

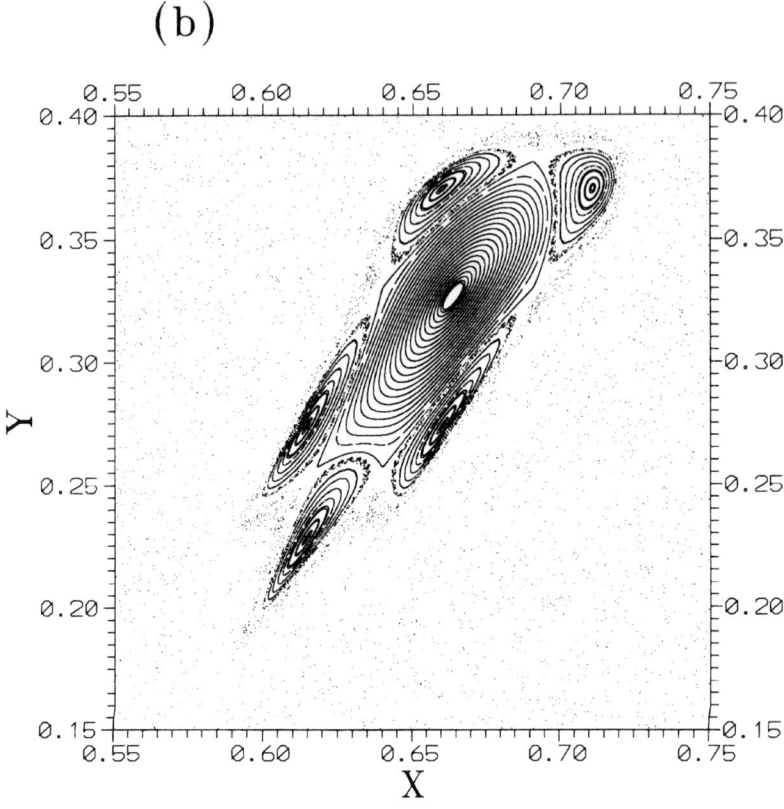

FIGURE 13b.

It is of interest to note that the last KAM curve to be destroyed has not the simplest noble rotation number [2, 1, 1, ...] in this area, but it is inside it. In other cases the last noble torus to be destroyed is outside a simpler noble torus (N. Voglis, private communication).

At any rate it seems that every "last KAM curve" is destroyed by an increase of chaos both from outside and inside, until the two chaotic domains join.

What is now of interest is how this joining of the inner and outer chaotic domains occurs. It is obvious that when the last KAM curve becomes a cantorus, a communication of the inner and outer domain can be realized by the passing of orbits through the holes of the cantorus. However, this effect cannot be observed in a map if we just take the iterates of a point inside the cantorus until one iterate is seen beyond the cantorus. This is because the map is discontinuous. Furthermore, the successive iterates of an orbit go many times outside and inside the cantorus but stay very close to the cantorus (sometimes for many iterations) before going well outside the cantorus.

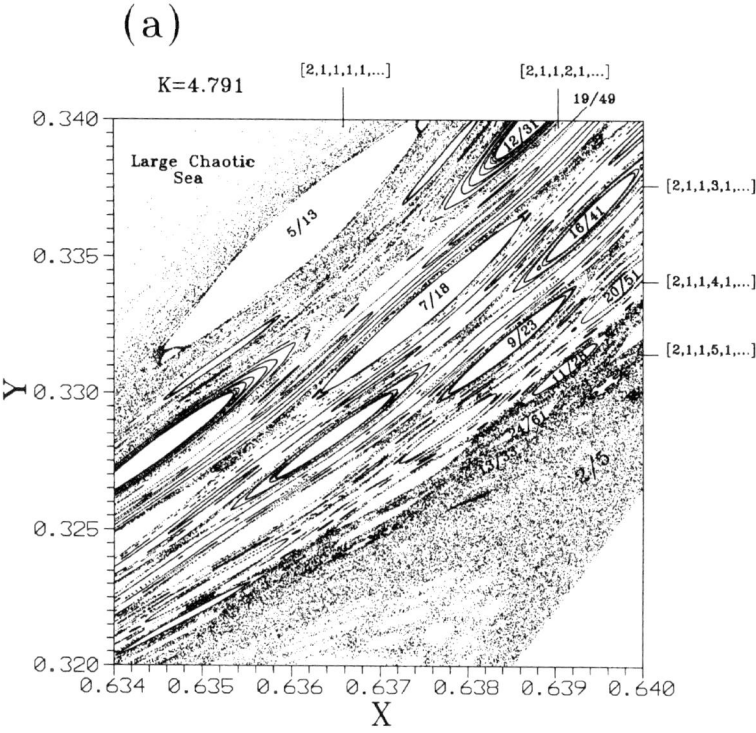

FIGURE 14. A set of noble invariant tori and cantori and chaos around the noble torus [2, 1, 1, 3, 1, ...], in the standard map. In case (**a**) $K = 4.791$. Then the noble tori [2, 1, 1, ...] (exterior) and [2, 1, 1, 5, 1, ...] (interior) are destroyed, but the noble tori [2, 1, 1, 2, 1, ...], [2, 1, 1, 3, 1, ...], [2, 1, 1, 4, 1, ...], still remain. In case (**b**) $K = 4.793$ and the tori [2, 1, 1, 2,1, ...] (exterior) and [2, 1, 1, 4, 1, ...] (interior) are also destroyed, but the torus [2, 1, 1, 3, 1, ...] still remains.

A much better technique is to take the unstable asymptotic curve of an unstable periodic orbit inside the cantorus until it passes in a continuous way through a hole of the cantorus.

In order to observe this phenomenon we have first to find the position of the cantorus and its holes.

CROSSING OF A CANTORUS

The position of a cantorus can be found by approaching it through periodic orbits. Namely, to each truncation of the continuous fraction representing a noble rotation number, correspond a couple of periodic orbits, and the sequence of such periodic

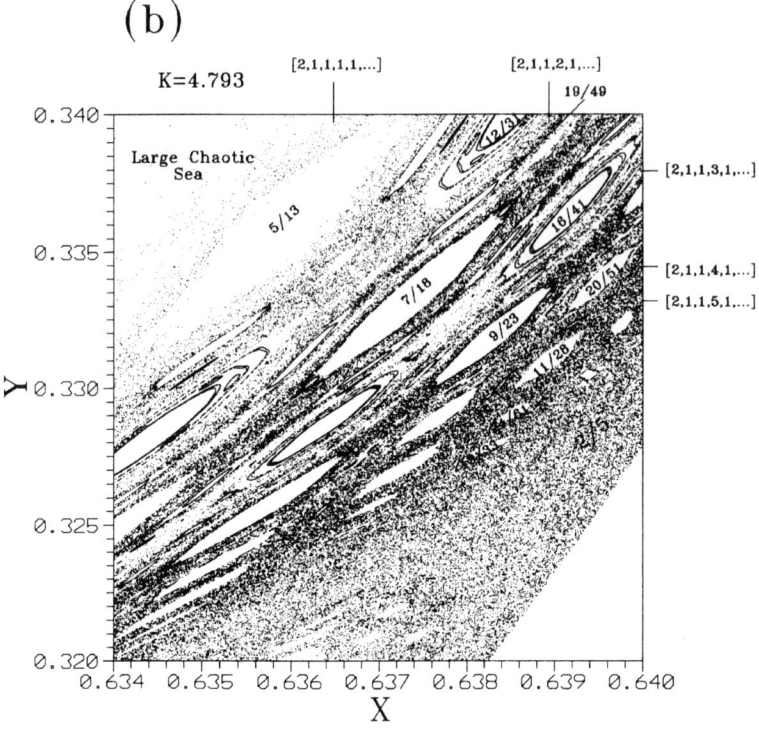

FIGURE 14b.

orbits tends to the cantorus. For example, the noble number $a = [2, 4, 1, 1, \ldots]$ has as successive truncations

$$\frac{1}{2}, \frac{4}{9}, \frac{5}{11}, \frac{9}{20}, \frac{37}{82}, \frac{60}{133}, \frac{97}{215}, \frac{157}{348}, \frac{254}{563}, \ldots \tag{10}$$

These numbers are successively larger and smaller than the noble number itself. For each truncation n/m there are two periodic orbits. The first is unstable, while the second is stable for relatively small K, becoming unstable as K increases beyond a critical value $K_{n/m}$. Greene [17] conjectured that the value of K_c, at which the last KAM curve is destroyed and a cantorus is formed, is such that all periodic orbits close to the cantorus have become unstable.

In a previous paper [19] we have used this conjecture in deriving the value of K_c by extrapolating the values of $K_{n/m}$, where n/m are the even or odd truncations of the noble number. Namely, we find that the values of $K_{n/m}$ for odd truncations represent periodic orbits inside the cantorus (in the present case the numbers $\frac{1}{2}, \frac{5}{11}, \frac{14}{31}$, becoming smaller at each successive approximation. Thus by extrapolating the numbers $K_{n/m}$ we find an approximate value of K_c which is smaller than all $K_{n/m}$. The same limit is reached if we extrapolate the values of $K_{n/m}$ for successive even truncations

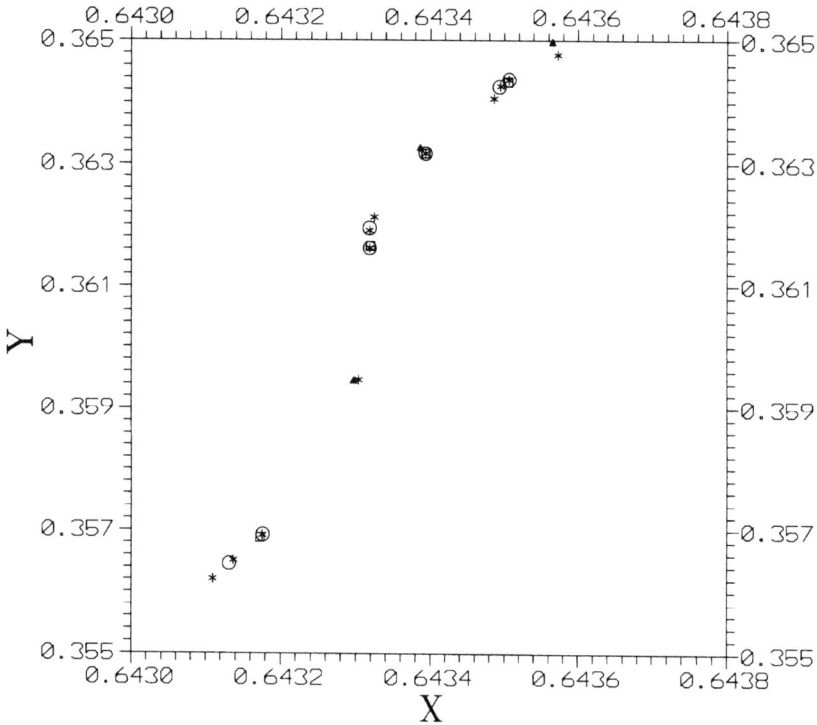

FIGURE 15. Part of the cantorus $a = [2, 4, 1, 1, ...]$ in the standard map for $K = 5$. The periodic orbits 97/215 (squares) and 254/563 (stars) are inside the cantorus, while the periodic orbits 60/133 (triangles) and 157/348 (circles) are outside the cantorus. Most of the points are so close to each other that they define approximately the cantorus.

of the noble number, representing periodic orbits outside the cantorus (in the present case the numbers $\frac{4}{9}$, $\frac{9}{20}$, $\frac{23}{51}$, ...).

For a value of K slightly larger than K_c the periodic orbits corresponding to high-order truncations of a have all become unstable, while orbits further away from the cantorus are still (partly) stable. As K increases, the orbits in a larger and larger zone around it, both outside and inside, are only unstable.

In order to find the approximate position of a cantorus we find the positions of the periodic orbits n/m inside and outside the cantorus. In this way we approach the cantorus, not only as regards its position, but also as regards its gaps. This is seen in FIGURE 15, where we see the positions of the points of the periodic orbits 97/215 (squares) and 254/563 (stars) inside the cantorus, and of the periodic orbits 60/133 (triangles) and 157/348 (circles) outside the cantorus. These points are very close to each other, thus they define approximately the gaps of the cantorus.

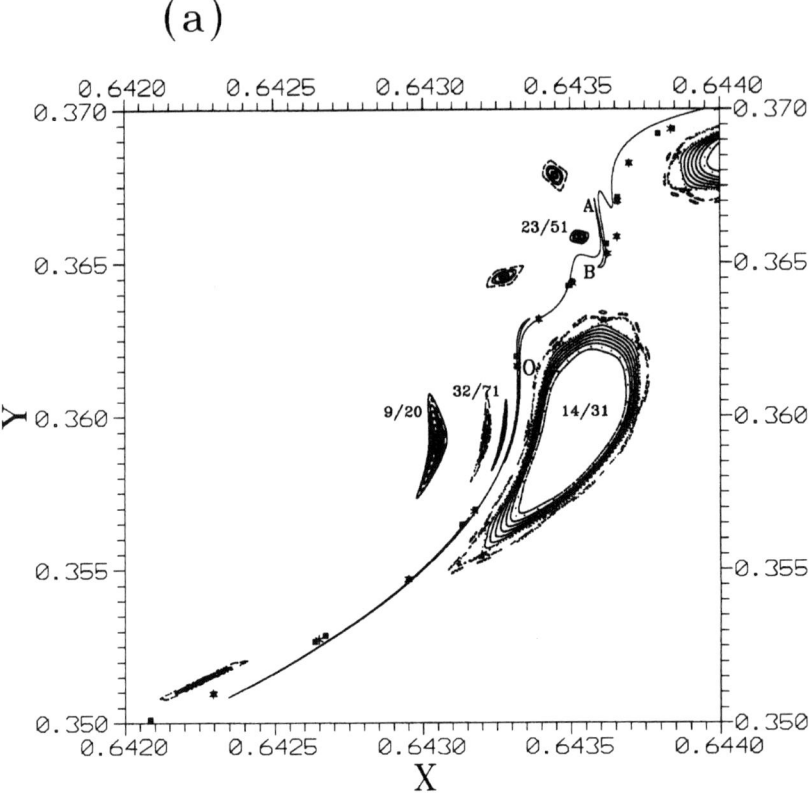

FIGURE 16. The crossing of the cantorus [2, 4, 1, 1, ...] by the unstable asymptotic curve of the periodic orbit 97/215 (**a**). The asymptotic curve after several iterations reaches large distances from the island (**b**).

Then we calculate an unstable asymptotic curve of one unstable periodic orbit inside the cantorus and follow its continuation until it crosses the cantorus, and even further, until it reaches any given distance outside the cantorus.

In FIGURE 16 we show the unstable asymptotic curve starting from an unstable periodic orbit (point O) of a periodic orbit of multiplicity 215, as it crosses the cantorus $a = [2, 4, 1, 1, ...]$ and goes to large distances in the large chaotic domain outside the main island of stability.

The asymptotic curve starts downward and to the left inside the cantorus (FIG. 16a). After making some oscillations in this region (two oscillations in the present case), it moves outward and crosses the cantorus just above and to the right of O. Then it makes a number of oscillations, going inside and outside the cantorus, but it continues with longer and longer oscillations outside the cantorus.

As we continue the asymptotic curve much further (FIG. 16b), we find that it extends to such large distances outside the cantorus that it reaches the boundary of the

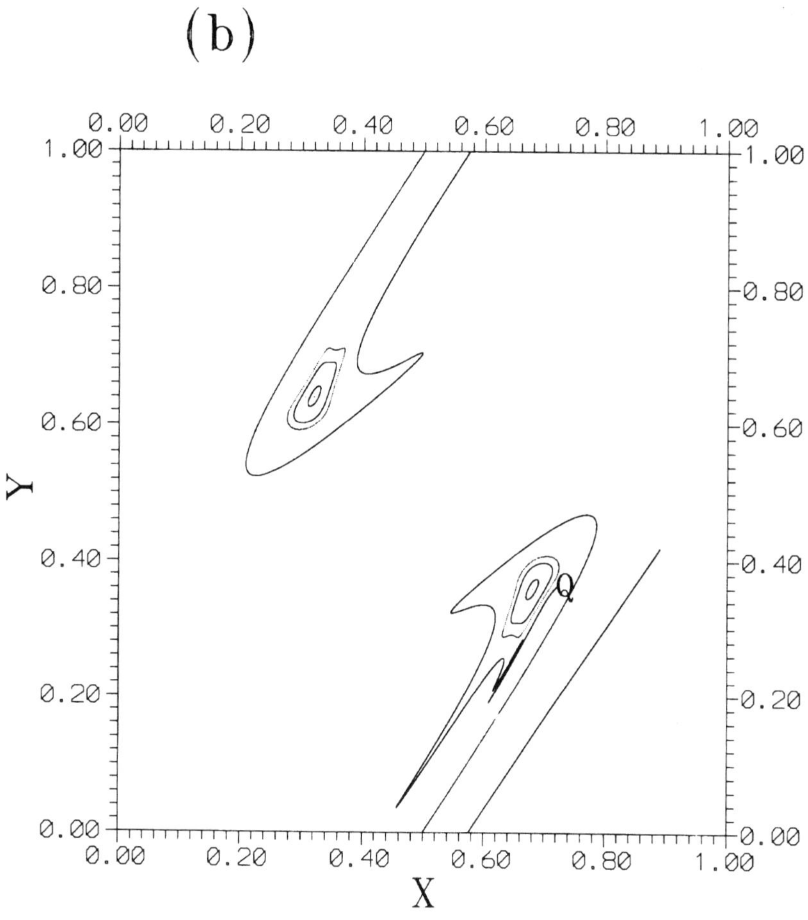

FIGURE 16b.

square ($0 < x < 1$, $0 < y < 1$). Later on the asymptotic curve fills densely the whole square except for the original island of stability and a dual island symmetric to the first with respect to the center of the figure.

However the asymptotic curve comes back to the sticky zone, close and inside the cantorus, an infinite number of times. This eventually leads to an equalization of the density of points in the sticky zone and in the outer chaotic domain, but this equalization requires a very long time.

The present method shows for the first time in a continuous way the successive crossings of a cantorus by an asymptotic curve.

The form of the lobes of the asymptotic curve gives much information about the diffusion through cantori. First, we see that the lobes are for a long time almost parallel to the cantorus. In fact, the lobes become longer and longer, surrounding the whole island before going to large distances from the island.

Second, the crossing of a cantorus is done many times outward and inward before the asymptotic orbit goes very far, in which case the successive crossings become more and more rare.

Third, it is well known that an unstable asymptotic curve cannot cross itself, or other unstable asymptotic curves. Thus the unstable asymptotic curves from other periodic orbits in the same chaotic zone inside the cantorus must follow the oscillations of the above asymptotic curve, before going far away from the cantorus. This implies that different asymptotic curves are very close to each other close to the cantorus and cross the cantorus together.

As a consequence the phenomenon of diffusion through cantori can be understood correctly only by following the lobes of one unstable asymptotic curve as in FIGURE 16. It is not a random diffusion of points outward from the cantorus, but it is governed by the forms of the lobes of an unstable asymptotic curve.

The lobes, in turn, avoid all islands of stability, both inside and outside the cantorus. Thus, they can go further outside the cantorus only by passing between islands of stability, and at the same time avoiding the unstable asymptotic curves of the unstable periodic orbits outside the cantorus.

The sizes and general forms of the lobes in a homoclinic tangle have been studied numerically in a particular model [20]. It was found that the low order intersections of the stable and unstable manifolds allow the prediction of the higher-order intersections. Thus a quantitative prediction of the form of the lobes can be made.

THREE DEGREES OF FREEDOM

The above methods can be extended to systems of three degrees of freedom.

In such cases the phase space of a conservative system has 5 dimensions. Then a surface of section has 4 dimensions. Thus the system can be reduced to a 4D map.

A typical example of a 4D map was given by Froeschlé and consists of two coupled standard maps [21].

$$x_1' = x_1 + y_1'$$

$$y_1' = y_1 + \frac{K}{2\pi}\sin 2\pi x_1 - \frac{\beta}{\pi}\sin 2\pi(x_2 - x_1)$$

$$x_2' = x_2 + y_2' \qquad \text{(mod1)} \qquad (11)$$

$$y_2' = y_2 + \frac{K}{2\pi}\sin 2\pi x_2 - \frac{\beta}{\pi}\sin 2\pi(x_1 - x_2),$$

If $\beta = 0$, the two maps (x_1, y_1) and (x_2, y_2) are independent. However, if $\beta \neq 0$ the two maps are coupled and a genuine 4D map is formed.

In such a case, we define a stretching number a, and three helicity angles, namely $\phi_{x1x2}, \phi_{x1y1}, \phi_{x1y2}$, giving the angles of the projections of the vector ξ on the planes x_1x_2, x_1y_1, x_1y_2 with the axis x_1.

The spectra of the stretching numbers and helicity angles are invariant with respect to the initial conditions along an orbit. In the case of chaotic orbits the spectra are invariant also with respect to initial conditions in the same chaotic domain, and

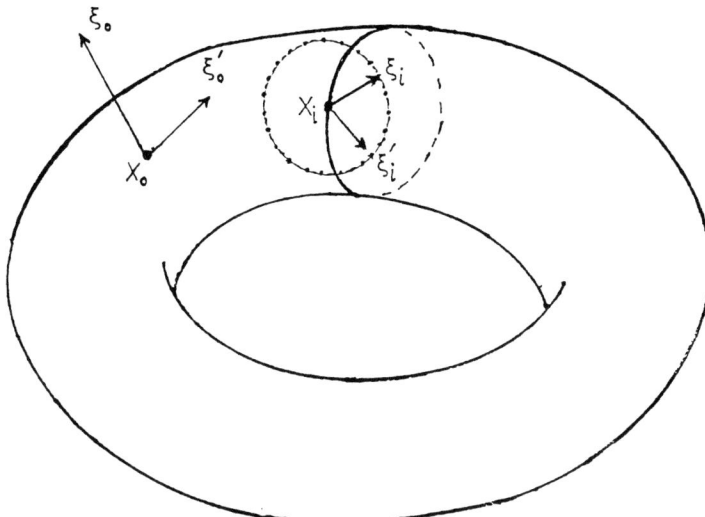

FIGURE 17. A torus on which lies a regular orbit in a 4D map (schematically). After some iterations the deviations ξ from an orbit become (almost) tangent to the torus, but two different initial deviations ξ_0 and ξ_0' lead to different tangents ξ_i and ξ_i' at the same point x_i on the tangent plane (circle).

with respect to the initial direction of the deviation ξ_0, but the spectra of regular orbits are not invariant with respect to the direction of the initial deviation ξ_0. In fact a regular orbit lies on an invariant torus (FIG. 17), and any initial deviation ξ_0 tends to become tangent to this torus after a few iterations. But two different initial deviations ξ_0 and ξ_0' lead in general to different tangents ξ and ξ'. Only in exceptional cases, if the deviations ξ, ξ' happen to be along the same tangent direction, the spectra are the same.

Thus one method to distinguish between chaotic and ordered orbits in 4D maps, or 3D systems, is to find whether two different initial deviations, ξ_0, ξ_0', taken at random from the same initial point of an orbit, lead to the same or different spectra.

This method requires a number of iterations sufficient to distinguish between the two spectra (at least about 10^3).

But if we want to distinguish chaotic and ordered domains we can use the method developed for 2D maps. Namely, we calculate only a few interations of an orbit, and the corresponding deviations ξ, and use the average value of a helicity angle, say $\langle \phi_{x1x2} \rangle_N$, or of the stretching number $\langle a \rangle_N$ [8]. For example, we get sufficient information by using only 10 iterations after the first 10 transients. We see that in the chaotic domain the average values of $\langle \phi_{x1x2} \rangle_{20-10}$ and $\langle a \rangle_{20-10}$ are almost constant, but with large variations of the individual values. On the other hand in the ordered domain the values of $\langle \phi_{x1x2} \rangle_{20-10}$ and $\langle a \rangle_{20-10}$ are very different from those of the chaotic domain. Their variations are still irregular, but much smaller than in the chaotic domain. Furthermore, the values of $\langle a \rangle_{20-10}$ are close to zero.

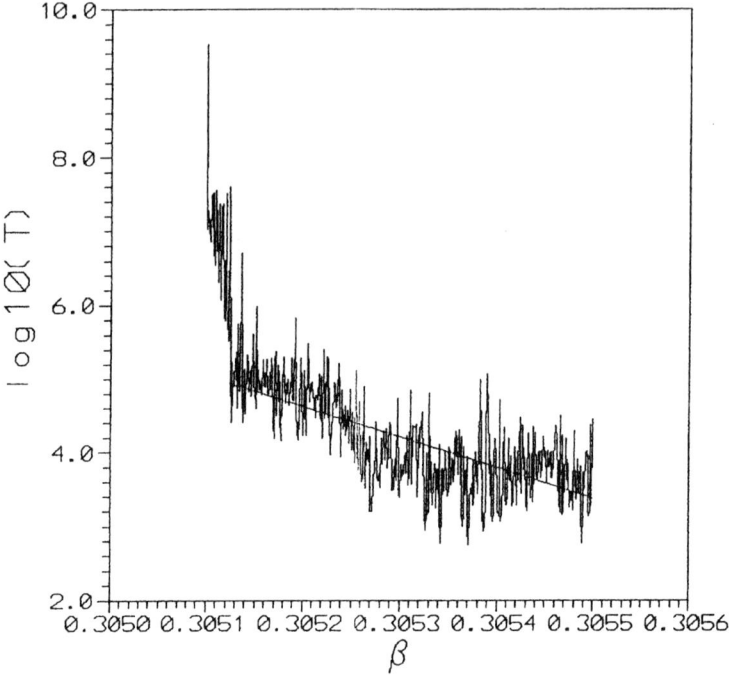

FIGURE 18. The escape time T from a regular region as a function of the coupling parameter β.

The change of the values of $\langle\phi_{x1x2}\rangle_{20-10}$ and $\langle a \rangle_{20-10}$ are rather abrupt at the limits of the islands of stability. Thus we can separate with a good approximation the chaotic and ordered domains.

At this point one may argue that in 4D maps and 3D systems any small error in the calculation of a regular orbit will lead to chaos, because of Arnold diffusion. Thus we may not find any regular orbit by numerical integrations. In order to check this point we have calculated the escape times for many orbits with the same initial conditions, $K = 3$, and various values of the coupling parameter β. The initial conditions correspond to an ordered orbit in the limiting case $\beta = 0$, when no Arnold diffusion exists, while for large values of β the diffusion is large.

The results of our calculations are shown in FIGURE 18. The individual values of T vary considerably, but we can clearly distinguish two very different regimes. When β is larger than $\beta_c = 0.305124$ the escape time decreases with increasing β approximately according to the exponential law

$$T = 10^{4.94 - 4160(\beta - \beta_c)}, \tag{12}$$

represented by a straight line in the logarithmic scale of FIGURE 18.

As expected, this law leads to a fast escape for large β. In fact for larger than $\beta = 0.3063$ the time T becomes of order 1. However, as β becomes smaller than β_c, the escape time increases superexponentially and becomes larger than $T = 10^{10}$ for slightly smaller β. Then we cannot continue our calculations for longer times.

The transition from exponential to superexponential escape time is rather smooth, although the change of the derivative of T is quite abrupt. We have tentatively called the exponential regime a "resonance overlap diffusion" regime, and the superexponential regime an "Arnold diffusion" regime.

The behavior of T has some similarities with the transition from exponential to superexponential stickiness time in 2D maps (compare our discussion of Eq. (8) above). In that case there is no Arnold diffusion and the diffusion is related to an interaction of resonances, inside and outside each cantorus.

But Arnold diffusion itself is also due to an interaction of resonances, that always occurs near the points of intersection of the resonant lines of the Arnold web. The only difference is that such interactions of resonances are rather localized in space, while resonance overlap usually implies relatively extended overlapping domains.

The only real difference between two-dimensional and higher-dimensional cases is that in the former case there are regions of infinite trapping time, while in the latter case the trapping time is finite.

Thus we expect that in our FIGURE 18 the trapping time will never become infinite, for any value of β different from zero, while for $\beta = 0$ the islands provide really infinite time of trapping. But this distinction may not be possible if the times required for the calculations are much larger than what any computer can provide.

ACKNOWLEDGMENTS

Most of this work was done in collaboration with N. Voglis and C. Efthymiopoulos. Several figures were prepared by C. Efthymiopoulos, M. Harsoula, and R. Dvorak. This paper was written while I was a visiting professor at the Florida State University, after an invitation by Dr. C. Hunter. I want to thank the Department of Astronomy of the University of Florida and the Organizing Committee (Drs. R. Buchler, S. Gottesman, J. Hunter and H. Kandrup) for preparing this Workshop for me.

REFERENCES

1. NIKOLIS, J.S., G. MEYER-KRESS & G. HAUBS. 1983. Z. Naturfosch. **38a:**1157.
2. FUJISAKA, H. 1983. Prog. Theor. Phys. **70:**1264.
3. UDRY, S. & D. PFENNIGER. 1998. Astron. Astrophys. **198:** 135.
4. FROESCHLE C., CH. FROESCHLE & E. LOHINGER. 1993. Cel. Mech. Dyn. Astron. **56:** 307.
5. VOGLIS, N. & G. CONTOPOULOS. 1994. J. Phys. A **27:** 4899.
6. KANDRUP, H.E. & M.E. MAHON. 1994. Astron. Astrophys. **290:** 762.
7. SMITH H. & G. CONTOPOULOS. 1996. Astron. Astrophys. **314:** 795.
8. CONTOPOULOS, G. & N. VOGLIS. 1997. Astron. Astrophys. **317:**73.
9. LASKAR, J. 1993. Physica D 67:257; LASKAR, J., C. FROESCHLE & A. CELLETTI. 1992. Physica D **56:** 253.

10. PATSIS, P.A., C. EFTHYMIOPOULOS, G. CONTOPOULOS & N. VOGLIS. 1997. Astron. Astrophys. **326:** 493.
11. CONTOPOULOS, G. 1971. Astron. J. **76:**147.
12. SHIRTS, R.B. & W.P. REINHARDT. 1982. J. Chem. Phys. **77:** 5204.
13. CONTOPOULOS, G., N. VOGLIS, C. EFTHYMIOPOULOS & E. GROUSOUSAKOU. 1995. In Waves in Astrophysics, J.H. Hunter, Jr. & R.E. Wilson, Eds. Ann. N. York Acad. Sci. **773:** 145.
14. CONTOPOULOS, G., N. VOGLIS, C. EFTHYMIOPOULOS, C. FROESCHLÉ, R. GONCZI, E. LEGA, R. DVORAK & E. LOHINGER. 1997. Cel. Mech. Dyn. Astron. **67:** 293.
15. MORBIDELLI, A. & A. GIORGILLI. 1995. J. Stat. Phys. **78:** 1607.
16. EFTHYMIOPOULOS, C., G. CONTOPOULOS, N. VOGLIS & R. DVORAK. 1997. J. Phys. A. **30:** 8167.
17. GREENE, J.M. 1979. J.Math. Phys. **20:** 1183.
18. CONTOPOULOS, G.1970. Astron. J. **75:** 96.
19. CONTOPOULOS, G., H. VARVOGLIS, & B. BARBANIS. 1982. Astron. Astrophys. 172:55.
20. CONTOPOULOS, G. & C. POLYMILIS. 1993. Phys. Rev. **47:** 1546.
21. FROESCHLE, C. 1972. Astron. Astrophys. **16:** 172.

Orbital Complexity, Short-Time Lyapunov Exponents, and Phase Space Transport in Time-Independent Hamiltonian Systems[a]

CHRISTOS SIOPIS, BARBARA L. ECKSTEIN, AND HENRY E. KANDRUP[b]

Department of Astronomy, University of Florida, Gainesville, Florida 32611

ABSTRACT: This paper compares two alternative characterizations of chaotic orbit segments, one based on the *complexity* of their Fourier spectra, as probed by the number of frequencies $n(k)$ required to capture a fixed fraction k of the total power, and the other based on the computed values of *short-time Lyapunov exponents* χ. An analysis of orbit ensembles evolved in several different two- and three-dimensional potentials reveals that there is a strong, roughly linear correlation between these alternative characterizations, and that computed distributions of complexities, $N[n(k)]$, and short-time χ, $N[\chi]$, often assume similar shapes. This corroborates the intuition that chaotic segments which are especially unstable should have Fourier spectra with particularly broad-band power. It follows that orbital complexities can be used as probes of phase space transport and other related phenomena in the same manner as can short-time Lyapunov exponents.

INTRODUCTION AND MOTIVATION

Viewed in an asymptotic $t \to \infty$ limit, orbits in a time-independent Hamiltonian system divide naturally into two classes, regular and chaotic. One possible definition of chaotic orbits is that they have one or more positive Lyapunov exponents [1]. For regular orbits, all the Lyapunov exponents vanish identically. Another possible definition relates to the form of the Fourier spectrum. Regular orbits are multiply periodic and, as such, have spectra where all the power is concentrated at a discrete set of frequencies. Alternatively, chaotic orbits have nonvanishing power for a continuous range of frequencies. (Strictly speaking, not every flow admitting one or more positive Lyapunov exponent must have nonzero power for a continuous range of frequencies, but one anticipates that, as a practical matter, positive Lyapunov exponent and broad-band power go hand in hand [2].)

Given the fact that, in a time-independent Hamiltonian system, Lyapunov exponents must come in pairs, $\pm\chi$, and that two of the exponents must vanish identically, it follows that for $t \to \infty$ every chaotic orbit in a D-dimensional system is character-

---

[a]H.E. Kandrup was supported in part by National Science Foundation Grant PHY92-03333. B.L. Eckstein and C. Siopis have been supported by NASA through the Florida Space Grant Consortium. Some of the computations were performed using computer time made available through the Research Computing Initiative at the Northeast Regional Data Center (Florida by arrangement with IBM.)

[b]Also with the Department of Physics and Institute for Fundamental Theory, University of Florida, Gainesville, Florida 32611.

ized completely by D-1 numbers, at least one of which must be positive. This is consistent with the idea that different chaotic segments in the same connected phase space region can all be viewed as pieces of the same chaotic orbit of infinite length. Similarly, if a chaotic initial condition is integrated into the future for a very long time, one finds that the Fourier spectrum for the resulting orbit will converge toward a time-independent invariant form [3].

This is all fine in principle, but in many, if not most, cases of physical interest one is concerned with the behavior of a piece of a chaotic orbit over a finite time interval. However, even when viewed for relatively long times, different segments of the same chaotic orbit can look extremely different. In some cases the segment may be trapped near a regular island and appear nearly indistinguishable from a regular orbit [4], [5]. In others, the segment may be far from any regular island and "wildly chaotic" in visual appearance. It would seem natural to search for characterizations of short-time chaotic behavior which reflect these basic features.

One possible tool is provided by short-time Lyapunov exponents which, unlike ordinary Lyapunov exponents [1], characterize the average exponential instability of a chaotic segment over a finite time interval [6]. Such short-time exponents have proven useful in a number of different settings as tools for tracking phase space transport [7]–[9], the important point being that chaotic segments which are trapped near regular regions and look more regular in appearance tend to be less unstable than wildly chaotic orbits which lie further from regular islands.

Another way in which to analyze a finite chaotic segment is by computing its Fourier spectrum. Given that the segment is of finite length, and that (in any numerical computation) the phase space coordinates are only sampled at discrete instants of time, one would expect in general that the computed spectrum will have nonzero power at all possible frequencies. However, one might also expect that segments which look wildly chaotic will have significant power spread over a larger number of frequencies than segments that are regular or nearly regular.

This leads naturally to the notion of the *complexity* $n(k)$ associated with some orbit segment. As formulated more carefully below, this $n(k)$ is defined as equalling the minimum number of frequencies in the Fourier spectrum needed to capture a given fraction k of the total power. Because "more regular" chaotic segments are less unstable than "wildly chaotic" segments, one might expect physically that there will be a strong positive correlation between the computed $n(k)$ and the largest short-time Lyapunov exponent χ. The principal purpose of this paper is to verify, and then quantify, this basic trend.

A second purpose is to provide two explicit examples of problems where the orbital complexity $n(k)$ can serve as a useful diagnostic. The first of these involves understanding the orbital building blocks associated with an elliptical galaxy, modeled (cf. Ref. [10]) as a self-consistent equilibrium. The construction of such equilibria using Schwarzschild's [11] method involves (1) specifying a mass distribution, and hence a gravitational potential, (2) evolving a large number of orbits in that potential, and then (3) selecting ensembles of orbit segments which (at least approximately) reproduce the assumed potential self-consistently. One obvious question here is: What types of orbits are incorporated in any given model and, in particular, what fraction of the orbits are chaotic? The relative abundance of chaos is important because models containing large numbers of chaotic segments may be especially susceptible to various sorts of perturbations, e.g., small amplitude irregularities associated with internal substructures and/or companion objects that destroy the strictly Hamiltonian character of the flow [12]–[14].

The other example involves understanding how the flow associated with a chaotic two-dimensional potential is impacted by allowing for the effects of weak friction and noise. It has long been known that, at least for simple maps [15], even very weak noise can significantly accelerate phase space transport through cantori [16], [17] by serving as a source of extrinsic diffusion. More recent work on continuous systems [12]–[14] has shown that, in some cases, even very weak perturbations, corresponding in natural units to a relaxation time $t_R \sim 10^6 - 10^9$ can have statistically significant effects within a time as short as $t \sim 100$. Consider, therefore, an orbit which, in the absence of friction and noise, is confined by one or more cantori near a regular island for some time $t > 100$. One might then like to know the probability that, within a time (say) $t = 100$, an orbit with the same initial condition but perturbed by friction and noise of specified amplitude will diffuse through the cantori and escape into the surrounding stochastic sea.

ORBITAL COMPLEXITIES AND SHORT-TIME LYAPUNOV EXPONENTS

The visual impression that wildly chaotic orbit segments are "more complex" than regular, or nearly regular, segments arises because regular segments are (multiply) periodic and nearly regular segments are nearly periodic. It thus seems natural to quantify the complexity of an orbital segment in terms of its Fourier spectrum. In particular, given a time series of orbital data, recorded at fixed times $t_i = i\Delta t$, ($i = 1, ..., N$), the complexity should hinge on the degree to which most of the power in the Fourier spectrum resides in a few special frequencies.

This intuition suggests the following as a possible definition of the *complexity* $n(k)$ *of an orbit segment at threshold k* in a D-dimensional Hamiltonian system. Consider separately the time series $\{r_a(t_i)\}$ for each configuration space coordinate r_a and compute the Fourier transformed $\{r_a(\omega_i)\}$. Next analyze the resulting Fourier series to determine the minimum number of frequencies $n_a(k)$ required to capture a fraction k of the total power, $\Sigma_i |r_a(\omega_i)|^2$. Finally, define the orbital complexity as

$$n(k) = \sum_{a=1}^{D} n_a(k). \qquad (1)$$

If this notion of complexity is to provide an effective orbital diagnostic, one must identify the "right" choice of threshold k. This depends on both the sampling interval Δt and the total integration time T. Suppose for specificity that $\Delta t \sim 0.1$ in natural units where a typical crossing time $t_{cr} \sim 1$, thus allowing one to obtain a reasonable visual representation of the orbit. Then suppose further that the total integration time $T \sim 50 - 500$, in many cases the typical time scale on which orbits exhibit interesting qualitative changes. In this case, experimentation with a number of different potentials has shown that a threshold $k \sim 0.9 - 0.95$ yields a classification in reasonable accord with visual impressions.

For choices of k that are significantly smaller, $n(k)$ does not serve as a sensitive diagnostic since the range of possible values of n is too small. Alternatively, much larger values of k are bad because, given the sampling limitations, even regular orbits will have a small amount of power at a large number of frequencies. Within the range $0.9 < k < 0.95$, the precise value of k is relatively unimportant. If, e.g., each orbit in some ensemble is ranked in terms of both $n(0.9)$ and $n(0.95)$, one finds typically that the rank correlation $\mathcal{R}[n(0.9), n(0.95)] > 0.98 - 0.99$.

This precise definition of complexity may seem *ad hoc*, and its ultimate justification resides in the fact that, as discussed below, it yields a classification of orbit segments in close agreement with classifications based on short-time Lyapunov exponents. However, certain aspects of the definition can be defended in and of themselves.

Why, for example, should one analyze each spatial coordinate individually by computing a separate n_a for each direction, rather than simply determining the number of frequencies required to capture a fraction of the total power in all directions? The answer here is simple: In certain cases, one can have elongated orbits where almost all the power is in a single direction, and if the other directions are not treated independently, the computed complexity will be insensitive to what the orbit is doing in these orthogonal directions. Similarly, one might ask: Why simply add the individual n_a? Why not consider instead a quantity like $\tilde{n} = \Pi_{a=1}^{D} n_a$? Again the answer is simple: If, e.g., in a three-dimensional system one considers an orbit restricted to some z = constant x-y plane, this $\tilde{n}$ will vanish identically even if the motion in the x-y plane is wildly chaotic.

More fundamentally, perhaps, why define the complexity in configuration space, rather than in momentum space or in the full phase space? The answer here is that complexities defined in terms of configuration space r_a, momentum space p_a, and phase space Z_a yield very similar classifications. Not surprisingly, orbit segments that look complicated in configuration space also look complicated when viewed in momentum space or in the full phase space. Given the availability of all the phase space information, one can choose to define a full phase space complexity $\tilde{n}(k) = \Sigma_{a=1}^{2D} n_a(k)$, but this $\tilde{n}$ will not yield significantly better classification.

Ordinary Lyapunov exponents can be defined as equalling the average exponential instability of a chaotic orbit in a $t \to \infty$ limit [1]. By analogy short-time Lyapunov exponents can be defined in terms of the average instability of a finite orbit segment [6]. Specifically, given some infinitesimal phase space perturbation $\delta Z(0)$,

$$\chi(t) \equiv \lim_{\delta Z(0) \to 0} \frac{1}{t} \left[\frac{\|\delta Z(t)\|}{\|\delta Z(0)\|} \right], \qquad (2)$$

where $\|\cdot\|$ denotes an appropriate norm. It is reasonable, albeit not obligatory, to choose the natural $2D$-dimensional Euclidean norm. All the results quoted here involved this choice of norm.

The computed value of χ will of course depend on the choice of initial perturbation. It is, however, clear that the magnitude of the growing $\delta Z(t)$ will eventually be dominated by its projection in the most unstable direction. It thus follows that, at sufficiently late times, the computed χ associated with a generic $\delta Z(0)$ will closely approximate the χ that would have been computed for a carefully chosen initial perturbation aligned in the most unstable direction. If one is only interested in an estimate of the largest χ and one is willing to integrate for a sufficiently long time, it thus suffices to track the evolution of a generic $\delta Z(0)$. Alternatively, one could use a more complicated (and computationally expensive) algorithm which [18] searches explicitly for the most unstable direction and, in principle, could permit the computation of all $2D$ exponents.

Extensive tests were performed to determine the degree to which, in three representative two-dimensional Hamiltonian systems, the values of orbital complexities correlate with the values of short-time Lyapunov exponents. In each case, the Hamiltonian was of the form

$$H(x, p_x, y, p_y) = \frac{1}{2}(p_x^2 + p_y^2) + V(x, y). \tag{3}$$

Two of the potentials are relatively well known. One of these,

$$V(x, y) = -(x^2 + y^2) + \frac{1}{4}(x^2 + y^2)^2 - \frac{1}{4}x^2 y^2, \tag{4}$$

is a special case of the so-called dihedral, or D-4, potential [19]. The other,

$$V(x, y) = \frac{1}{2}(x^2 + y^2) + x^2 y - \frac{1}{3}y^3 + \frac{1}{2}x^4 + x^2 y^2 + \frac{1}{2}y^4 \\ + x^4 y + \frac{2}{3}x^2 y^3 - \frac{1}{3}y^5 + \frac{1}{5}x^6 + x^4 y^2 + \frac{1}{3}x^2 y^4 + \frac{11}{45}y^6, \tag{5}$$

constitutes the sixth-order truncation of the two-particle Toda lattice potential [20]. The third potential,

$$V(x, y) = -\frac{1}{(1 + x^2 + y^2)^{1/2}} - \frac{0.3}{(1 + x^2 + 0.1 y^2)^{1/2}}, \tag{6}$$

is the sum of a softened Kepler potential and a softened anisotropic Kepler potential [9]. These three potentials manifest very different symmetries. The fact that the same correlations between χ and $n(k)$ were observed in all three cases thus suggests strongly that the basic conclusions are robust.

For a number of different energies in each potential, surfaces of section were sampled to generate ensembles of initial conditions corresponding to as many as 5120 chaotic orbits with the same energy E. These ensembles were evolved into the future for times between $T \sim 50$ and $T \sim 2500$, with orbital data recorded at regular intervals varying between $\Delta t \sim 0.05$ and $\Delta t \sim 0.25$. An estimate of the largest Lyapunov exponent was obtained simultaneously by tracking the evolution of an initial perturbation of amplitude $\delta Z \sim 10^{-10}$–10^{-8}, periodically renormalized in the usual way [1] at intervals $\Delta T = 10$ or less. The orbital data were Fourier transformed using a standard FFT program [21], and the resulting spectra analyzed to compute complexities $n(k)$ for a variety of values of k between 0.5 and 0.99. The computed χ and n were analyzed in two different ways. First they were analyzed to extract rank correlations between χ and $n(k)$. Then they were binned to extract distributions $N[\chi]$ and $N[n(k)]$, and the resulting distributions compared in a search for both qualitative and quantitative similarities.

When, over the integration interval T, (almost) all the segments exhibit comparable values of χ, they will also exhibit comparable values of $n(k)$. Thus neither $n(k)$ nor χ is useful for drawing distinctions between different segments in the ensemble. However, what is still true is that if the ensemble contains a few exceptional segments where χ is significantly higher (lower) than the typical χ, those exceptional segments will also assume values of $n(k)$ that are significantly higher (lower) than the typical $n(k)$.

If the ensemble exhibits a broad range of χ, significantly more interesting results obtain. In this case, comparisons of chaotic segments in ensembles of fixed energy E reveal a strong positive correlation between the computed values of χ and $n(k)$, the rank correlation typically assuming a value $\mathcal{R}(n, \chi) \sim 0.75 - 0.95$. One example is provided in FIGURE 1(a), which exhibits a plot of χ versus $n(0.9)$. The data for this plot were generated from 3135 initial conditions with energy $E = 10$ evolved in the D-4 potential. Orbital data were recorded at intervals $\Delta t = 0.25$ for a total time $T = 256$, corresponding to $\sim 100 t_{cr}$. This particular case illustrates a relatively weak correlation, with $\mathcal{R}[n(0.9), \chi] = 0.770$. Another example is shown in FIGURE 1(b). This analogous plot was generated from an ensemble of 2560 initial conditions with $E = -0.6$ evolved in the potential (6). Here $T = 2048$, which corresponds to $\sim 200 t_{cr}$ and the sampling interval $\Delta t = 1.0$. In this case the correlation is manifestly stronger, the rank correlation assuming a value $\mathcal{R}[n(0.9), \chi] = 0.918$.

FIGURES 1(a) and 1(b) exhibit a roughly linear correlation. However, it is also clear visually that there is at least some curvature. Thus, for example, in FIGURE 1(b) it is tempting to fit the data points to a relation $\chi \propto n^p$ for some constant $p \ne 1$. Such fits can in fact be done, both here and in other cases. However, residuals for power law fits with $p \ne 1$ are not significantly smaller than for linear fits with $p = 1$.

The observed roughly linear correlation between $n(k)$ and χ would suggest that binned distributions of complexities and short-time Lyapunov exponents, $N[n]$ and

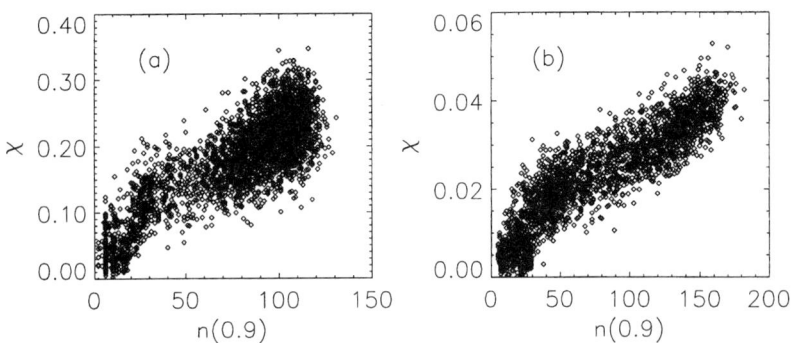

FIGURE 1. (a) The short-time Lyapunov exponent χ plotted as a function of the complexity $n(0.9)$ for 3135 segments with energy $E = 10$ evolved in the D-4 potential for a total time $T = 256$. (b) χ plotted as a function of $n(0.9)$ for 2560 segments with $E = -0.6$ evolved in the potential (6) for $T = 2048$.

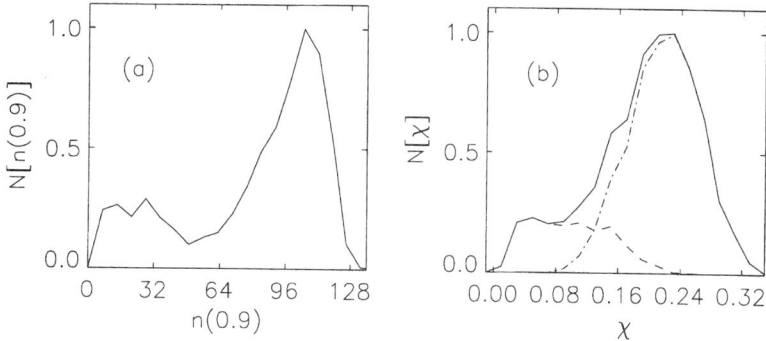

FIGURE 2. (a) A binned distribution of complexities, $N[n(0.9)]$, for the segments plotted in FIGURE 1a. (b) The solid curve gives the corresponding distribution of short-time Lyapunov exponents, $N[\chi]$, for the same segments. The dashed and dot-dashed curves overlay distributions $N[\chi]$ computed for those segments with complexities $n \leq 50$ and $n > 50$.

$N[\chi]$, will exhibit similar features. However, the significant spreads and the residual curvatures might also suggest that plots of $N[n]$ and $N[\chi]$ will manifest some interesting differences. Both these expectations are in fact correct.

FIGURES 2(a) and (b) exhibit distributions $N[n]$ and $N[\chi]$ generated from the orbit segments analyzed in FIGURE 1(a), in each case allowing for 20 bins. Both curves have a distinct peak at a large value of n (or χ) and a substantially weaker peak at a lower value. However, it is evident that $N[n]$ manifests a sharper distinction between segments with low and high values than does $N[\chi]$. The simplest interpretation is that the total collection of segments is comprised of two separate populations, one with low n and χ, the other with higher n and χ, and that, in both cases, the total N is a sum of two distinct, singly peaked distributions. The obvious inference then is that the two subdistributions of n are relatively distinct, whereas the subdistributions of χ have significant overlap.

To test this hypothesis, the initial ensemble was divided into two separate components, namely segments with $n(0.9) \leq 50$ and $n(0.9) > 50$, and distributions $N[\chi]$ were then computed individually for each component. The resulting curves are superimposed in FIGURE 2(b) as dashed and dot-dashed curves. It is apparent that the large n component yields a nearly Gaussian distribution of χ centered below the high χ peak, and that the lower n component yields a more irregular distribution which includes both the region under the secondary peak and regions well toward the right of the peak. This is in accord with expectation. Chaotic segments in two-dimensional potentials often divide naturally into two subclasses, confined chaotic segments trapped near regular islands by cantori and unconfined chaotic segments. Ensembles of unconfined chaotic segments tend typically to be characterized by a Gaussian distribution of χ, whereas ensembles of chaotic segments can exhibit more complicated distributions of χ reflecting the substructures associated with cantori [9], [22].

FIGURES 3(a) and 3(b) show analogous plots for the data analyzed in FIGURE 1(b). It can be seen from FIGURE 3(a) that the distribution of complexities, $N[n]$, has peaks at both low and high values of n, indicative of two distinct populations. There is also

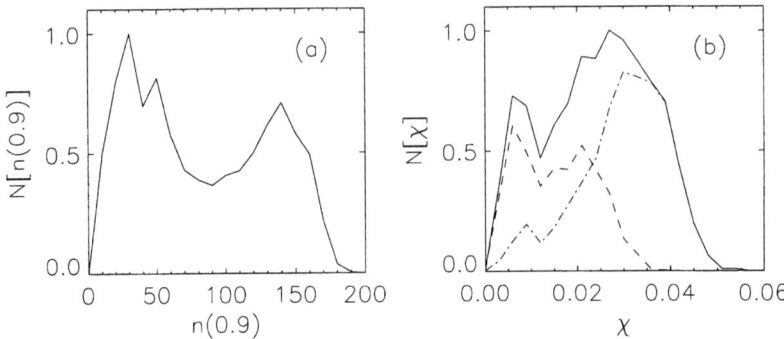

FIGURE 3. (a) $N[n(0.9)]$ for the segments plotted in FIGURE 1b. (b) The solid curve gives $N[\chi]$ for the same segments. The dashed and dot-dashed curves give $N[\chi]$ for segments with $n < 80$ and $n > 80$.

a less unambiguous, but still statistically significant, indication that there are really two lower-n peaks. The existence of at least two peaks is also apparent from FIGURE 3(b) but, as in FIGURE 2, the low and high n populations are less distinct. It is thus natural once again to divide the segments into low and high n populations, and to compute separate distributions $N[\chi]$ for each population. The results of such a calculation are indicated by the dashed and dot-dashed curves in FIGURE 3(b) which exhibit, respectively, distributions for segments with $n(0.9) \leq 80$ and $n(0.9) > 80$. The distribution $N[\chi]$ appropriate for the high n orbits is again roughly Gaussian, albeit with an extended low χ tail, whereas the $N[\chi]$ for the lower n orbits is a bimodal distribution which can be well approximated as a sum of two other Gaussians. This reinforces the idea that there are really two low χ, low n populations.

Similar conclusions also obtain for three-dimensional Hamiltonian systems, including, for example, three-dimensional generalizations of the D-4 potential like

$$V(x, y, z) = -(x^2 + y^2 + z^2) + \frac{1}{4}(x^2 + y^2 + z^2)^2 - \frac{1}{4}(x^2y^2 + ay^2z^2 + bz^2x^2), \quad (7)$$

with a and b constants. When $a = b = 1$, this potential is symmetric in all three directions. Other choices break the symmetry. Suppose, for example, that $a = b = 1$ and consider again ensembles of segments with $E = 10$ integrated for a total time $t = 256$. One then finds that $\mathcal{R}[n(0.9), \chi] = 0.797$, $\mathcal{R}[n(0.95), \chi] = 0.792$, and $\mathcal{R}[n(0.9), n(0.95)] = 0.990$. However, as is evident from FIGURE 4, despite this strong correlation the shapes of the distributions $N[n]$ and $N[\chi]$ exhibit appreciable differences. The form of the distribution of complexities $N[n(0.9)]$ suggests strongly the existence of reasonably distinct low and high complexity populations, whereas the distribution of short-time exponent $N[\chi]$ is less conclusive.

For some different choices of a and b, the relative importance of a low χ, low n population is significantly reduced. This is illustrated in FIGURES 5(a) and 5(b), which again exhibit $N[n(0.9)]$ and $N[\chi]$ for segments with $E = 10$ evolved for $t = 256$, but now with $a = 0$ and $b = 2$. Here both distributions seem characterized at least ap-

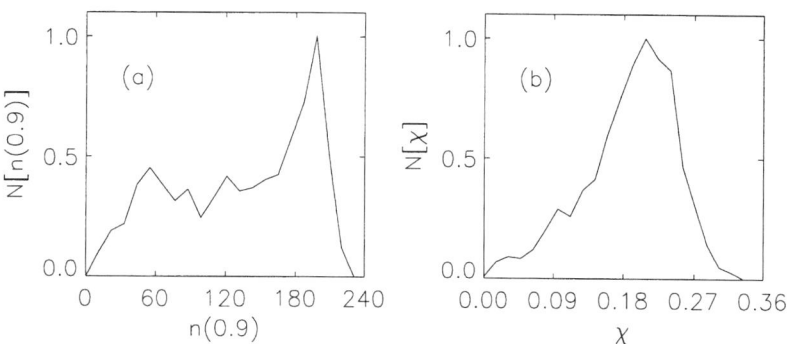

FIGURE 4. (a) $N[n(0.9)]$ for a collection of 3200 segments with energy $E = 10$ evolved in the generalized D-4 potential (7) with $a = b = 1$ for a total time $t = 256$. (b) The corresponding $N[\chi]$ for the same segments.

proximately by a singly peaked population which, modulo a low χ, low n tail, is reasonably well fit by a Gaussian. Alternatively, other choices of a and b result in significantly more complicated distributions. This is illustrated in FIGURES 5(c) and 5(d), which exhibit analogous data generated with $a = 1$ and $b = 3/2$. In this case, both distributions demonstrate the existence of three distinct populations, one with low values of χ and n, and two with significantly higher values.

SELF-CONSISTENT GRAVITATIONAL CHAOS

There is growing observational evidence that many galaxies may be genuinely triaxial, i.e., neither spherical nor axisymmetric, and that they may contain high density central cusps, the density increasing seemingly without bound down to scales as small as a few lightyears (see Ref. [23] and citations therein). For this reason, there is an obvious need to construct models of cuspy triaxial galaxies, which is typically done numerically using Schwarzschild's method. Unlike most spherical and axisymmetric models, especially those without cusps, such models seem to require large measures of chaotic orbits. Any attempt to understand the overall orbital structure of a cuspy triaxial model thus require that one quantify the "degree" of chaos exhibited by different chaotic segments, determining in particular what fraction of the segments are regular or trapped near regular regions and what fraction are more wildly chaotic.

As a specific example, consider orbits generated in the gravitational potential $V(x, y, z)$ associated via the Poisson equation,

$$\nabla^2 V(x, y, z) = 4\pi G \rho(x, y, z), \tag{8}$$

with a density distribution of the form [24]

$$\rho(m) = \rho_0 m^{-\gamma}(1 + m)^{-(4-\gamma)}, \quad m^2 = \frac{x^2}{a^2} + \frac{y^2}{b^2} + \frac{z^2}{c^2}. \tag{9}$$

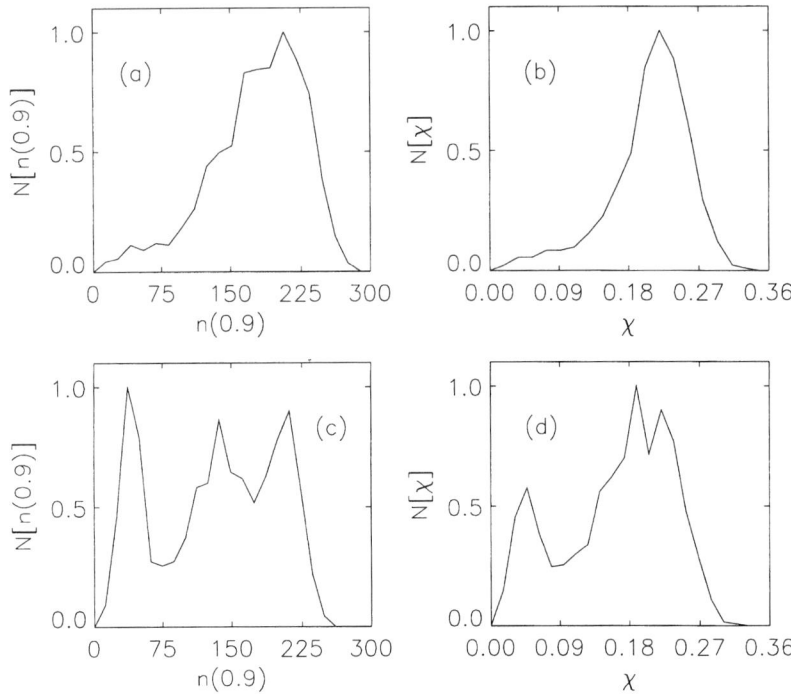

FIGURE 5. (a) $N[n(0.9)]$ for a collection of 3200 segments with $E = 10$ evolved for $t = 256$ in the potential (7) with $a = 0$ and $b = 2$. (b) $N[\chi]$ for the same segments. (c) $N[n(0.9)]$ for a collection of 3200 segments with $E = 10$ evolved for $t = 256$ in the potential (7) with a = 1 and $b = 3/2$. (d) $N[\chi]$ for the same segments.

Here the axis ratios $a:b:c$ satisfy $1: \sqrt{5/8} : 1/2$, and $-\gamma = 1$ or $-\gamma = 2$. As is often the case in galactic astronomy, where it is ρ rather than V which is motivated from observations, this potential cannot be written in a simple analytic form, so that the calculation of orbits and Lyapunov exponents is comparatively expensive computationally.

Model building with this potential has involved [24], [25] selecting 20 well-spaced energies and, for each energy, evolving an ensemble of about 500 representative initial conditions into the future for a total time $T = 200 t_{cr}$ where $t_{cr}(E)$ is an energy-dependent characteristic crossing time. Orbital data were sampled at intervals $\Delta t = 0.1 t_{cr}$, leading to a total of 2000 data points for each segment. These integrated ensembles can then be viewed either (a) as interesting entities in their own right and/or (b) as a library which can be sampled (using, e.g., a quadratic programming algorithm [21]) to generate a self-consistent galactic model. The restriction to such relatively small ensembles arises both because the integrations are expensive computationally and, more importantly, because it is expensive to apply programming algorithms to very large orbital ensembles. The choice of an integration time $T = 200 t_{cr}$ is motivated by the observation that real galaxies are comparatively young objects, only ~$100 t_{cr}$, in age. Given these ensembles, several natural questions arise:

(1) What fraction of the computed segments are wildly chaotic, as opposed to regular or nearly regular? In other words, how ubiquitous is chaos in such a potential?
(2) How much "variety" is observed in the wildly chaotic orbits? If one focuses on time intervals as short as ~$100 t_{cr}$, how much do different segments in the same connected phase space region differ from one another? Do they, e.g., come with a broad range of n and χ?
(3) To what extent do the integrated chaotic segments sample an invariant distribution? If a galactic model is to constitute a statistical equilibrium, its building blocks must be time-independent so that, for example, if the chaotic segments are evolved for longer times their statistical properties do not change.

As for the potentials described in Section II, one finds once again a strong, often nearly linear, correlation between the computed values of the short-time Lyapunov exponent χ and the complexity n. Plots of χ as a function of the configuration space complexity $n(0.9)$ are presented in FIGURES 6(a) and 6(b) for two representative cases with $\gamma = 2$, namely $E = -2.884$, the second lowest energy, and $E = -1.114$, the eighth lowest energy. Here, unlike the corresponding plots in Section II, one sees a dense accumulation of points at very low values of χ since the ensembles contain both regular and chaotic segments (in total, about 40% of the computed segments correspond to wildly chaotic orbits).

The degree to which, overall, n and χ are correlated can be quantified by computing rank correlations for each of the 20 energies in a given model and then averaging over the different energies to compute mean values and dispersions (individual energies have too few segments to yield statistically significant results). For a total integration time $T = 200 t_{cr}$, the $\gamma = 2$ model yields $\mathcal{R}[n(0.9), \chi] = 0.796 \pm 0.021$ and $\mathcal{R}[n(0.95), \chi] = 0.797 \pm 0.019$. These numbers were computed by including all the segments at each energy, both regular and chaotic. However, neglecting the dense accumulation of points at low χ and $n(k)$, changes things very little. In this case, $\mathcal{R}[n(0.9), \chi] = 0.786 \pm 0.071$ and $\mathcal{R}[n(0.95), \chi] = 0.786 \pm 0.069$.

The results also remain essentially unchanged if instead one defines complexity in terms of the orbit as viewed in momentum space or in the full six-dimensional phase space. Thus, for example, a momentum space analysis applied to all the segments yields $\mathcal{R}[n(0.9), \chi] = 0.789 \pm 0.021$ and $\mathcal{R}[n(0.95), \chi] = 0.788 \pm 0.021$, and an analysis of phase space complexities gives $\mathcal{R}[n(0.9), \chi] = 0.797 \pm 0.021$ and $\mathcal{R}[n(0.95), \chi] = 0.795 \pm 0.020$. It is also of interest to determine the effect of rotating the coordinate system so that the principal planes of the mass distribution are not all aligned with the x, y, and z axes. Suppose, for example, that the x-y plane is rotated by $\pi/4$. In this case, the original and rotated configuration space complexities, n and n_R, satisfy $\mathcal{R}[n(0.9), n_R(0.9)] = 0.924 \pm 0.015$ and $\mathcal{R}[n(0.95), n_R(0.95)] = 0.924 \pm 0.014$. Rotations change things, but not all that much.

The strong correlation between n and χ suggests that, once again, the distributions $N[n]$ and $N[\chi]$ should be relatively similar. The degree to which this is true is illustrated in FIGURE 7, which exhibits $N[n]$ and $N[\chi]$ for the orbit segments analyzed in FIGURES 6(a) and 6(b). The highest peaks in these distributions correspond to a

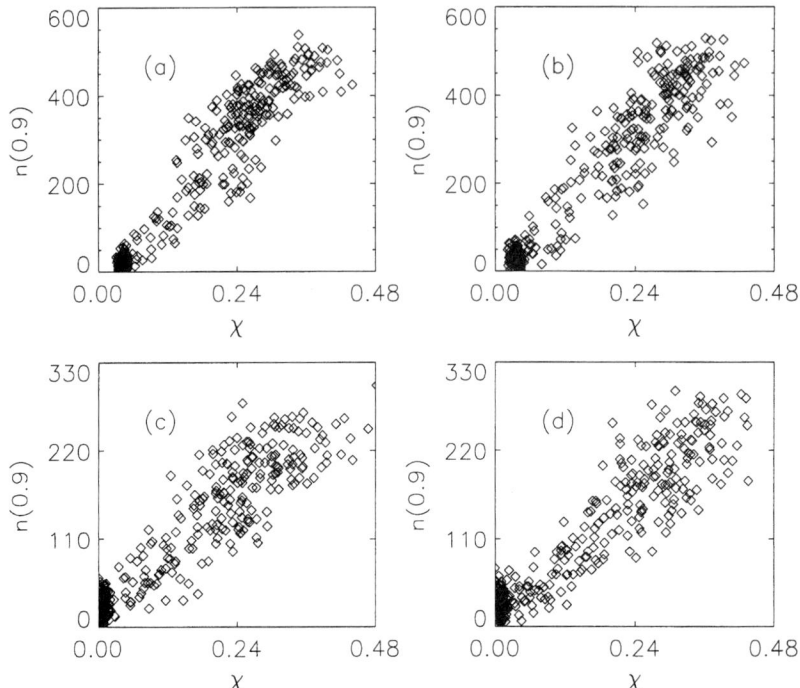

FIGURE 6. (a) The short-time Lyapunov exponent χ plotted as a function of $n(0.9)$ for 496 orbits with energy $E = -2.884$ evolved in the potential (9) with $\gamma = 2$ for a total time $T = 200 t_{\rm cr}$. (b) The same as (a) for a different energy $E = -1.114$. (c) χ plotted as a function of $n(0.9)$ for the same initial conditions as (a), but now analyzing only the interval $100 t_{\rm cr} < T < 200 t_{\rm cr}$. (d) The same as (c) for the initial conditions with $E = -1.114$.

large number of (relatively uninteresting) regular orbits. For this reason, the plots were so normalized that the peak in the distribution of wildly chaotic orbits has height $N = 1$ and the y-axis is truncated at a value $N = 2$ (the actual data extend up to $N \sim 4$). The data are extremely poor compared with those exhibited in Section II since one is dealing with a much smaller number of orbits. However, the spike in FIGURE 7(a) at $n \sim 200$ is statistically significant, leading once again to the interpretation that the chaotic population includes a subpopulation for which n and (presumably) χ are especially small.

Similar results also obtain if one focuses separately on the early and late portions of the integration, namely $0 < T < 100 t_{\rm cr}$ and $100 t_{\rm cr} < T < 200 t_{\rm cr}$, identifying a short-time exponent for the latter interval via the obvious formula [22]

$$\chi(t_1 < t < t_2) = \frac{1}{(t_2 - t_1)}[(t_2 - t_0)\chi(t_0 < t < t_2) - (t_1 - t_0)\chi(t_0 < t < t_1)]. \qquad (10)$$

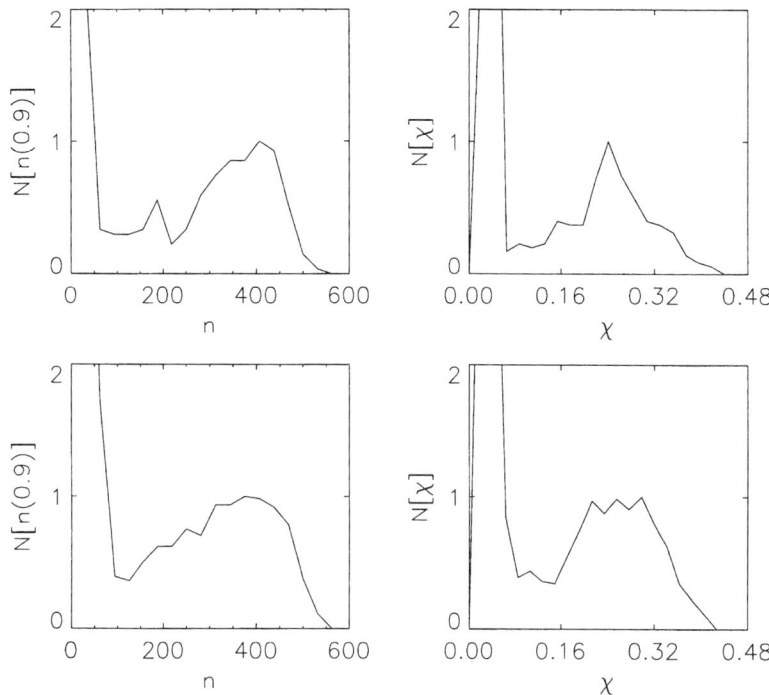

FIGURE 7. (a) $N[n(0.9)]$ for the segments plotted in FIGURE 5a. (b) $N[\chi]$ for the same segments. (c) $N[n(0.9)]$ for the segments plotted in FIGURE 5b. (d) $N[\chi]$ for the same segments.

Analysis of the 20 ensembles with $\gamma = 2$ for $0 < T < 100 t_{cr}$ yields $\mathcal{R}[n(0.9), \chi] = 0.774 \pm 0.026$ and $\mathcal{R}[n(0.9), \chi] = 0.773 \pm 0.025$, whereas the interval $100 t_{cr} < T < 200 t_{cr}$ gives $\mathcal{R}[n(0.9), \chi] = 0.802 \pm 0.026$ and $\mathcal{R}[n(0.9), \chi] = 0.804 \pm 0.025$. FIGURES 6(c) and 6(d) exhibit representative points generated for $100 t_{cr} < T < 200 t_{cr}$ for the same initial conditions used to generate FIGURES 6(a) and 6(b).

Considering separately the two halves of the integration also provides a useful check on whether the chaotic initial conditions have succeeded in sampling an invariant or nearinvariant, distribution. If both halves sample an invariant distribution, the distributions of complexities for the intervals $0 < T < 100 t_{cr}$ and $100 t_{cr} < T < 200 t_{cr}$ must be identical modulo statistical uncertainties. FIGURE 8(a) overlays the distributions $N[n(0.9)]$ generated for the two halves of the integration with the initial conditions exhibited in FIGURE 6(a). Here the raw data in 20 separate bins are too sparse to yield a statistically significant comparison, so that they have been smoothed by a box-car averaging over adjacent cells. With this averaging, one concludes that, to within statistical uncertainties, the early and late time distributions do in fact coincide, although there are hints of systematic differences for orbits with $n \sim 80\text{--}120$.

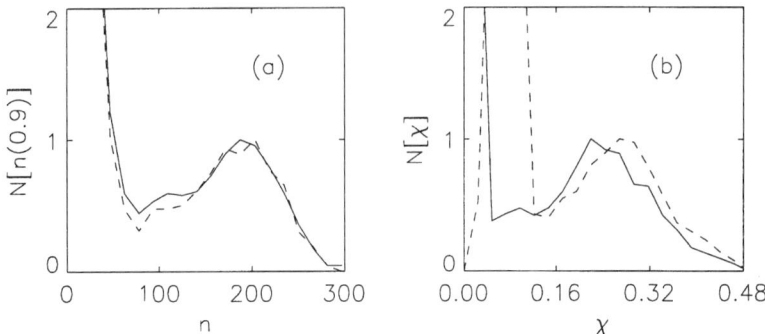

FIGURE 8. (a) The distributions $N[n(0.9)]$ for the initial conditions plotted in FIGURE 5a, analyzing separately the intervals $0 < T < 100 t_{cr}$ (dashed line) and $100 t_{cr} < T < 200 t_{cr}$ (solid line). (b) The corresponding distributions $N[\chi]$.

Short-time Lyapunov exponents computed by tracking the evolution of a generic δZ cannot be used to reach a similar conclusion. The computed short-time exponents for regular segments decrease systematically in time as they eventually asymptote toward the late time value $\chi = 0$; and, even for chaotic segments it takes a finite time for the computed χ to find the most unstable direction. Only after a time $T \sim 30-40 t_{cr}$ does the computed $N[\chi]$ yield an unambiguous distinction between regular/nearly regular and wildly chaotic segments, and, as is evident from FIGURE 8(b), even for the manifestly chaotic segments the distributions $N[\chi]$ for $0 < t < 100 t_{cr}$ and $100 t_{cr} < t < 200 t_{cr}$ exhibit systematic differences. By contrast, using complexities a time as short as $\sim 25-50 t_{cr}$ suffices to make a relatively clean distinction between wildly chaotic and regular/nearly regular orbits.

EXTRINSIC DIFFUSION THROUGH CANTORI

For many energies, the chaotic sea in the potentials (4)–(6), and many other two-dimensional potentials, is connected in the sense that a single chaotic orbit can, and presumably will eventually, pass arbitrarily close to every point in the sea. However, cantori [16], [17] can trap a chaotic orbit near a regular island for comparatively long intervals, $100 t_{cr}$ or more, during which time the chaotic orbit behaves very much like a regular orbit. This phenomenon allows one to distinguish between confined chaotic segments, that are stuck near a regular island, and unconfined chaotic segments. However, even relatively small non-Hamiltonian perturbations, modeled as low amplitude friction and noise, can facilitate extrinsic diffusion through cantori, allowing confined segments to become unconfined (or unconfined segments to become confined) on a time scale $\ll 100 t_{cr}$ [12]–[14].

There is an obvious need to quantify this phenomenon, both for individual initial conditions and for ensembles of initial conditions. For a single initial condition corresponding to an unperturbed confined segment, this can be done by (1) computing multiple noisy integrations of the same initial condition, generating thereby large

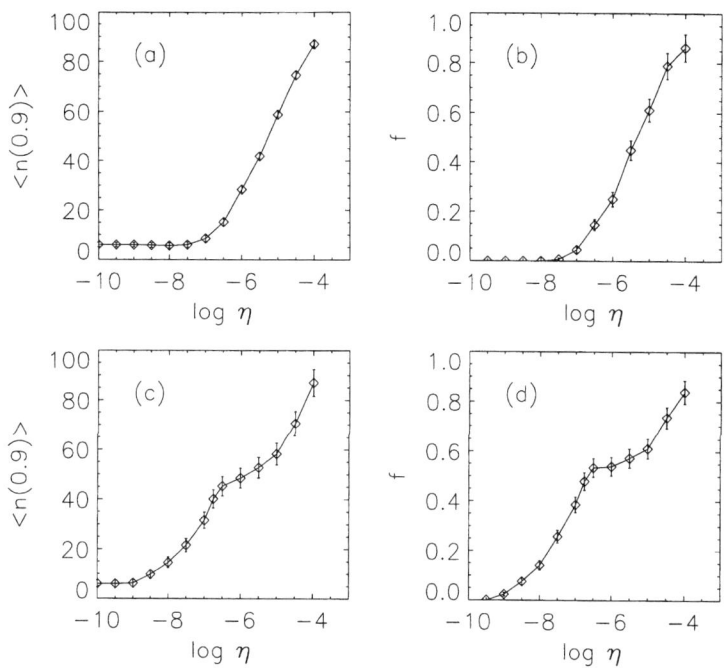

FIGURE 9. (a) The mean complexity, $\langle n(0.9)\rangle$, plotted as a function of η, each point representing 1024 noisy integrations of the same initial condition which, in the absence of friction and noise, would correspond to a confined chaotic segment with $E = 10$ in the D-4 potential. The total integration time was $T = 256$. (b) The fraction f of the orbits in (a) which escaped to become unconfined. (c) and (d) The same as (a) and (b) for a different initial condition.

collections of noisy chaotic segments, and then (2) studying, as a function of the amplitude of the perturbations, the fraction of the noisy segments that escape to become unconfined within a given time interval.

Conceptually, the most straightforward way to determine whether some chaotic segment has become unconfined is to define a masked region in configuration space, corresponding to the region where the chaotic segment is confined, and to determine whether, in any given time interval, the segment has left this region. The construction of such a mask is comparatively easy when the confined region has a relatively simple geometry (even though determining the precise location of the last important cantorus can be tedious). However, identifying a suitable mask becomes more complicated if the shape of the confined region is highly irregular (or, especially, if one tries to address similar issues in three- or higher-dimensional systems, where things become much harder to visualize). One might therefore hope to use complexities as probes of phase space transport, the obvious point being that chaotic segments that have become unconfined should be more complex than orbits that remain confined near regular regions.

That complexities can be used in this way is well illustrated by a study of perturbations of unperturbed orbits with $E = 10$ evolved in the D-4 potential for a total time $T = 256 \sim 100 t_{cr}$. The initial conditions selected all corresponded to orbits which, in the absence of any perturbations, remained stuck near a regular island for a time $T > 500$. The equations to be solved took the form

$$\frac{dr^a}{dt} = p_a \quad \text{and} \quad \frac{dp_a}{dt} = -\frac{\partial V}{\partial r^a} - \eta p_a + F_a, \quad (a = x, y) \quad (11)$$

with $V(x, y)$ the D-4 potential, η a constant coefficient of dynamical friction, and F_a Gaussian white noise with zero mean. The force F_a was treated as a random variable, characterized completely by its first two moments, which were taken to be

$$\langle F_a(t) \rangle \equiv 0 \quad \text{and} \quad \langle F_a(t_1) F_b(t_2) \rangle = 2\eta \Theta \delta_{ab} \delta_D(t_1 - t_2). \quad (12)$$

Here δ_{ab} and δ_D denote, respectively, Kronecker and Dirac deltas, and $\langle \ \rangle$ denotes an ensemble average. The second moment condition implies that the force is delta-correlated in both space and time and related to η by a Fluctuation-Dissipation Theorem [26] in terms of some temperature Θ. The integrations were effected using an algorithm described in Ref. [27].

Initial conditions were so chosen that the unperturbed segments correspond to confined chaotic orbits which are trapped in configuration space in a spherical annulus and look nearly indistinguishable from regular loop orbits. Identifying the exact locations of the cantori which confine the segment to this annulus is very difficult but, fortunately, not necessary: Once the initially confined orbit has breached the last important cantorus it will typically move far away from the confined regions very quickly. A sensible choice of masked region was made on the basis of visual inspection, and it was verified that allowing for small changes in the inner and outer radii of the masked annulus had only minimal effects on the quantitative conclusions.

Different ensembles of perturbed orbits were generated by specifying an initial condition, fixing the temperature at a value $\Theta \sim E$, and, for various values of η, performing 1024 different perturbed integrations. Each integration involved random forces generated using different seeds.

Given an appropriate mask, it is straightforward to compute, for fixed Θ and η, the fraction f of the perturbed chaotic segments that become unconfined within a time $T = 256$. To establish a correlation between this f and the complexity of the perturbed orbits, the quantity $n(0.9)$ was computed for each of the 1024 noisy segments and the resulting complexities were combined to obtain the mean complexity $\langle n(0.9) \rangle$ for the ensemble. It was found that there is a strong correlation between f and $\langle n(0.9) \rangle$.

For $\eta < 10^{-9}$, friction and noise have only minimal effects on confined chaotic orbits over times as short as $T = 256$. Fewer than 1% of the perturbed orbits become unconfined, and the complexities of the perturbed orbits are virtually identical to the complexity n_u of the unperturbed orbit, so that $\langle n \rangle \approx n_u$. However, for larger values of η, one begins to see appreciable numbers of transitions from confined to unconfined chaos and a corresponding increase in the mean complexity $\langle n \rangle$.

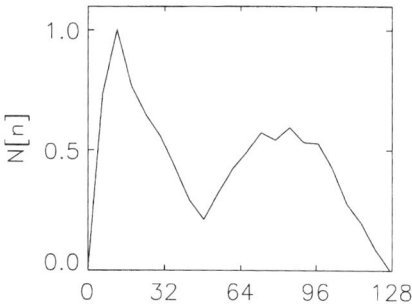

FIGURE 10. $N[n(0.9)]$ for an ensemble of 512 noisy segments with $\eta = 10^{-6}$ and $k_B T = 6$, integrated with the initial condition used to generate FIGURE 5.

FIGURE 9 exhibits data generated from two different representative initial conditions with $E = 10$ and $\Theta = 6$. The two left panels each exhibit the mean complexity $\langle n(0.9) \rangle$ for the 1024 segment ensemble; the right panels plot the fraction f of the segments that have become unconfined by the end of the integration. For the first initial condition, one observes a reasonably smooth growth for $\langle n \rangle$ and f, the first appreciable increases in both quantities arising for $\eta \sim 10^{-7}$. For the other, one observes a more complex structure. Both $\langle n \rangle$ and f exhibit a smooth initial increase which begins at a substantially lower $\eta \sim 10^{-8.5}$–10^{-9}, an apparent plateau near $\eta = 10^{-6.5}$, and another period of relatively rapid increase at values $\eta > 10^{-6}$. It is clear that the perturbing friction and noise affect the two initial conditions somewhat differently, and that these differences are manifested in both $\langle n \rangle$ and f.

It is not hard to understand what is happening. For sufficiently weak perturbations, $\eta < 10^{-6}$ or so, the perturbed orbits have nearly the same energy as the unperturbed orbits, so that one is considering extrinsic diffusion on a nearly constant energy hypersurface. In this setting, it is apparent that the first initial condition in FIGURE 9 is "more stuck" to the regular region than the second initial condition, so that it typically requires a larger perturbation to become unconfined. (In the absence of friction and noise, it remains trapped almost twice as long!) However, for $\eta > 10^{-6}$ or so, the energies of the perturbed segments begin to exhibit significant changes, so that one can no longer visualize the flow as entailing diffusion on a very nearly constant energy hypersurface. In this case, the detailed choice of initial condition become less important, so that the computed values of f and $\langle n \rangle$ for these two different (and other) initial conditions are the same within statistical uncertainties.

That some segments have become unconfined but others have not, would suggest that the distribution of complexities, $N[n(0.9)]$, should be bimodal. However, one would anticipate that there should not be a completely sharp break between the low- and high-n segments since, in at least some cases, a segment will become unconfined only very near the end of the integration, so that its overall complexity remains quite small. That this is in fact the case is illustrated in FIGURE 10, which exhibits the distribution of complexities for an ensemble of 2048 orbits with $\eta = 10^{-6}$ for the first initial condition in FIGURE 9.

CONCLUSIONS

The work described here demonstrates that there is a precise, quantitative sense in which chaotic orbit segments with an especially large short-time Lyapunov exponent χ tend to have a more complex Fourier spectrum, with substantial power at a larger number of frequencies, than do chaotic segments for the same energy in the same potential for which χ is smaller. This involved introducing the quantity $n(k)$, the orbital complexity at threshold k, defined in terms of the minimum number of frequencies in the Fourier spectrum of the segment required to capture a fixed fraction k of the total power. Individual orbit segments in an ensemble of fixed energy E typically exhibit a strong, often roughly linear correlation between the computed values of χ and $n(k)$. Moreover, if the computed values of χ and n are binned to generate distributions of short-time Lyapunov exponents and complexities, the resulting $N[\chi]$ and $N[n]$ tend to be similar in shape, although significant differences can exist which make one distribution easier to interpret than the other.

Two concrete examples were also used to demonstrate that the orbital complexity $n(k)$ can be used as a diagnostic for characterizing orbital segments and/or tracking phase space transport in much the same way as various authors [7]–[9] have used short-time Lyapunov exponents. Indeed, there are several reasons why, in certain cases, orbital complexities may be preferable to short-time Lyapunov exponents in providing such characterizations.

(1) Typically it is both easier and less expensive computationally to compute the complexity of an orbit segment than the largest short-time Lyapunov exponent. Computing $n(k)$ requires little more than evaluating a fast Fourier transform, which is cheap and easy, whereas computing a short-time χ requires two integrations, namely the evolution of the unperturbed initial condition and the evolution of some initial perturbation. If one estimates the largest χ by introducing an arbitrary initial $\delta Z(0)$, one must integrate long enough to converge toward the most unstable direction. This difficulty can be avoided by working with the full tangent dynamics, implementing a computational scheme which at each instant searches explicitly for the most unstable direction, but this leads to a more complicated, and expensive, computation.

(2) In some cases, the distribution of complexities, $N[n(k)]$, can prove a more discerning diagnostic than the corresponding distribution of short-time Lyapunov exponents, $N[\chi]$. Indeed, when considering an ensemble of chaotic segments that contains both confined and unconfined orbits, one discovers oftentimes (1) that an examination of $N[n]$ permits a more clean cut distinction between different types of chaotic segments and/or (2) that considering both $N[n]$ and $N[\chi]$ can allow one to decipher effects which would not be completely evident from a consideration of only one of these two distributions.

(3) Orbital complexities $n(k)$ can be computed, and interpreted, trivially in settings where the meaning of short-time Lyapunov exponents is not completely clear, e.g., when considering a Hamiltonian flow perturbed by weak friction and noise. For example, $n(k)$ can provide a useful diagnostic in determining the efficiency of extrinsic diffusion as a mechanism in allowing a chaotic segment originally confined near some regular island to diffuse through one or more cantori so as to move unimpeded throughout a larger stochastic sea.

Finally, it should be noted that complexities are not the only way in which Fourier transforms have been used to distinguish between regular and chaotic orbits or between different "degrees" of chaos. If one is interested simply in deciding whether a given orbit segment is regular or chaotic, one can use the frequency map techniques developed by Laskar and his collaborators (see, e.g., Ref. [28]), which involves implementing a Hanning filter [21] to generate highly accurate (five or more significant figures) estimates of the principal frequencies of an orbit. The important point, then, is that, for regular orbits, the computed ω will fit along sequences in frequency space, whereas the principal frequencies for chaotic orbits will be distributed more erratically. Moreover, if the principal frequencies are computed for successive segments of even a very nearly regular, confined chaotic orbit, one finds that they will "drift" through frequency space in a way that true regular orbits do not.

This latter fact suggests in turn another way to distinguish between different classes of chaotic orbits, namely by estimating the rate at which the principal frequencies change. Recent work by Valluri and Merritt [29] seems to indicate that nearly regular, confined chaotic orbits will diffuse through frequency space much more slowly than will wildly irregular, unconfined chaotic orbits. However, this would suggest that there might exist a strong correlation between the rate at which chaotic segments diffuse through frequency space and their overall complexity.

ACKNOWLEDGMENTS

Work on this manuscript was completed while H.E. Kandrup was a visitor at the Aspen Center for Physics, the hospitality of which is acknowledged gratefully.

REFERENCES

1. LICHTENBERG, A.J. & M.A. LIEBERMAN. 1992. Regular and Chaotic Dynamics. Springer-Verlag. Berlin.
2. TABOR, M. 1989. Chaos and Nonintegrability in Nonlineax Dynamics. Wiley. New York.
3. KANDRUP, H.E. & B.O. BRADLEY. 1995. University of Florida preprint.
4. CONTOPOULOS, G. 1971. Astron. J. **76**: 147.
5. SHIRTS, R.S. & W.P. REINHART. 1982. J. Chem. Phys. **77**: 5204.
6. GRASSBERGER, P., R. BADII & A. POLITI. 1988. J. Stat. Phys. **51**: 135.
7. VAROSI, F., T.M. ANTONSEN & E. OTT. 1991. Phys. Fluids A **3**: 1017.
8. FINN, J. M., J.D. HANSON, I. KAN & E. OTT. 1991. Phys. Fluids B **3**: 1250.
9. MAHON, M.E., R.A. ABERNATHY, B.O. BRADLEY, & H.E. KANDRUP. 1995. Mon. Not. R. Astr. Soc. **275**: 443.
10. BINNEY, J. & S. TREMAINE. 1987. Galactic Dynamics. Princeton University Press. Princeton.
11. SCHWARZSCHILD, M. 1979. Astrophys. J. **232**: 236.
12. HABIB, S., H.E. KANDRUP, & M.E. MAHON. 1996. Phys. Rev. E **53**: 5473.
13. HABIB, S., H.E. KANDRUP, & M.E. MAHON. 1997. Astrophys. J. **480**: 155.
14. KANDRUP, H.E. 1998. Ann. N. Y. Acad. Sci. Vol. 867. This issue.
15. LIEBERMAN, M. A. & A. J. LICHTENBERG. 1972. Phys. Rev. A **5**: 1852.
16. AUBRY, S. & G. ANDRE. 1978. In Solitons and Condensed Matter Physics. A. J. Bishop and T. R. Schneider, Eds.: 264. Springer-Verlag, Berlin.
17. MATHER, J. 1982. Topology 21: 457.

18. WOLF, A., J. SWIFT, H. SWINNEY & J. VASTANO. 1985. Physica D **16**: 285.
19. ARMBRUSTER, D., J. GUCKENHEIMER & S. KIM. 1989. Phys. Lett. A **140**: 416.
20. TODA, M. 1967. J. Phys. Soc. Japan **32**: 431.
21. PRESS, W.H., B.P. FLANNERY, S.A. TEUKOLSKY & W.T. VETTERLING. Numerical Recipes in C. 2nd edit. Cambridge University Press. Cambridge.
22. KANDRUP, H.E. & M.E. MAHON. 1994. Astron. Astrophys. **290**: 762.
23. LAUER, T.R., E.A. AHJAR, Y.-L. BYUN, A. DRESSLER, S.M. FABER, C. GRILLMAIR, J. KORMENDY, D. RICHSTONE & S. TREMAINE. 1995. Astron. J. **110**: 2622.
24. MERRITT, D. & T. FRIDMAN. 1996. Astrophys. J. **460**: 136.
25. SIOPIS, C. 1998. University of Florida Ph. D. Dissertation.
26. VAN KAMPEN, N.G. 1981. Stochastic Processes in Physics and Chemistry. North Holland, Amsterdam.
27. GRINER, A., W. STRITTMEYER & J. HONERKAMP. 1988. J. Stat. Phys. **51**: 95.
28. PAPAPHILIPPOU, Y. & J. LASKAR. 1998. Astron. Astrophys. 329: 451.
29. VALLURI, M. & D. MERRITT. 1998. Astrophys. J. **506**: 686.

Bifurcations of Periodic Orbits in Axisymmetric Scalefree Potentials[a]

C. HUNTER,[b] BALŠA TERZIĆ,[b] AMY M. BURNS,[c] DONALD PORCHIA,[b] AND CHRIS ZINK[b]

[b]*Department of Mathematics, Florida State University, Tallahassee, Florida 32306-4510*
[c]*Department of Engineering Physics, Embry Riddle Aeronautical University, Daytona Beach, Florida 32114*

ABSTRACT: We study orbits in potentials with central cusps, emphasizing the spheriodal equidensity (SED) potentials generated by mass distributions with spheroidal equidensity surfaces. The most prominent bifurcations are those related to 1:1 and 4:3 resonances between radial motions and motions perpendicular to the central plane. We find that 1:1 resonances can cause the thin tube orbit, as well as the equatorial plane orbit, to become unstable. We concentrate on period-tripling bifurcations because they appear to be the least understood. We study them via a class of analytic maps. This study suggests that stable period-three orbits generally arise *de novo* in stable and unstable pairs via a turning-point bifurcation, and not through a bifurcation from the thin tube at a 120° rotation angle. The stable period-three orbits typically have only a short span of existence before becoming unstable to a period-doubling instability through a supercritical pitchfork bifurcation.

1. INTRODUCTION

An understanding of orbits is essential to an understanding of the dynamical structure of elliptical galaxies. Elliptical galaxies are believed to be composed primarily of stars moving collisionlessly, because two-body encounters are rare, in their own gravitational fields. The orbits of these stars therefore are the building blocks from which a dynamical model for a galaxy is constructed.

This paper is concerned with orbits in a special class of such gravitational fields, those arising from scalefree potentials for which the potential varies as some power $-\beta$ of distance from the center. The dynamics is then simplified because the range of orbits that is possible at any one energy is geometrically similar to the range that is possible at any other energy [1]. This feature, which obviously simplifies the task of surveying the full range of orbits, has been one reason for studying dynamics in scalefree potentials. Another reason is that scalefree potentials allow naturally for the presence of a central density cusp. Hubble Space Telescope (HST) observations have shown that cusps are a common feature of early-type elliptical galaxies, and that some appear even to have nearly pure power-law profiles [2].

A wide variety of potentials remain even with the simplifying restriction to scalefree. That is because the scalefree property places no restriction on the manner in

---

[a]This work has been supported in part by National Science Foundation Grants DMS-9304012 and 9704615, and by the Florida Space Grant Consortium.

which the potential varies on a spherical surface centered on the cusp; it requires only that this variation be the same at all radii. We restrict this variation very considerably in focusing on SEP (spheroidal equipotential) and SED (spheroidal equidensity) potentials. SEP and SED are mnemonics for spheroidal equipotential and spheroidal equidensity. Because equipotentials are always rounder than the equidensity surfaces of the underlying mass distribution that gives rise to the potential, that mass distribution can become quite extreme with spheroidal equipotentials[1]. Specifically regions of negative density arise on the axis of symmetry if

$$q^2 < \frac{1}{2}(1+\beta),\tag{1}$$

and in the equatorial plane if

$$\frac{1}{q^2} < \beta.\tag{2}$$

Here q is the ratio of the axis of the equipotential spheroid along the axis of symmetry to that perpendicular to it. These spheroids are oblate if $q < 1$, prolate if $q > 1$, and spherical if $q = 1$. Inequalities (1) and (2) allow a range of possibilities with everywhere positive densities for $\beta < 1$, that is, for potentials less singular than that of a point mass. Evans's [1] FIGURE 1(a) displays a more restricted region of parameter space. That is, because he has imposed the extra requirement of a positive phase-space density of the unique two-integral distribution function. Earlier surveys, as well as the present one, show considerable evidence for the existence of a third integral of motion on which the distribution function can, and probably does, depend. Hence the region of parameter space for which there are distribution functions with everywhere positive phase-space densities is likely larger.

By construction, SED potentials arise from everywhere positive mass densities. As FIGURE 6 of Qian et al. [3] shows, the extra requirement of positive phase-space density of the two-integral distribution function restricts their parameter space too. Our survey of these models shows that they too generally have third integrals and hence a much greater variety of possible distribution functions. We carry out our survey for each family by selecting values of the parameters β and q, while choosing the representative value of the energy to be that for which the maximum magnitude of the angular momentum L_z about the axis of symmetry is $1/\sqrt{e} = .6065$. This choice gives the same range of L_z as in Richstone's pioneering investigation of orbits in the logarithmic SEP [4]. We then study the dynamics of the orbits at lesser values of L_z by constructing surfaces of section. Our main interest here is to determine the types of orbits that occur, the constraints to which they are subject in phase space, and the formation of new classes of orbits that arise through bifurcations.

Section 2 gives basic definitions and equations. Section 3 describes some of the more significant findings of our survey. The bifurcations that are least understood, though among the most prominent, are period-tripling ones. Section 4 describes a class of maps constructed specifically to model period-tripling bifurcations, and which are helpful in understanding their occurrence in this and other work. Section 5 summarizes our results and draws conclusions.

2. SCALEFREE AXISYMMETRIC POTENTIALS

2.1. Potentials

We follow Binney and Tremaine [5] in using $r = \sqrt{x^2 + y^2 + z^2}$ to denote distance in spherical coordinates, and $R = \sqrt{x^2 + y^2}$ to denote distance from the z-axis of symmetry. Any axisymmetric potential of the form

$$\Phi = \frac{F(\theta)}{r^\beta}, \tag{3}$$

where θ is a colatitude angle measured from the axis of symmetry and F is an arbitrary function, is scalefree. We restrict attention to potentials with reflective symmetry with respect to the central $z = 0$ plane so that

$$\Phi(R, -z) = \Phi(R, z), \tag{4}$$

and hence the function $F(\theta)$ of Eq. (3) has the reflective symmetry

$$F(\theta) = F(\pi - \theta), \tag{5}$$

about $\theta = \pi/2$. We normalize the spheroidal equipotential (SEP) to the form

$$\Phi(R, z) = \frac{-1}{\beta[R^2 + z^2/q^2]^{\beta/2}} = -\frac{1}{\beta r^\beta}\left[\sin^2\theta + \frac{\cos^2\theta}{q^2}\right]^{-\beta/2}, \tag{6}$$

The logarithmic SEP potential

$$\Phi(R, z) = \frac{1}{2}\ln\left[R^2 + \frac{z^2}{q^2}\right], \tag{7}$$

studied by Richstone [4] is obtained as the $\beta \to 0$ limit of $(\Phi + \beta^{-1})$ for the Φ of (6), the added constant, which has no effect on the derivatives of Φ which give the forces, is needed to obtain a finite limit as $\beta \to 0$. The more general $\beta \neq 0$ cases of (6) were studied by Evans [1].

The SED scalefree potentials that arise from a mass distribution with spheroidal equidensity surfaces are defined implicitly by Poisson's equation

$$\nabla^2\Phi = \frac{4\pi G \rho_0}{[R^2 + z^2/q^2]^{1+\beta/2}}, \tag{8}$$

where ρ_0 is the density at $R = 1$, $z = 0$. The function $F(\theta)$ now has a more complicated form that involves elliptic integrals [6]. Because it is time consuming and computationally inefficient to have to evaluate these integrals at each step of an orbit, we first derive Fourier series representations, which are of the form

$$F(\theta) = \frac{1}{2}B_0 + \sum_{j=1}^{\infty} B_j \cos 2j\theta, \tag{9}$$

because of the symmetry (5). The B_j coefficients depend on β and q, and must be computed anew for each potential. That is a simple task because the integrals that define the coefficients integrate a periodic integrand over a complete period. Hence a simple repeated trapezoidal rule suffices [7]. Few Fourier coefficients are needed to achieve high accuracy, except as $q \to 0$, because $F(\theta)$ is analytic and the $B_j(q, \beta)$ consequently decay exponentially with j for large j [8]. To aid comparisons, we choose ρ_0 in (8) such that $F(\pi/2) = -1/\beta$. Then the values of both SED and SEP potentials are

$$\Phi(R, 0) = -\frac{1}{\beta B^\beta}, \tag{10}$$

in the central plane $z = 0$.

2.2. Dynamics of Orbits

The angular momentum

$$L_z = R^2 \dot{z}, \tag{11}$$

of any orbit about the z-axis of symmetry is conserved, because the potential is axisymmetric. The equations of motion for the orbit are

$$\ddot{R} = \frac{L_z^2}{R^3} - \frac{\partial \Phi}{\partial R}, \quad \ddot{z} = -\frac{\partial \Phi}{\partial z}, \tag{12}$$

and can be solved without explicitly considering motion in the longitudinal angle. The orbit has an energy integral

$$\frac{1}{2}\left(\dot{R}^2 + \dot{z}^2 + \frac{L_z^2}{R^2}\right) + \Phi(R, z) = E. \tag{13}$$

The orbit that lies always in the equatorial plane $z = 0$ is found by integrating the $z \equiv 0$ form of the energy integral:

$$\dot{R}^2 = 2E + \frac{2}{\beta R^\beta} - \frac{L_z^2}{R^2}. \tag{14}$$

This expression for $\dot{R}^2$ has a unique maximum at $R^{2-\beta} = L_z^2$. When this maximum is zero, that is, when

$$E = \left(\frac{1}{2} - \frac{1}{\beta}\right) L_z^{-2\beta/(2-\beta)},$$

then the only possible orbit is a circular orbit in $z = 0$. We use the freedom given by the scalefree nature of the potential to fix the energy at

$$E = \left(\frac{1}{2} - \frac{1}{\beta}\right) e^{\frac{\beta}{2-\beta}}, \tag{15}$$

so that all other orbits occur only in the range

$$0 \le L_z^2 < \frac{1}{e} = .3679. \tag{16}$$

2.3. Surfaces of Section

We study $z = 0$ surfaces of section of orbits, marking crossings in the $(R, \dot{R})$-plane at which the section is crossed with $\dot{z} > 0$. The value of $\dot{z}$ at a crossing is known to be

$$\dot{z} = \sqrt{\left(1 - \frac{2}{\beta}\right) e^{\frac{\beta}{2-\beta}} - \dot{R}^2 + \frac{2}{\beta R^\beta} - \frac{L_z^2}{R^2}} \tag{17}$$

from the energy integral (13). Each surface of section is finite and is bounded by the $\dot{z} = 0$ curve that is the locus (14) of the equatorial plane orbit. We explore the surface of section by integrating, for sufficiently long times, orbits that start from a selected range of feasible positions in the $(R, \dot{R})$-plane with the initial value (17).

3. ORBITAL SURVEYS AND BIFURCATIONS

Our survey has included both oblate and prolate SEP and SED potentials, which we study for decreasing values of L_z^2 in the range (16). The power β has largely been restricted to the range $-.5 \le \beta \le .5$, which includes both rising ($\beta < 0$) and falling ($\beta > 0$) rotation curves. All potentials in this β-range arise from density distributions that have an $R^{-1-\beta}$ cusp when seen in projection.

3.1. Orbits at High Angular Momenta

Surfaces of section can be understood analytically when L_z^2 is only slightly less than its maximum $1/e$. Orbits then depart little from circular. The equations of motion (12), linearized about an $R = R_0 = L_z^{2/(2-\beta)}$, $z = 0$, circular orbit are

$$\ddot{R} + \kappa^2(R - R_0) = 0,$$

$$\ddot{z} + \nu^2 z = 0.$$

Here the constants κ and ν, defined via the equations

$$\kappa^2 = \left.\frac{\partial^2 \Phi}{\partial R^2} + \frac{3}{R}\frac{\partial \Phi}{\partial R}\right|_{\substack{R = R_0 \\ z = 0}} = \frac{2-\beta}{R_0^{\beta+2}},$$

$$\nu^2 = \left.\frac{\partial^2 \Phi}{\partial z^2}\right|_{\substack{R = R_0 \\ z = 0}} = \frac{F''(\pi/2) - \beta F(\pi/2)}{R_0^{\beta+2}},$$

(18)

are, respectively, the epicyclic and perpendicular out-of-plane frequency for the potential in the equatorial plane. The generic orbit crosses the surface of section at

$$R - R_0 = \frac{C}{\kappa}\sin\left[\frac{2n\pi\kappa}{\nu} + \chi\right], \quad \dot{R} = C\cos\left[\frac{2n\pi\kappa}{\nu} + \chi\right],$$

(19)

for integers $n \geq 0$ and constants C and χ; that is, at a sequence of points that lie on an ellipse. The surface of section is bounded by a largest ellipse for which $C = [2E + ((1/\beta) - 1)R_0^{-\beta}]^{1/2}$ and there is a single point at its center at which the orbit whose perturbation from circular is solely in z-motion crosses repeatedly. If the ratio κ/ν is irrational, then the succession of crossing points generated by a single orbit does not repeat, and continues to fill in the outline of the ellipse. The crossing points do repeat when κ/ν is rational and the orbit is periodic.

FIGURE 1(a) shows a surface of section that is typical of relatively high angular momenta. It is qualitatively similar, indeed topologically equivalent, to that given by equations (19) for nearly circular orbits, even though $L_z^2 = 0.1$ here is well below the maximum of e^{-1}. The periodic orbit represented by the fixed point of the surface of section at $R = 0.54$, $\dot{R} = 0$ is a thin tube orbit. As FIGURE 1(b) shows, it is confined to a curve in (R, z)-space, and oscillates in R as well as z. The loci of crossings of individual orbits lose their elliptical shape away from the central fixed point, but they are still familially related to the thin tube orbit. They are now confined to fat tubes, with nonzero cross-sectional area in (R, z)-space, that encircle the z-axis of symmetry.

3.2. Bifurcations

3.2.1. Instability of the Equatorial Plane Orbit

The most prominent bifurcation seen in previous work on SEP potentials [1], [4], [9] is that due to a 1:1 resonance between the nonlinear oscillation in R of the equatorial plane orbit and a small oscillation in z. It causes that previously stable orbit to become unstable at the critical value of $L_z^2 = 0.0485$ for the SEP potential of FIGURE 2, and to be replaced by the pair of stable banana orbits. FIGURE 2(a) shows the surface of section at the lower value of $L_z^2 = 0.01$, and FIGURE 2(b) shows the banana

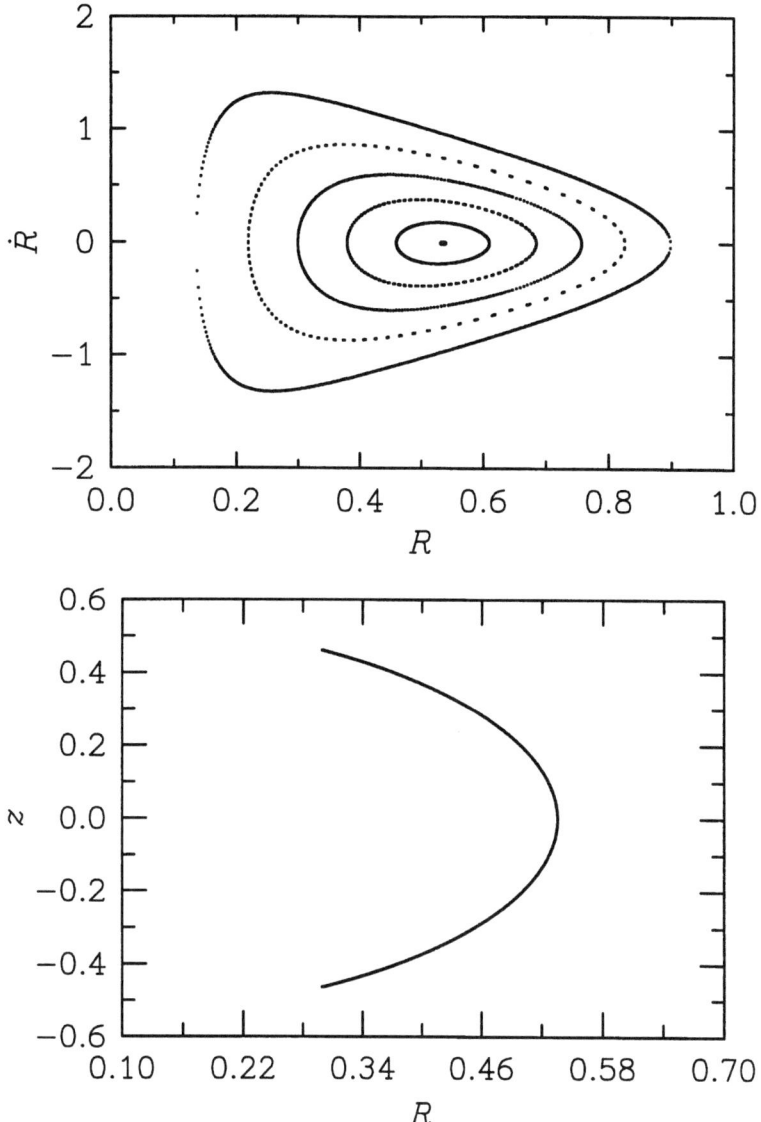

FIGURE 1. (a, top) The $(R, \dot{R})$ surface of section for an oblate SED potential ($q = 0.8$, $\beta = 0.3$) with $L_z^2 = 0.1$. **(b, bottom)** The path in (R, z) space of the thin tube orbit represented by the central point of the surface of section.

orbits for it. Though a $z = 0$ surface of section is not the easiest way to view the development of this instability, it is a simple case of a supercritical pitchfork bifurcation. It is more common in SEP than in SED potentials in our survey because of the stronger z-dependence of the former for the same values of the flattening parameter q.

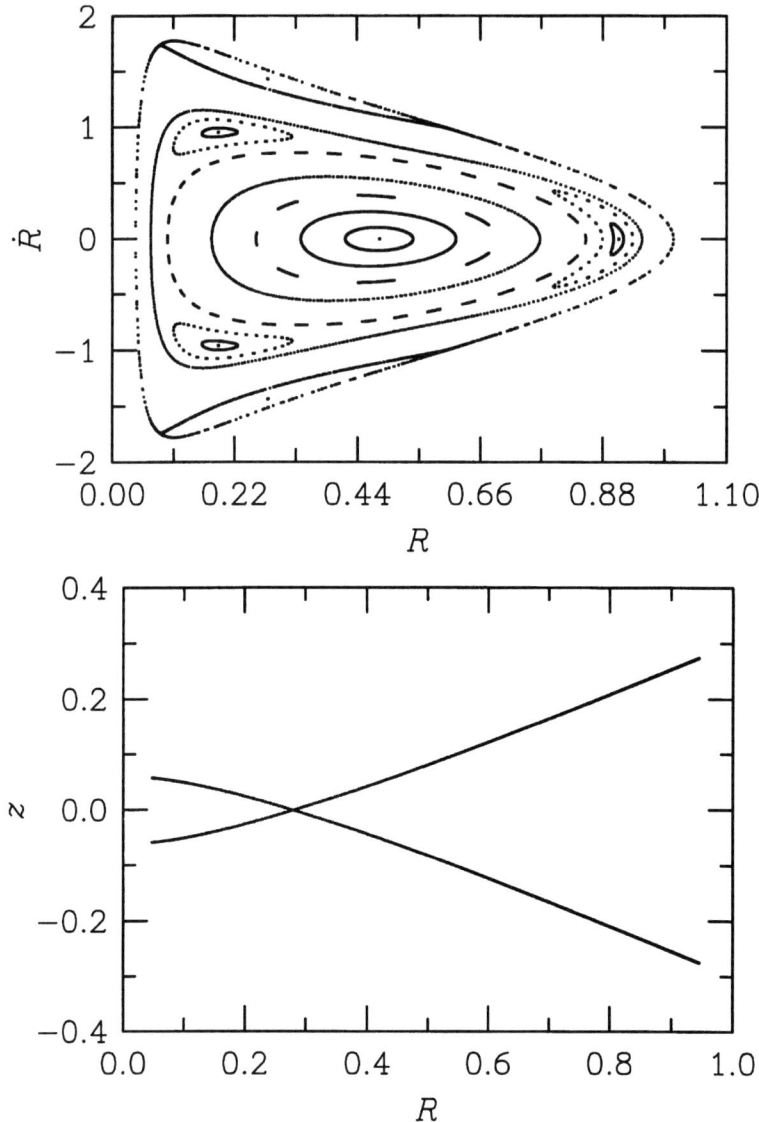

FIGURE 2. (a, top) The $(R, \dot{R})$ surface of section of an oblate SEP potential with $q = 0.8$, $\beta = -0.1$ at angular momentum $L_z^2 = 0.01$. It shows two bifurcations that have occurred; the equatorial plane orbit has become unstable, giving rise to the upper and lower islands, and a three-island chain has appeared. **(b, bottom)** The (R, z) paths of the pair of banana orbits that have bifurcated symmetrically from the equatorial plane orbit, and are represented by the central points of the upper and lower islands in FIGURE 2(a). **(c)** The (R, z) paths of the periodic 4:3 reflected fish orbit represented by the points at the centers of the three-island chain.

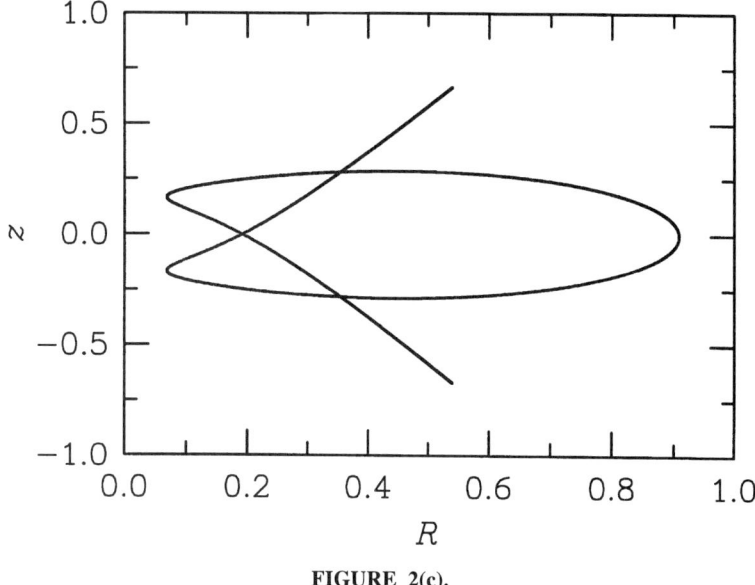

FIGURE 2(c).

3.2.2. Period-Tripling Bifurcations

FIGURE 2(a) shows the additional feature of a three-island chain. A new family of orbits has appeared because the three points at the center of the islands represent the successive crossings of the surface of section by the "reflected fish" orbit shown in FIGURE 2(c). This is a periodic orbit in which four oscillations in R are performed in the time that it takes to accomplish three oscillations in z. The other orbits of the family, represented by curves lying in the islands, fill a fatter region in $R - z$ space of the same characteristic shape.

It has been suggested [1], [9] that the three-island chains in a surface of section arise through a period-tripling bifurcation of the central thin tube orbit when there is a 4:3 resonance between motions in R and those in z. We expended considerable effort trying to make this connection. Sometimes we seemed to be successful, as in the case illustrated in FIGURE 3(a). Note that the periodic orbit corresponding to the centers of the islands shown in FIGURE 3(b) differs in one respect from the reflected fish of FIGURE 2(c). It adds a fourfold oscillation in R and a threefold one in z to a thin tube like that of FIGURE 1(b) which has two in-and-out oscillations in R for every one in z. The net result is the six oscillations in R for every three in z in FIGURE 3(b).

The fact that three-island chains can occur through some mechanism other than a bifurcation of the central thin tube orbit was made clear when careful and detailed integrations by Balša Terzić [10] caught not one, but two three-island chains, shortly after their creation at $L_z^2 = 0.022$ in an SED potential with $q = 0.8$, $\beta = 0.3$. That surface of section is shown in FIGURE 3 of his paper in this volume. No islands are visible in surfaces of section for slightly larger values of L_z^2, and it is apparent that two pairs of period-three orbits, one of each pair stable and the other unstable, are created

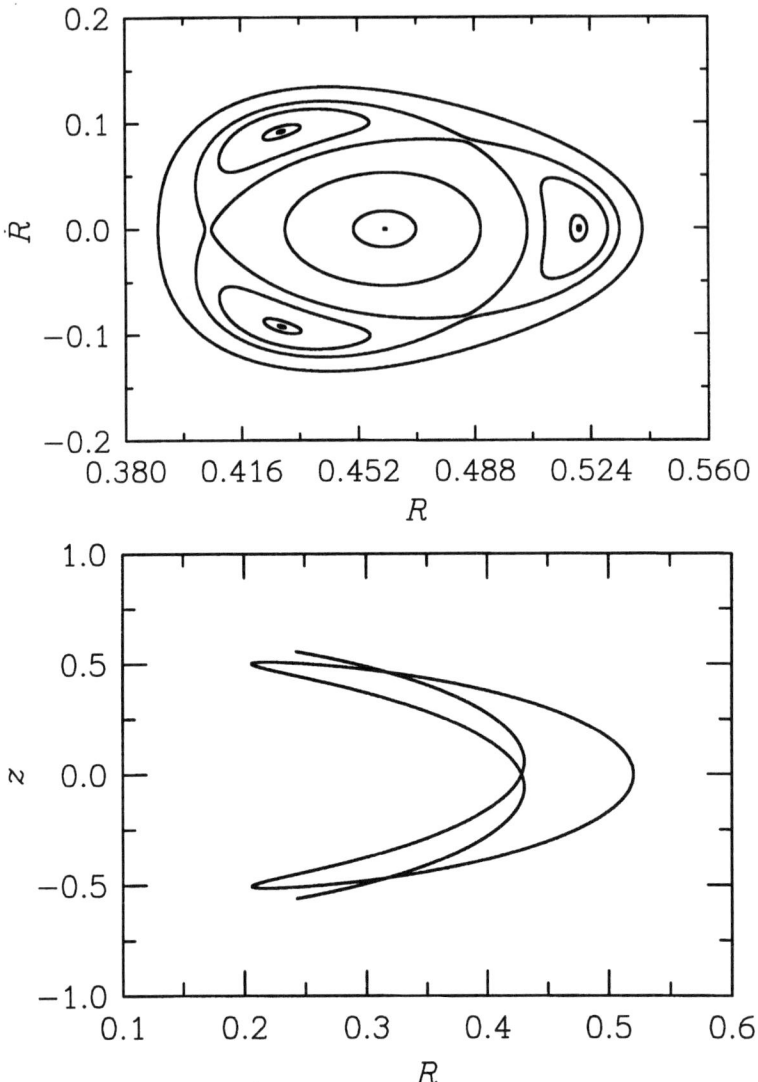

FIGURE 3. (a, top) The central part of the $(R, \dot{R})$ surface of section for a $q = 0.75$, $\beta = 0$, logarithmic SEP potential (7), in which a pair of period-three orbits has appeared near the center. The rotation angle, as defined in our Section 4, is $\alpha = 120.0123°$. **(b, bottom)** The (R, z) path of the stable period-three orbit represented by the centers of the three surrounding islands.

in the outer regions of the surface of section and not at its central point. FIGURE 4 shows the similar shapes of one of these pairs, the pretzels. The other pair are reflected fish similar to that shown in FIGURE 2(c).

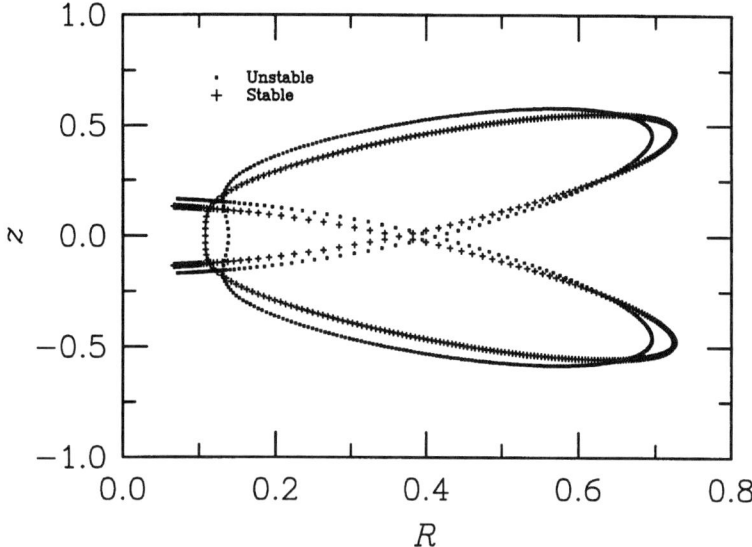

FIGURE 4. The (R, z) paths of the pretzel pair of period-three orbits, one stable and the other unstable, that occur at $L_z^2 = 0.222$ in an SED potential with $q = 0.8$, $\beta = 0.3$. The $(R, \dot{R})$ surface of section for this potential is that shown in Terzić, FIGURE 3 [10].

Similar bifurcation phenomena are found with prolate SED potentials. The basic bifurcation-free surfaces of section are now more complex because there are two families of tube orbits, inner and outer tubes, characterized by whether their parent orbit crosses $z = 0$ when closest to, or furthest from, the z-axis. Both families are apparent in FIGURE 5(a), which also shows that a period-three orbit has arisen in the midst of the outer tube region. Its (R, z) path is shown in FIGURE 5(b); the tail of this reflected fish is bent back so that it has the same six oscillations in R to three in z as FIGURE 3(b). Period-three orbits also occur in the inner tube region. FIGURE 6 shows an instance from a different surface of section, for which the bifurcated orbit has a shape akin to that of an inner tube.

3.2.3. Instability of the Thin Tube Orbit

The thin tube orbit can become unstable through a 1:1 resonance between radial and perpendicular motions, and this leads to the distinctive sequence of surfaces of section displayed in FIGURE 7. This phenomenon has been observed only in strongly flattened potentials; the $q = 0.4$ SED potential of FIGURE 7 corresponding to an E6 elliptical galaxy. There is first a supercritical pitchfork bifurcation as the thin tube, which is symmetric in z, is replaced with a pair of stable thin tubes which are not. The symmetric thin tube orbit soon regains its stability at a lower angular momentum through a subcritical pitchfork bifurcation, when an unstable period-one orbit is shed. This orbit follows a simple circuit in (R, z) space which is symmetric in z and which crosses the $z = 0$ plane at the two unstable fixed points in the surface of section; the outer one corresponding to a counterclockwise description of the circuit,

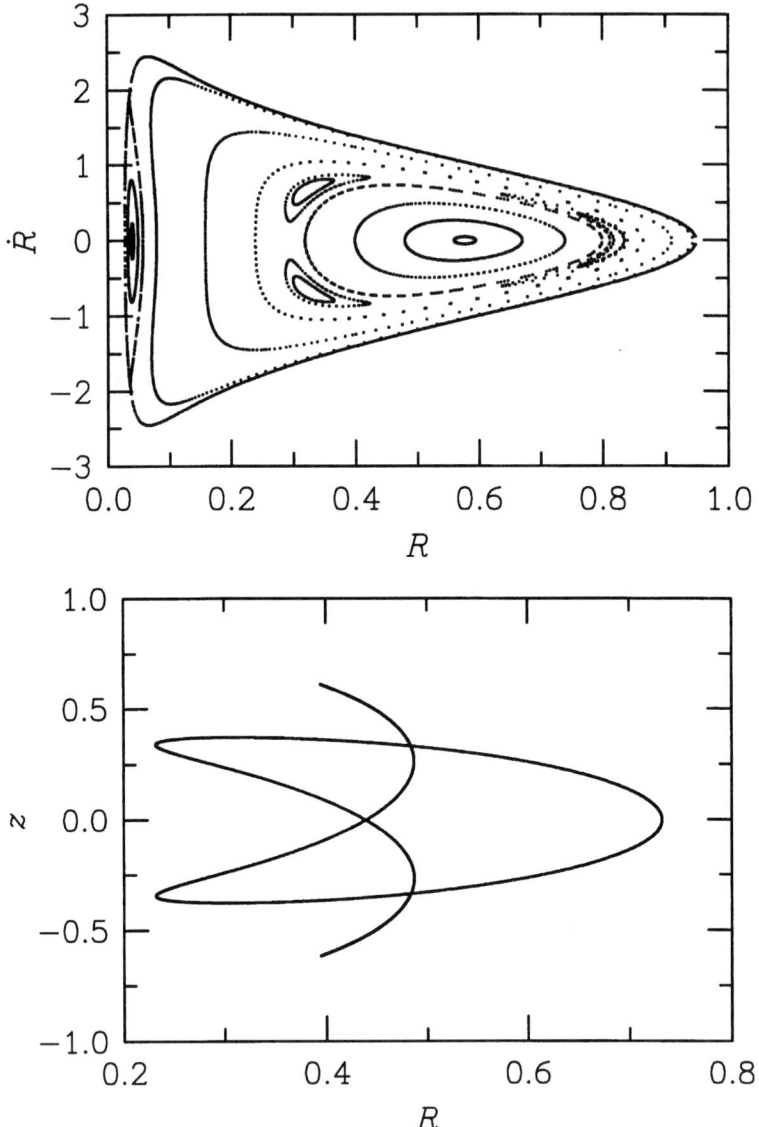

FIGURE 5. (a, top) The $(R, \dot{R})$ surface of section for the prolate SED potential with $q = 1.2$, $\beta = 0.1$, and $L_z^2 = 0.01$. The elongated narrow region to the left is occupied by inner tubes, while the larger region to the right is that of the outer tubes. A stable period-three orbit has appeared in the outer long axis tube region; its (R, z) path is shown in **(b, botttom)**.

and the inner one to a clockwise description. The stable pair of unsymmetric thin tube orbits which is formed at the first bifurcation from the symmetric thin tube per-

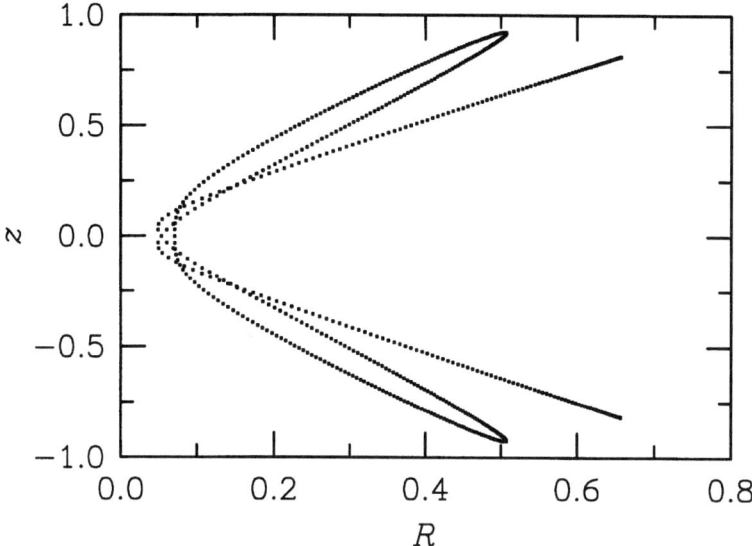

FIGURE 6. The (R, z) path of the stable period-three orbit which corresponds to the small three-island chain that closely surrounds the center of the inner tube region of the surface of section for the prolate SED potential with $q = 1.2$, $\beta = -0.1$, $L_z^2 = 0.004$. The surface of section is like that depicted in FIGURE 3(a), while the bifurcated orbit retains the characteristic shape of an inner tube.

sists unaffected by the second bifurcation. The families of the three stable orbits then grow to dominate the surface of section at yet lower angular momenta.

3.2.4. Weakly Cuspy Stäckel Potentials

A family of scalefree potentials, with quite different orbital populations, are the weakly cuspy Stäckel potentials studied by Sridhar and Touma [11]. They form a one-parameter family with

$$\Phi = -\frac{1}{2\beta r^\beta}[(1 + \cos\theta)^{1-\beta} + (1 - \cos\theta)^{1-\beta}]. \tag{20}$$

There is now no second parameter q because the angular variation is fully determined by the parameter β. These potentials are only mildly cuspy because the parameter β here is limited to the range $-2 < \beta < -1$, well removed from that of our other models, by the requirements that density be everywhere positive and singular only at $r = 0$.

Like all Stäckel potentials, (20) has an exact third integral I_3 that is quadratic in the velocities. Hence $z = 0$ surfaces of section can be constructed analytically as contours of constant I_3. There are no tube orbits at any angular momentum, only banana orbits, so that the surface of section consists of two sets of concentric rings, one above and one below the R-axis, separated by this axis which represents the single

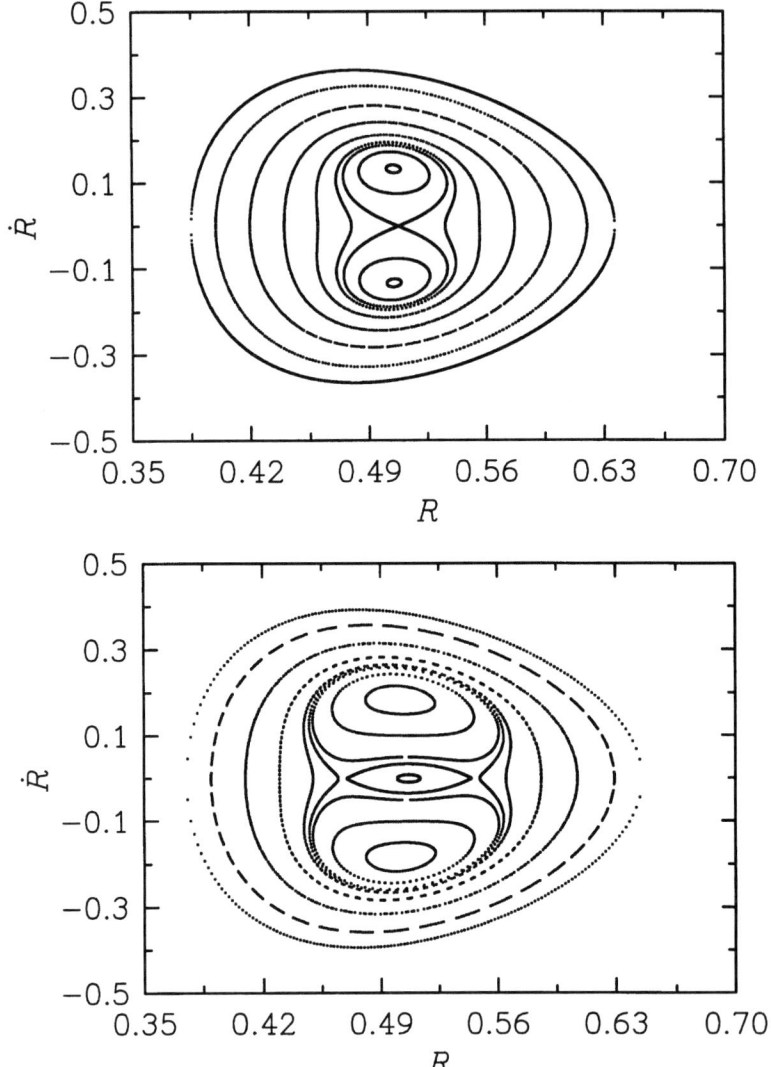

FIGURE 7. A sequence of $(R, \dot{R})$ surfaces of section at different angular momenta for the strongly flattened oblate SED potential with $q = 0.4$ and $\beta = 0.5$. The thin tube orbit becomes unstable to a 1:1 resonance and is replaced with a pair of stable orbits before **(a)** $L_z^2 = 0.335$. Then, before the stage **(b)** $L_z^2 = 0.33$ is attained even, the central thin tube has regained its stability. Subsequent surfaces of section at lower angular momenta, such as **(c)** $L_z^2 = 0.1$, have the same topology, but the families of the three stable orbits grow to dominate the surface of section. Note that this topology differs from those in which a pair of banana orbits has simply bifurcated from the equatorial plane orbit because of the connection between the upper and lower regions through the two unstable points on the R-axis.

unstable orbit in the plane, and surrounded by the bounding equatorial plane orbit. The banana orbits arise, as usual, from a 1:1 resonance because $\kappa = \nu$ for the potential (20).

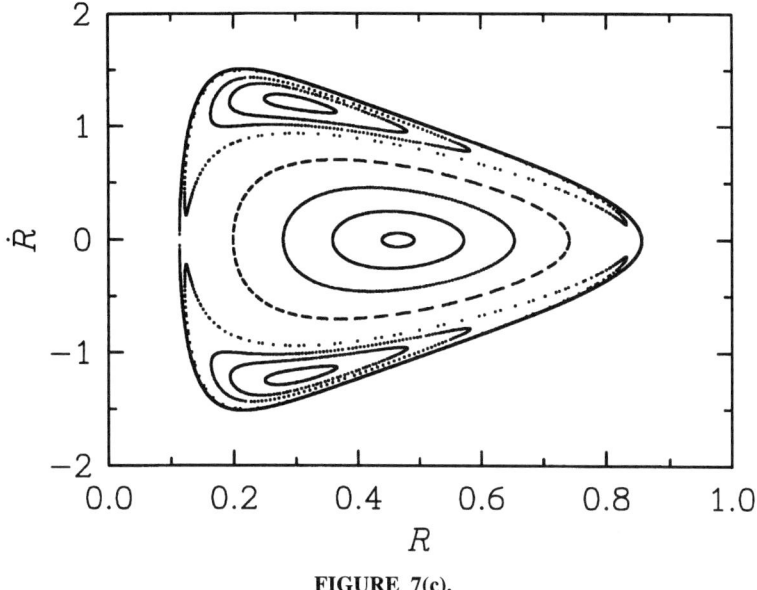

FIGURE 7(c).

Another remarkable property of the potential (20) is that it remains Stäckel when a central point mass $-GM/r$ term is added to it, though it is then no longer scalefree. In the limit of a dominant central mass, the third integral I_3 becomes the z-component of the Laplace-Runge-Lenz vector [12].

4. MAPS AND THEIR BIFURCATIONS

4.1. Elliptic Fixed Points

Like many before us, we have sought to understand bifurcations through a study of maps that model the area-preserving map that takes one crossing of a surface of section to the next. We write this map in the form

$$\begin{bmatrix} x \\ y \end{bmatrix} \mapsto \begin{bmatrix} x' \\ y' \end{bmatrix} = \begin{bmatrix} F(x, y) \\ G(x, y) \end{bmatrix} \quad (21)$$

for some functions F and G, aligning the x- and y-axes with the R and $\dot{R}$ axes of our surfaces of section respectively. Our goal is to study the process through which period-three orbits and their accompanying three-island chains arise.

To study the behavior in the neighborhood of the elliptic fixed point that represents the thin tube orbit, we place our origin there. We expand the map, which we assume to be locally analytic, about the fixed point in the form

$$\begin{bmatrix} x \\ y \end{bmatrix} \mapsto \begin{bmatrix} x' \\ y' \end{bmatrix} = \begin{bmatrix} x\cos\alpha + y\sin\alpha + \ldots \\ -\sin\alpha + y\cos\alpha + \ldots \end{bmatrix}, \quad (22)$$

with the dots signifying higher-order terms. A rescaling to make the invariant ellipses circular, such as $x = R - R_0$, $y = \dot{R}/\kappa$ in (19), is the only transformation necessary in this case to reduce the linear part of the map (21) to the simple rotation form in (22). Our rotational angle α, which is in the opposite sense from that usually chosen, is a natural choice here with $\alpha = 2\pi\kappa/\nu$ for (19).

A compact and convenient way of representing the expanded form of (22) is the complex form

$$z' = x' + iy' = e^{-i\alpha}z + az^2 + bz\bar{z} + c\bar{z}^2 + \ldots . \quad (23)$$

for some constants a, b, and c. Siegel and Moser [13] prove that, if $e^{-3i\alpha} = 1$, as it is when $\kappa/\nu = 4/3$ and there is a 4:3 resonance between radial and perpendicular motions, *and* $c \neq 0$, then the fixed point at (0,0) is unstable. In our survey we study a sequence of surfaces of section that change as a parameter such as L_z^2 is varied. The angle α then varies continuously and this instability is momentary. A more significant result therefore is that of Meyer and Hall [14] who prove that the orbit of period three, which, with varying α, bifurcates from the fixed point at $\mathcal{O}$ when $e^{-3i\alpha} = 1$, consists of three hyperbolic points. FIGURE A.6 of their Chapter VIII illustrates this process schematically, and Hale and Koçak's [15] FIGURE 15.18 shows it in detail for an area-preserving Cremona map. No three-island chains, such as we see in our FIGURE 3(a) are formed.

4.2. Generating Functions

Maps are guaranteed to be exactly area-preserving if they are derived from generating functions. As in mechanics [12], the generating function combines an old and a new variable. We shall use a generating function of the form $W(x, x')$ to define an area-preserving map according to relations

$$y = \frac{\partial W}{\partial x}, \quad y' = -\frac{\partial W}{\partial x'}. \quad (24)$$

We also require that the map incorporate the following symmetry that is present in our orbital dynamics. Orbits can be reversed in time and run backwards because our potentials are time-independent. Reversing time changes $\dot{z} > 0$ crossings of $z = 0$ into $\dot{z} < 0$ ones, which are not the crossings that our surfaces of section record. However, every reversed orbit has a mirror image twin (reflected in the $z = 0$ plane) because our potentials are even in z, and this has $\dot{z} > 0$ in place of $\dot{z} < 0$ crossings of $z = 0$. The mathematical expression of this combination of time reversibility and symmetry in z is the requirement that the map satisfy the condition

$$\begin{bmatrix} x' \\ -y' \end{bmatrix} \mapsto \begin{bmatrix} x \\ -y \end{bmatrix}.$$

This condition implies that $W(x, x')$ is symmetric in its arguments. We now define $Q(x, x') = \partial W/\partial x$, and write the map as

$$y = Q(x, x'), \quad y' = -Q(x', x). \tag{25}$$

4.3. Orbits of Period Three

An orbit of period three

$$\begin{bmatrix} x_0 \\ y_0 \end{bmatrix} \mapsto \begin{bmatrix} x_1 \\ y_1 \end{bmatrix} \mapsto \begin{bmatrix} x_2 \\ y_2 \end{bmatrix} \mapsto \begin{bmatrix} x_0 \\ y_0 \end{bmatrix},$$

of the map (25) requires a solution of the system

$$\begin{aligned} y_0 &= Q(x_0, x_1), \quad y_1 = Q(x_1, x_2), \quad y_2 = Q(x_2, x_0), \\ y_1 &= -Q(x_1, x_0), \quad y_2 = -Q(x_2, x_1), \quad y_0 = -Q(x_0, x_2). \end{aligned} \tag{26}$$

A nontrivial and generic possibility is

$$y_0 = 0, \quad x_2 = x_1, \quad y_2 = -y_1, \tag{27}$$

which is achieved if some solution, other than $x_0 = x_1$, can be found of the pair of simultaneous equations

$$Q(x_0, x_1) = 0, \quad Q(x_1, x_1) + Q(x_1, x_0) = 0. \tag{28}$$

The $x_0 = x_1$ case can be eliminated by combining Eqs. (28) in the form

$$Q(x_1, x_1) + Q(x_1, x_0) - 2Q(x_0, x_1) = 0, \tag{29}$$

and then explicitly eliminating the $(x_0 - x_1)$ factor.

4.4. A Cubic Generating Function

We now select a class of generating functions for studying how and when orbits of period three arise. We study the general symmetric cubic W,

$$W(x,x') = \frac{xx' - \frac{1}{2}(x^2 + x'^2)\cos\alpha}{\sin\alpha} + [I_1(x^3 + x'^3) + I_2 xx'(x + x')]. \tag{30}$$

Its quadratic terms give the linear part of (22), while the additional cubic terms introduce nonlinear effects. This form of generating function is inappropriate for analyzing the bifurcations related to $\alpha = 0$ or $\alpha = 180°$, but that is not our concern. We restrict attention to $0 < \alpha < 180°$ since maps for $0 > \alpha > -180°$ are simply reflections of ones we consider. Because scaling transformations do not alter the map in any significant way, the constants multiplying the cubic terms in (30) can be scaled so that they depend on a single parameter. We therefore introduce a polar representation and write

$$I_1 = \cos\gamma, \quad I_2 = \sin\gamma. \tag{31}$$

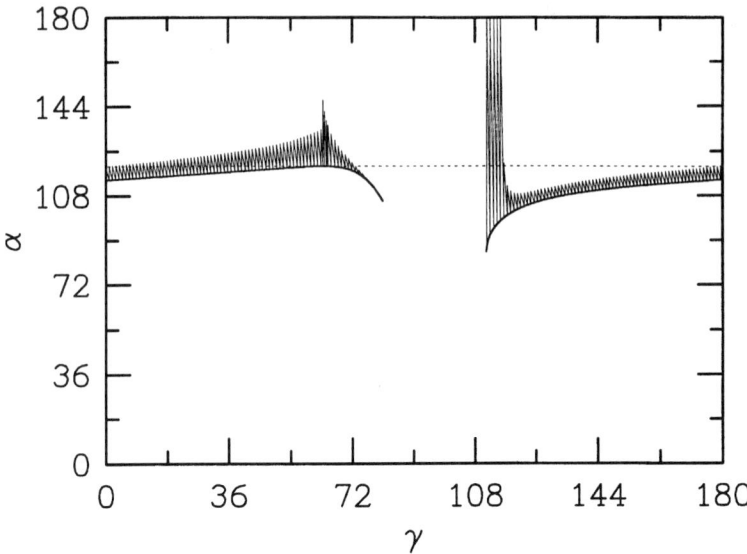

FIGURE 8. The two shaded regions display the portions of (γ, α) space in which the map generated by (30) has stable period-three orbits. The regions are bounded below by the solid curve which corresponds to the critical rotation angle α_{crit} at which a pair of period-three orbits, one stable and the other unstable, is created. They are bounded above by the occurrence of a destabilizing period-doubling bifurcation.

Because of symmetries, we can restrict γ to the range $0 \leq \gamma \leq 180°$, which leaves us with a $180° \times 180°$ box of (α,γ) parameter space to survey. Our map is explicit only in the special case of $I_2 = \sin\gamma = 0$. Otherwise we solve $y = \partial W/\partial x$, which is a quadratic for x', by selecting the root that $\to 0$ as $(x, y) \to 0$. The algebra is elementary (see Appendix A.2 for some details) with (29), which is the only possibility in this case, reducing to a linear relation between x_0 and x_1. The results confirm that, generically, orbits of period three are created as a pair, one stable and the other unstable, at a triad of distinct points (27). This happens through a turning-point bifurcation, as captured in Terzić's [10] FIGURE 3, except that two pairs form there whereas only one pair ever forms with the present model. When γ is held fixed and α is increased, a pair of period-three orbits is created at values $\alpha = \alpha_{crit}$ which are included in the plot shown in FIGURE 8.

Only unstable period-three orbits are found for the range $80.54° < \gamma < 111.43°$ for which values of α_{crit} are missing. The critical rotation angle α_{crit} never exceeds $120°$, and is $120°$ only for $\gamma = \tan^{-1}2 = 63.43°$, when the constant $c = 0$ in the expansion of (23) so that Siegel and Moser's [13] instability result does not apply. Otherwise α_{crit} is close to $120°$ over a wide range. Stable period-three orbits are rare because they exist in only a small portion of the parameter space of FIGURE 8. Except in a narrow band to the immediate right of $\gamma = 111.43°$, the stable period-three orbit has only a short span of existence, a few degrees in α, before it becomes unstable through a supercritical pitchfork bifurcation corresponding to a period-doubling instability. The

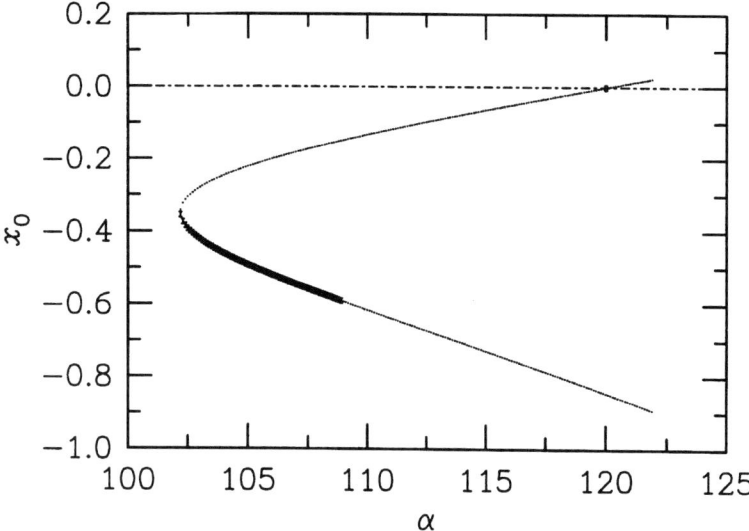

FIGURE 9. The generic bifurcation diagram, for fixed γ and varying α, for period-three orbits, showing how x_0, the x-coordinate of the point of the orbit on the y-axis, varies with α. The specific case plotted is that of $\gamma = 120°$. The crosses, which form the thicker segment, mark the range of stable period-three orbits, while the dot–dash line marks the period-one fixed point. The diagram is reversed, and the values of x_0 are positive for stable period-three orbits as in FIGURE 3(a), when $\gamma < \tan^{-1} 2$.

generic bifurcation diagram for fixed γ and varying α is shown in FIGURE 9. Note that it is the unstable period-three orbit that passes through the fixed point at $(0, 0)$ at $\alpha = 120°$, which conforms with the result of Meyer and Hall [14]. FIGURE 10(a) shows the $\alpha = 120°$ map for $\gamma = 65°$ for which $\alpha_{\text{crit}} = 119.97°$. The stable period-three orbit is evident there at the stage that the unstable one has momentarily destabilized the elliptic fixed point at $\mathcal{O}$. FIGURE 10(b) at $\alpha = 122°$ shows the stability of the elliptic fixed point at $\mathcal{O}$ restored, and stable and unstable period-three orbits around it, qualitatively similar to the surface of section of FIGURE 3(a).

Hénon [16] studied the most general quadratic map that is area-preserving and explicit,

$$x' = x\cos\alpha + (y - x^2)\sin\alpha,$$
$$y' = x\sin\alpha + (y - x^2)\cos\alpha.$$

(We have changed the sign of his α to conform with our usage.) Although not explicitly time-reversible, Hénon's map can be obtained by transformation from a time-reversible DeVogelaere map [17]; that is, a map with a generating function $W = W_1(x) + W_1(x')$ in our notation Hénon found that as α increased, a pair of orbits of period three are created together at $\alpha = \cos^{-1}(1 - \sqrt{2}) = 114.47°$. The unstable orbit passes through $\mathcal{O}$ at $\alpha = 120°$, at which stage the stable orbit loses its stability.

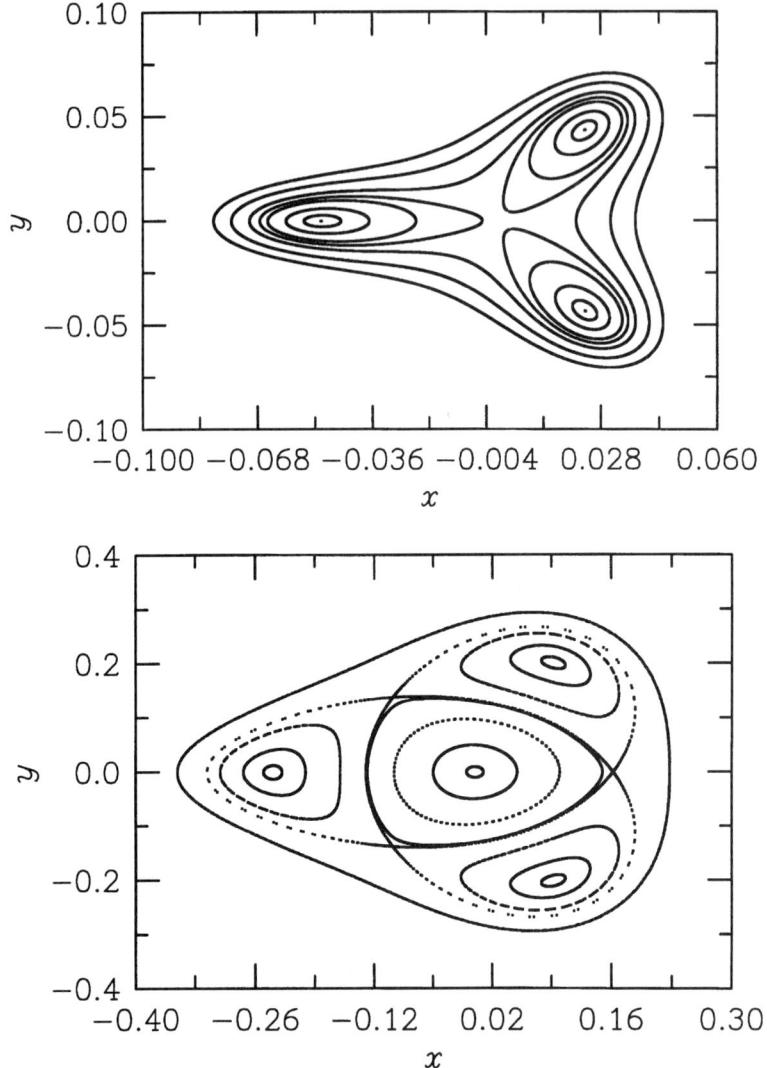

FIGURE 10. Two cases of the map generated by (30) for $\gamma = 65°$. **(a, top)** At $\alpha = 120°$, the unstable period-three orbit has collapsed to the fixed point at $\mathcal{O}$ which becomes momentarily unstable. There is a stable period-three orbit present too. **(b, bottom)** At $\alpha = 122°$, the fixed point at $\mathcal{O}$ has restabilized, the unstable period-three orbit has reemerged, and the stable period-three orbit is still present.

Hénon's map is equivalent to the special $I_2 = 0$ case for which our map becomes explicit; the requirement of explicitness greatly reducing the variety of time-reversible maps that can be obtained from a generating function.

5. DISCUSSION AND CONCLUSIONS

This has been a selective and incomplete study of bifurcations of periodic orbits in axisymmetric scalefree potentials with central cusps. It has focused on the most prominent bifurcations, and especially on the less understood orbits of period three in their crossing of the central plane. The figures we display are selected to highlight instances of bifurcations. Many surfaces of section, especially ones for spheroidal equidensity (SED) potentials, are duller, showing only tube orbits and no prominent bifurcations.

The family of maps introduced in Section 4 gives some significant insights. It is derived from a generating function which is of general type for problems with the same symmetries as ours, and which is truncated consistently at cubic terms. Hence it should encompass all phenomena that generally arise sufficiently close to an elliptic fixed point. Our analysis shows that its stable period-three orbits generically arise away from an elliptic fixed point through a turning-point bifurcation. Only exceptionally does a stable period-three orbit arise from a bifurcation of an elliptic fixed point when its rotation angle passes through 120°. However, there is a wide parameter range in which the turning-point bifurcation occurs when the rotation angle is close to 120°, which can explain why stable period-three orbits are often found quite close to the elliptic fixed point. On the other hand, there are none to be found in other parameter ranges. That, plus the generally short range between the stages at which a stable period-three orbit arises and that at which it becomes unstable, is consistent with the scarcity of three-island chains found in surfaces of section. All of these period-three orbits belong to what Contopoulos [18] classifies as a regular family in that they are all connected, as in the bifurcation diagram of FIGURE 9, to the basic thin tube orbit. A more elaborate and higher-order map than our equations (24) and (30) allows is needed to account for the creation of the two pairs of irregular period-three orbits that Terzić's [10] remarkable FIGURE 3 shows.

ACKNOWLEDGMENTS

It is no surprise that many of the phenomena described in this work are already known to George Contopoulos, this meeting's honoree. In addition to his earlier classification of families of periodic orbits as either regular or irregular, Contopoulos and Michaelidis [19] studied the specific problem of the bifurcations of triple-periodic orbits in the context of the two-dimensional potential $\Phi(x, y) = \frac{1}{2}(\omega_1^2 x^2 + \omega_2^2 y^2) - \varepsilon x y^2$ as the parameter ε is increased. Their FIGURE 2(a) plots what at first appears to be a stable and an unstable period-three orbit bifurcating from a stable period-one orbit. The closer analysis shown in their magnified FIGURE 2(b) reveals that this is not so, and that the two period-three orbits are in fact created as a pair slightly away from the period-one orbit. The unstable period-three orbit then passes through the period-one orbit as in our FIGURE 9. These instances serve to remind us of some of the many ways in which George has been a pioneer in the understanding of dynamical phenomena. We appreciate the privilege of being invited to this meeting and of being able to contribute this paper in his honor.

APPENDIX

A.1. Computing Surfaces of Section

Here we present the simple method which we have used for computing crossings of a surface of section without any need for interpolation, and without losing any of the accuracy of a high-order ordinary differential equation (ODE) solver. Like Hénon [20], we accomplish this by using an independent variable whose value at the surface of section is known. Unlike Hénon, we use that same independent variable throughout the integration. To be specific, we introduce polar coordinates A and ψ in the $(z, \dot{z})$-plane, via the transformations

$$z = A \sin\psi, \quad \dot{z} = A \cos\psi, \tag{32}$$

where the amplitude A is to be always positive. We use ψ, rather than time, as the variable of integration. Then the crossings of the $z = 0$ surface-of-section occur at $\psi = n\pi$ for integer n, and the crossings with $\dot{z} > 0$, which are the ones we record, occur for even values of n.

Differentiation of equations (32) gives

$$\dot{z} = A\dot{\psi}\cos\psi + \dot{A}\sin\psi = A\cos\psi,$$
$$\ddot{z} = -A\dot{\psi}\sin\psi + \dot{A}\cos\psi = -\nu^2 z = -\nu^2 A\sin\psi, \tag{33}$$

where ν is now defined as the local frequency in the z-direction

$$\nu^2(R, z) = \frac{1}{z}\frac{\partial\Phi}{\partial z}. \tag{34}$$

This reduces to the earlier definition of (18) in the limit of $z \to 0$. Equations (33) can be solved to give the equations

$$\dot{A} = A(1 - \nu^2)\sin\psi\cos\psi, \quad \dot{\psi} = \cos^2\psi + \nu^2\sin^2\psi. \tag{35}$$

The first of these equations shows that an initially positive A cannot become zero in finite time and hence remains positive if it is so initially. The second shows that ψ is always an increasing function of t, and so can be used as the independent variable of integration. The integration needed is that of the following first-order system of three equations for the three unknowns A, R, and $\dot{R}$:

$$\frac{dA}{d\psi} = \frac{\dot{A}}{\dot{\psi}}, \quad \frac{dR}{d\psi} = \frac{\dot{R}}{\dot{\psi}}, \quad \frac{d\dot{R}}{d\psi} = \frac{\ddot{R}}{\dot{\psi}}. \tag{36}$$

Their right sides are all known functions of A, ψ, R, and $\dot{R}$ because $z = A\sin\psi$. We integrate from a starting point (R_0, V_0) in an $(R, \dot{R})$ surface of section using the initial conditions

$$\psi = 0, \quad A = \sqrt{2E - \frac{L_z^2}{R_0^2} - 2\Phi(R_0, 0) - V_0^2}, \quad R = R_0, \quad \dot{R} = V_0. \tag{37}$$

A.2. Period-Three Orbits of the Map Generated by Equation (30)

The Q function for the map (25) is

$$Q(x, x') = \frac{x' - x\cos\alpha}{\sin\alpha} + 3x^2\cos\gamma + x'(2x + x')\sin\gamma. \tag{38}$$

Elimination of the y_i between equations (26) then shows that one pair of x_i must coincide, and hence that (27) is now the only possibility for period-three orbits. Eliminating the $(x_0 - x_1)$ factor from equation (29) gives the condition

$$(6\cos\gamma - \sin\gamma)x_0 + (6\cos\gamma + \sin\gamma)x_1 = 2\cos\alpha + \frac{2\cos\alpha + 1}{\sin\alpha}.$$

Using this condition to eliminate x_1 from the equation $Q(x_0, x_1) = 0$ gives a quadratic for x_0, and hence two *possible* period-three orbits. They are not *valid* period-three orbits unless each step of the map satisfies the condition

$$x' =$$

$$\frac{-(1 + 2x\sin\alpha\sin\gamma) + \sqrt{[1 + 2x\sin\alpha\sin\gamma]^2 - 4\sin\alpha\sin\gamma[(3x^2\cos\gamma - y)\sin\alpha - x\cos\alpha]}}{2\sin\alpha\sin\gamma} \tag{39}$$

that is necessary for $(0, 0)$ is to be a fixed point.

Orbits of period three are created, where none are present before, at a stage at which the two roots of the quadratic for x_0 cease to be a complex conjugate pair and become equal. That stage occurs for

$$(2\lambda^2 - 1)\cos^2\alpha_{\text{crit}} - 2\cos\alpha_{\text{crit}} + \tfrac{1}{2}(\lambda + 1)^2 - 1 = 0 \tag{40}$$

where

$$\lambda = \frac{\tan\gamma - 6}{\tan\gamma + 6}.$$

The roots of the quadratic (40) for $\cos\alpha_{\text{crit}}$ are complex when $\lambda > 1$ and when $\lambda < -2.47569$, the real root of the cubic $2\lambda^3 + 6\lambda^2 + 3\lambda + 1 = 0$. Otherwise, there is just one feasible real root for $\cos\alpha_{\text{crit}}$ which lies in the range $[-\tfrac{1}{2}, 1)$ for $-1.4 < \lambda < 1$, but two in the range $(-\tfrac{1}{2}, 0)$ for $-2.47569 < \lambda < -1.4$.

We now relate these properties to FIGURE 8. As γ increases from 0 on its left branch, λ increases from -1 and we are in the region of one real root for $\cos\alpha_{\text{crit}}$. The maximum value $\alpha_{\text{crit}} = 120°$, $\cos\alpha_{\text{crit}} = -\tfrac{1}{2}$, is attained only for $\lambda = -\tfrac{1}{2}$, that is, for $\tan\gamma = 2$. Otherwise α_{crit} is less than $120°$ because the left side of (40) evaluated for $\cos\alpha_{\text{crit}} = -\tfrac{1}{2}$ is $(\lambda + \tfrac{1}{2})^2$ and so vanishes for $\lambda = -\tfrac{1}{2}$ only. The left branch of FIGURE 8 ends at $\gamma = \tan^{-1}6$ where $\lambda = 0$ and $\alpha_{\text{crit}} = \cos^{-1}(2^{-1/2} - 1) = 107.03°$. This is a stage at which the term in the square root of equation (39) vanishes when evaluated for $(x, y) = (x_0, 0)$. It marks a transition beyond which the roots for α_{crit} found for yet larger values of γ no longer give valid period-three orbits because the condition (39) is not satisfied.

The left end of the right branch of α_{crit} values in FIGURE 8 lies at $\gamma = 111.43°$ where $\lambda = -2.47569$. The value of x_0 is infinite at this starting-point. Moving to its right, λ is again negative and increasing, and the quadratic (40) has two feasible roots for $\cos\alpha_{crit}$ as γ increases up until $\gamma = 135°$ when $\lambda = -1.4$. However, only the smaller value of these with the larger value of α_{crit} gives a valid period-three orbit. It continues to do so up to $\gamma = 180°$.

REFERENCES

1. EVANS, N.W. 1994. Mon. Not. R. Astron. Soc. **267**: 333–360.
2. LAUER, T.R. *et al.* 1995. Astron. J. **110**: 2622–2654.
3. QIAN, E.E., P.T. DE ZEEUW, R.P. VAN DER MAREL & C. HUNTER. 1995. Mon. Not. R. Astron. Soc. **274**: 602–622.
4. RICHSTONE, D.O.. 1982. Astrophys. J. **252**: 496–507.
5. BINNEY, J. & S. TREMAINE. 1987. Galactic Dynamics. Chapter 3. Princeton Univ. Press. Princeton, NJ.
6. CHANDRASEKHAR, S. 1969. Ellipsoidal Figures of Equilibrium, Chapter 3. Yale Univ Press. New Haven, CT.
7. ISAACSON, E. & H.B. KELLER. 1966. Analysis of Numerical Methods, Chapter 7. John Wiley. New York.
8. DAVIS, P.J. 1975. Interpolation and Approximation, Chapter 12. Dover. New York, NY.
9. LEES, J.F. & M. SCHWARZSCHILD. 1992. Astrophys. J. **384**: 491–501.
10. TERZIĆ, B. 1998. Ann. New York Acad. Sci. **867**: this volume.
11. SRIDHAR, S. & J. TOUMA. 1997. Mon. Not. R. Astron. Soc. **292**: 657–661.
12. GOLDSTEIN, H. 1980. Classical Mechanics. 2nd edit. Chapters 3 and 9. Addison-Wesley. Reading, MA.
13. SIEGEL, C.L. & J.K. MOSER. 1971. Lectures on Celestial Mechanics. Section 31. Springer Verlag. New York.
14. MEYER, K.R. & G.R. HALL. 1992. Introduction to Hamiltonian Dynamical Systems and the N-Body Problem. Chapter VIII. Springer Verlag. New York.
15. HALE, J. & H. KOÇAK. 1991. Dynamics and Bifurcations. Springer Verlag. New York.
16. HÉNON, M. 1969. Quart. App. Math. **37**: 291–312.
17. GREENE, J.M., R.S. MACKAY, F. VIVALDI, & M.J. FEIGENBAUM. 1981. Physica D. **3**: 468–486.
18. CONTOPOULOS, G. 1970. Astron. J. **75**: 96–107.
19. CONTOPOULOS, G. & P. MICHAELIDIS. 1980. Celest. Mech. **22**: 403–413.
20. HÉNON, M. 1982. Physica D. **5**: 412–414.

Irregular Period-Tripling Bifurcations in Axisymmetric Scalefree Potentials

BALŠA TERZIĆ[a]

Department of Mathematics, Florida State University, Tallahassee, Florida 32306-4510

ABSTRACT: We investigate the phenomenon of irregular bifurcations in axisymmetric scalefree potentials, in which a tube orbit gives rise to a new family of orbits of period 3. A proper understanding of the formation of this irregular family of orbits is important because the family occupies a fair portion of the phase space in models where it occurs. We show that these bifurcations are related to the transition of the rotation number of the phase-space torus through 1/3. This type of bifurcation is closely related to the behavior of nontwist maps.

1. INTRODUCTION

We have examined a range of orbits in axisymmetric potentials with central cusps, focusing on SED potentials [1]. The latter arise from mass distribution with spheroidal equidensity (SED) surfaces [2]. They are characterized by two parameters — the power $-\beta$ of dependence on distance, and the flatness parameter (axis ratio) q. Our primary focus is on understanding period-tripling bifurcations in these potentials, which is related to 4:3 resonances between radial and vertical frequencies.

The period-3 orbits which bifurcate in stable/unstable pairs via pitchfork bifurcation from the stable thin tube orbit at a 120° rotation angle at some critical L_z^2 are called *regular* [3]. However, of particular interest to us are *irregular* orbits which bifurcate away from the thin tube, via a turning-point bifurcation [3].

In our study of orbits in scale-free potentials, the most prevalent bifurcation we encountered has been the regular, pitchfork bifurcation in which a stable/unstable period-3 orbit pair bifurcates near, or in some special cases, exactly from the stable period-1 orbit (FIGURE 1). This paper focuses on investigating the dynamics of period-3 orbits that arise away from the stable period-1 orbit and away from the orbit in the central plane, the last invariant curve in the phase space (FIGURE 2). It is important to note that the difference between the number of stable and the number of unstable orbits is a conserved quantity in both regular and irregular bifurcations.

We restrict our attention to the axisymmetric, scalefree SED potential when the axis ratio $q = 0.8$ and $\beta = 0.3$. This is an oblate case, as the axis ratio is less then unity.

2. IRREGULAR PERIOD-TRIPLING BIFURCATIONS

We first observed these stable/unstable pairs of period-3 orbits in our surface of section (SOS) graphs that we generated using the integration routines (FIGURE 3).

---

[a]Address for telecommunication: e-mail, bterzic@math.fsu.edu

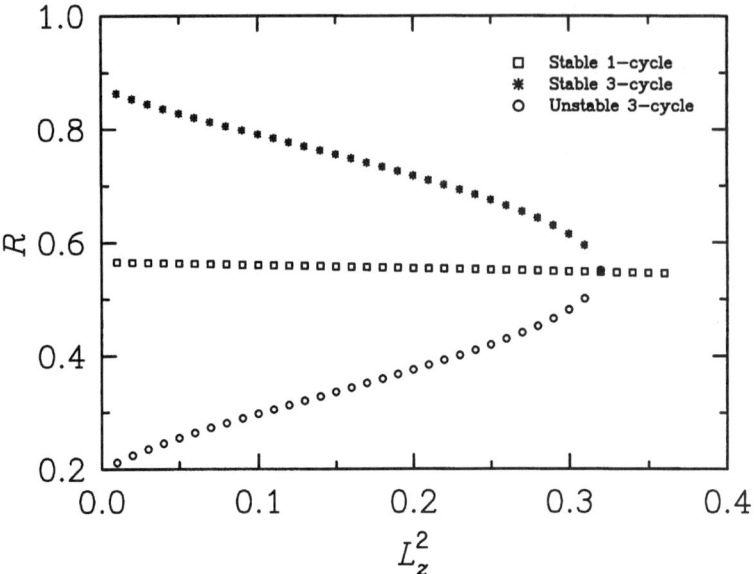

FIGURE 1. Regular pitchfork bifurcation. SED potential $q = 1.2$, $\beta = 0.35$.

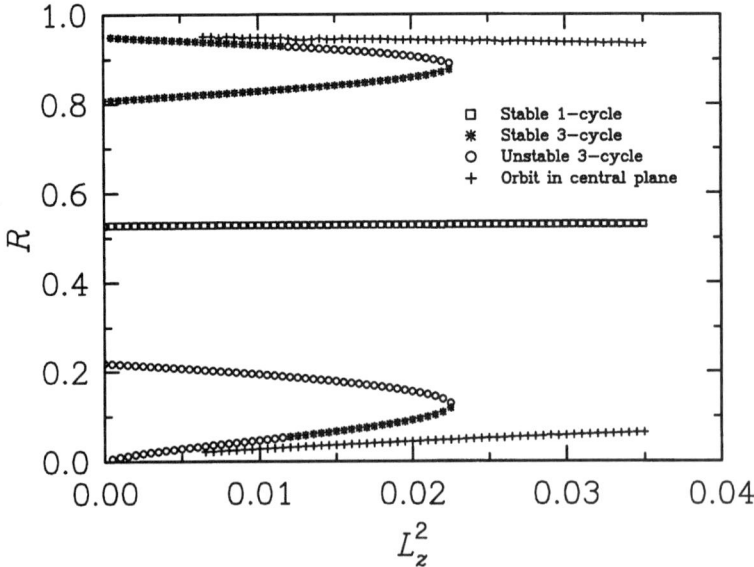

FIGURE 2. Irregular turning-point bifurcation. SED model $q = 0.8$, $\beta = 0.3$.

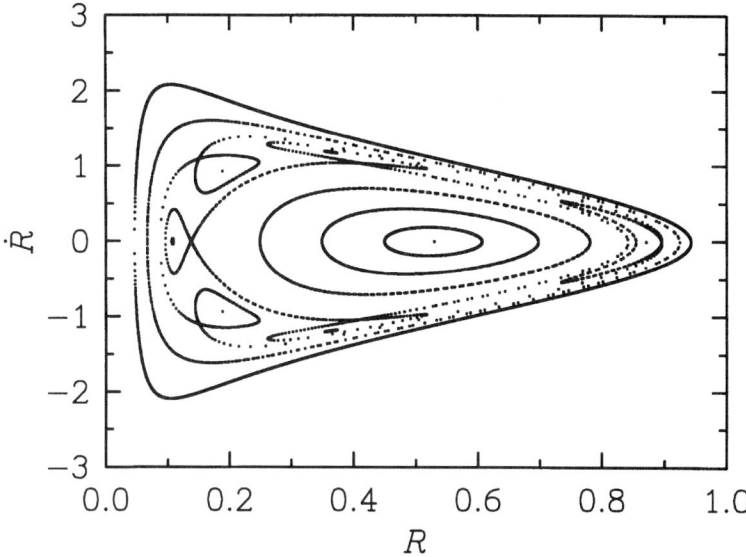

FIGURE 3. Surface of section $(R, \dot{R})$ for SED potential $q = 0.8$, $\beta = 0.3$, $L_z^2 = 0.022$ (oblate case). Two stable/unstable pairs of period-3 orbits have just bifurcated in the outer regions.

These are graphs of the phase space or Poincaré space, the standard tool we use for analysis of the dynamics. This surface of section graph shows two stable/unstable pairs of period-3 orbits, shortly after they bifurcated. We see that they are originating in the region away from both the stable period-1 orbit at the center and the orbit in plane.

2.1. Average Rotation Number

In order to understand this phenomenon of irregular period-tripling bifurcations, we have analyzed the rotation numbers of invariant curves in $(R, \dot{R})$ Poincaré space. The rotation number of an invariant curve is defined to be a fraction of a full revolution around the period-1 orbit that a point traverses in one iteration. Hence, the rotation number is the mapping angle divided by 2π. If we rotate any point around the stable period-1 orbit by an angle α, the resulting point, is the same one we would have obtained if the rotation angle was $(2\pi - \alpha)$ in the opposite direcion of rotation. Therefore, we have two different rotation numbers for the same mapping. To avoid this ambiguity, we use Floquet perturbation analysis [4] for the linearized mapping in the immediate vicinity of the period-1 orbit. The analysis shows that mapping angles in our case are around 120°, rather then 240° (FIGURE 4). Therefore, the rotation number of these sets of period-3 orbits is 1/3 rather then −2/3 or any other multiple of 1/3.

We calculated the average rotation number of invariant curves for different values of L_z^2, the square of the z-component of the angular momentum, which is an integral of motion due to the axisymmetric nature of the potential. In FIGURE 5, we

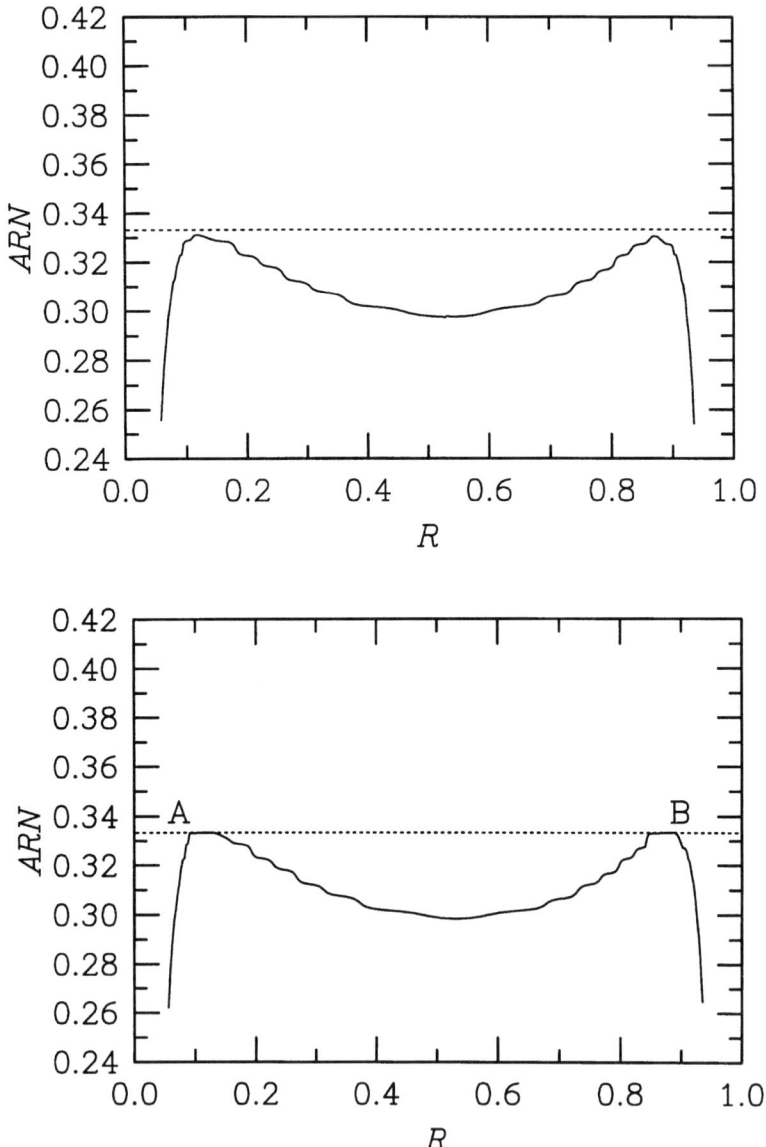

FIGURE 4. Average rotation number (ARN) as a function of R. SED potential $q = 0.8$, $\beta = 0.3$, and (**a, top**) $L_z^2 = 0.025$, (**b, bottom**) $L_z^2 = 0.0225$.

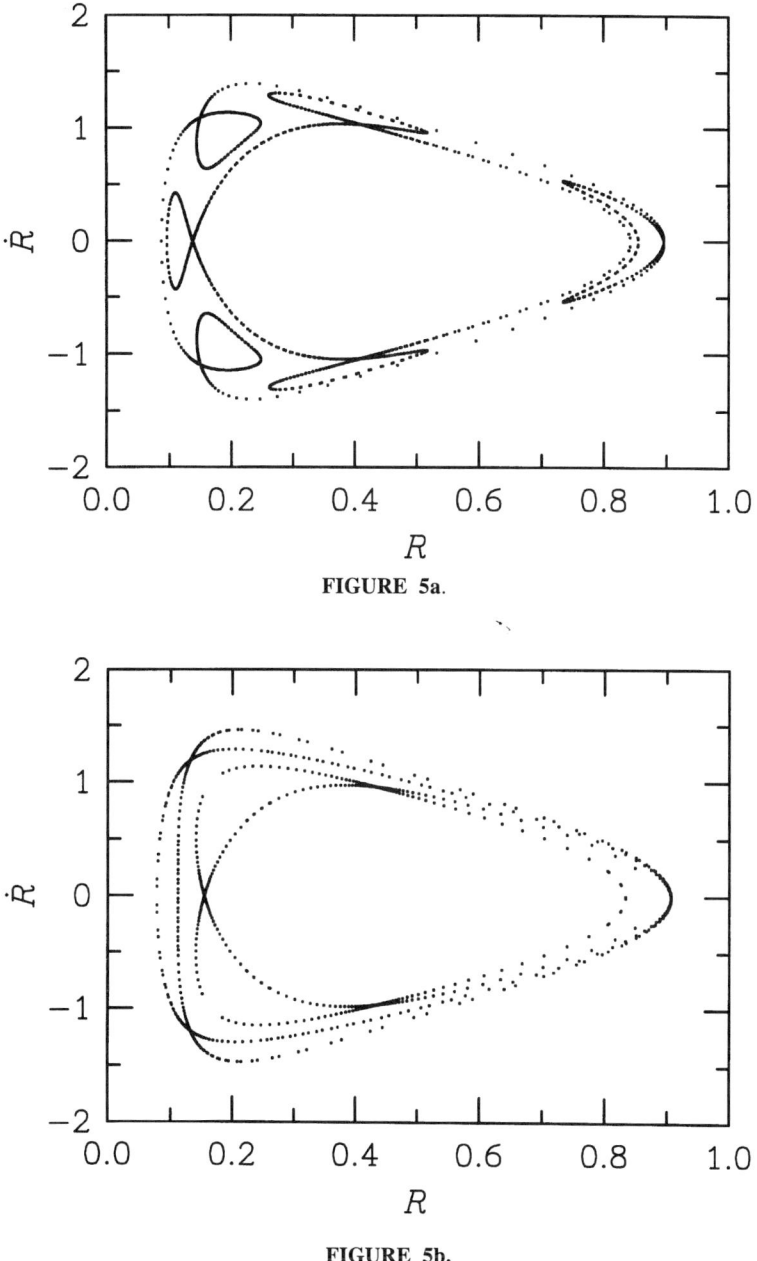

FIGURE 5. Separatrix reconnection occurs as L_z^2 decreases (between (**a**) and (**b**)). The stability of the outer pair of stable/unstable period-3 orbits reverses as L_z^2 decreases (between (**b**) and (**c**)). SED potential $q = 0.8$, $\beta = 0.3$, and (**a**) $L_z^2 = 0.022$, (**b**) $L_z^2 = 0.02$, (**c**) $L_z^2 = 0.01$.

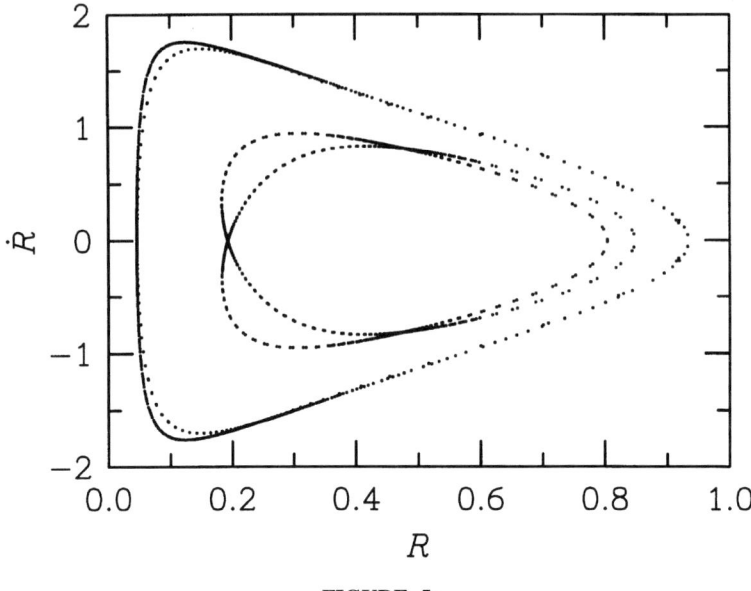

FIGURE 5c.

follow the transition of the rotation curve as L_z^2 is decreased through the bifurcation point. In the SOS graphs, where the irregular period-3 orbits were initially observed, bifurcations occur at $L_z^2 = 0.0225$. At that point, as can be seen in FIGURE 5b, the rotation curve touches the 1/3 line. It is important to note that it is the same invariant curve in the SOS graphs that touches 1/3 line at both A and B, as these are its two crossings of the $\dot{R} = 0$ in the phase space. Therefore, these two pairs of stable/unstable 3-cyles bifurcate simultaneously. This analysis shows that the irregular period-tripling bifurcations occur in the phase space when the average rotation number of one of the invariant curves equals 1/3.

In the SOS graphs, the stable period-3 orbits form small islands. These are invariant curves which also have the rotation number equal to 1/3 around the stable 1-cycle (FIGURE 5). This is manifested by the flattenings of the rotation curve at 1/3 near the stable period-3 orbits.

As L_z^2 decreases, the maxima of the rotation curve increase, while the flattenings around the stable period-3 orbits on the left get smaller, due to shrinking of the islands in the phase space. The rotation curve will cross the 1/3 line at both unstable period-3 orbits, but no flattening will be present as no islands around the unstable point exist in the phase space. In the outer parts of the rotation curve smaller flattenings are observed. These flattenings correspond to higher resonances responsible for generation of orbits of higher periodicity.

2.2. Separatrix Reconnection and Exchange of Stability of Outer Period-3 Orbit

Another important feature of these irregular bifurcations is that of separatrix reconnection (FIGURE 6). The *separatrix* is the line separating the regions of different dynamics. We observe that in FIGURE 6b, separatrices delimiting the inner and the outer sets of period-3 orbits reconnect so that their position reverses; that is, the inner set moves out, and the outer moves in. In this process of reconnection, only the stable period-3 orbits reverse their positions in reference to one another, while the unstable ones remain at their previous locations. As L_z^2 further decreases, the invariant curves around the outer stable period-3 orbit become more elongated (FIGURE 7a, 7b) until, finally, the stability of the outer pair of stable/unstable period-3 orbits reverses — the unstable period-3 orbit becomes stable and the stable one destabilizes (FIGURE 7c, 7d).

2.3. Nontwist Maps

The phenomena that we encountered, periodic orbit formation and separatrix reconnection, are important features of the so-called *nontwist* maps that violate the twist condition. For detailed treatment of these nontwist maps and their dynamics, see del-Castillo-Negrete *et al.* [5]. This clearly implies that the dynamics of this system cannot be described by the standard map. Our future efforts will be directed toward developing a nontwist map which will be able to qualitatively describe the dynamics of this kind of systems; this will give us a dramatic improvement in speed over traditional integration methods. Also, we will work on generalizing the conclusions reached in order to describe the broader range of cases.

3. SUMMARY

In summary, it is important to realize that at all times the difference between the number of stable and unstable points is constant (in our case 1), that the irregular period-tripling bifurcations occur when the average rotation number of an invariant curve in the phase space is 1/3 and that the creation as well as the transformation (change of stability) of these irregular period-3 orbit is simultaneous. The orbits in SED potentials seem to obey an effective third integral of motion in addition to energy and the z-component of the angular momentum, as we did not find any chaotic regions in SOS graphs. Thus, we conclude that this is at least a near-integrable, if not fully integrable, dynamical system. Further work is required for a more definitive resolution of this problem.

ACKNOWLEDGMENTS

I would like to thank the organizers of the workshop for giving me this opportunity to present this paper. I also would like to thank Dr. George Contopoulos for his guidance and insight in exploring these irregular bifurcations, Dr. Philip Morrison of the University of Texas at Austin, and Dr. James Meiss of the University of Colorado for their idea of investigating rotation numbers of the invariant curves in SOS. Last, but certainly not least, I would like to thank my advisor Dr. Chris Hunter of Florida State University for his careful mentorship and valuable instruction over the years.

REFERENCES

1. HUNTER, C., B. TERZIĆ, A.M. BURNS, D. PORCHIA & C. ZINK. 1998. Ann. New York Acad. Sci. **867**. This volume.
2. QIAN, E.E., P.T. DE ZEEUW, R.P. VAN DER MAREL & C. HUNTER. 1995. Mon. Not. R. Astron. Soc. **274:** 602.
3. CONTOPOULOS, G. 1970. Astron. J. **75:** 96.
4. FLOQUET, G. 1883. Annales de l'Ecole Normale Supérieur. **12:** 47.
5. DEL-CASTILLO-NEGRETE, D., J.M. GREENE & P.J. MORRISON. 1996. Physica D **91:** 1.

Negative Energy Modes and Gravitational Instability of Interpenetrating Fluids

A.R.R. CASTI,[a,d] P.J. MORRISON,[b] AND E.A. SPIEGEL[c]

[a]*Department of Applied Physics, Division of Applied Mathematics, Columbia University, New York, New York 10027*
[b]*Department of Physics and Institute for Fusion Studies, The University of Texas at Austin, Austin, Texas 78712*
[c]*Department of Astronomy, Columbia University, New York, New York 10027*

> "A negative energy electron will have less energy the faster it moves and will have to absorb energy in order to be brought to rest. No particles of this nature have ever been observed." — P. A. M. Dirac, 1930

ABSTRACT: We study the longitudinal instabilities of two interpenetrating fluids interacting only through gravity. When one of the constituents is of relatively low density, it is possible to have a band of unstable wavenumbers well separated from those involved in the usual Jeans instability. If the initial streaming is large enough, and there is no linear instability, the indefinite sign of the free energy has the possible consequence of explosive interactions between positive and negative energy modes in the nonlinear regime. The effect of dissipation on the negative energy modes is also examined.

1. BACKGROUND

It used to be thought that the stellar components of two spiral galaxies would pass right through each other in the event of a collision and that only the gaseous components would merge. However, simulations over the past twenty years or so [1], [2] have shown that the macroscopic energy of such a collision is quickly converted into internal energy and that merger of the stellar systems is a common natural outcome of a collision. How is this conversion effected?

An answer to this may lie in the physics of streaming instabilities. In the context of plasma physics, interpenetrating electron beams produce the two-stream instability [3] whose gravitational analog has long been recognized, beginning with the investigations of Sweet [4] and Lynden-Bell [5]. Of course, in the case of galaxy collisions, which occur quickly, the conventional two-stream instability may operate too slowly to be effective. However, it is known that even when two streams of interacting particles are not linearly unstable, they may collectively produce negative energy modes that lead to an explosive nonlinear growth of perturbations for arbitrarily small disturbances. There are well-developed criteria for the occurrence of this explosive growth in plasma physics [6] and, as Lovelace *et al.* [7] have suggested

---

[d]Present address: Laboratory of Applied Mathematics, Mount Sinai School of Medicine, Box 1012, One Gustav L. Levy Place, New York, NY 10029.

in their analysis of counterrotating galaxies, we may expect something analogous in the gravitational setting. Our aim here is to briefly develop this topic of negative energy modes for the case of gravitational interaction in the expectation that the phenomena involved will be found significant in a variety of astrophysical processes.

Even in the event of linear instability, the case of counterstreaming populations is significantly different from the standard gravitational instability, which occurs only for perturbation scales greater than the Jeans length [8]. When there are two interpenetrating fluids such as stars and gas, modes of arbitrary wavelength can be rendered unstable. Numerous investigators have reported on these issues. Most of them (such as Ikeuchi et al. [9], Fridman and Polyachenko [10], and Araki [11]) focused primarily on the symmetric situation of identical fluids in counterstreaming motion. In that case one finds that the spectrum of any instabilities arising from the relative motion is wholly contained within the Jeans instability band, and this blurs the distinction between the two processes. This need not be true when this symmetry is broken, and indeed not all authors restricted themselves entirely to the symmetric case. The present venture into the asymmetric problem is intended to focus on the possibility of well-separated instability bands, which has not been elucidated in the gravitational context, as far as we are aware.

2. EQUATIONS OF MOTION

To see the problem in its simplest version, it is useful to have a uniform medium as the unperturbed state. Rather than formulate the problem inconsistently to achieve this end, as Jeans did, we prefer the Einstein device of introducing a cosmological repulsion term. In the Newtonian setting we readily see how to redefine the gravitational potential so that, instead of introducing such a repulsion term, we fill space with a fluid of *negative* gravitational mass of density ρ_Λ. As in the one-fluid plasma model, we treat this density as a constant since its purpose is to allow a gravitationally neutral background state. One may also contemplate the analog of the two-fluid plasma model in which this background antigravitational fluid has its own dynamics, but we do not do that here. The two dynamically active fluids we consider are gravitationally ordinary and polytropic. Thus, the Poisson equation is written here as,

$$\nabla^2 V = 4\pi G(\rho_1 + \rho_2 - \rho_\Lambda), \qquad (1)$$

where V is the gravitational potential, ρ_1 and ρ_2 are the source densities of the conventional fluids, and $-\rho_\Lambda$ is the cosmological background density.

The equations of motion for the two fluids ($j = 1, 2$) are,

$$\rho_j(\partial_t + \mathbf{u}_j \cdot \nabla)\mathbf{u}_j = -\nabla p_j - \rho_j \nabla V \qquad (2)$$

$$\partial_t \rho_j + \nabla \cdot (\rho_j \mathbf{u}_j) = 0, \qquad (3)$$

where we do not sum over repeated indices. Each fluid has a sound speed, $c_j^2 = \partial p_j / \partial \rho_j$, and a Jeans wavenumber, $k_{Jj}^2 = 8\pi G \hat{\rho}_j / c_j^2$, where the hat signifies the equilibrium value and an uncustomary factor of 2 appears in the definition of the Jeans wavenumbers.

We shall use natural units with k_{J1}^{-1} as the length scale and $(k_{J1}c_1)^{-1}$ as the time scale. We further simplify the description by considering only longitudinal motions in one-dimension, so each single-component velocity field may be expressed as the gradient of a velocity potential: $u_j = \partial_x \phi_j$. The fundamental equations (1)–(3) take the dimensionless form,

$$M_1 \partial_t \phi_1 + \frac{1}{2} M_1^2 (\partial_x \phi_1)^2 + \frac{\rho_1^{\gamma_1 - 1}}{\gamma_1 - 1} + V = \beta_1 \tag{4}$$

$$cM_2 \partial_t \phi_2 + \frac{1}{2} c^2 M_2^2 (\partial_x \phi_2)^2 + \frac{c^2 \rho_2^{\gamma_2 - 1}}{\gamma_2 - 1} + V = \beta_2 \tag{5}$$

$$\partial_t \rho_1 + M_1 \partial_x (\rho_1 \partial_x \phi_1) = 0 \tag{6}$$

$$\partial_t \rho_2 + cM_2 \partial_x (\rho_2 \partial_x \phi_2) = 0 \tag{7}$$

$$\partial_x^2 V - \frac{1}{2} \rho_1 - \frac{1}{2} \beta \rho_2 + \frac{\rho_\Lambda}{2\hat{\rho}_1} = 0 \tag{8}$$

where the $M_j = \mathcal{U}_j/c_j$ are Mach numbers, $\mathcal{U}_j$ measures the initial streaming velocities, $\beta = \hat{\rho}_2/\hat{\rho}_1$, $c = c_2/c_1$, and the Bernoulli constants B_j are chosen to balance the basic state.

3. LINEAR THEORY

We perturb from the state of uniform densities and constant velocities by setting $\phi_j = (-1)^{j+1}x + \delta\phi_j$, $\rho_j = 1 + \delta\rho_j$, and $V = \hat{V} + \delta V$. The density terms of the Poisson equation (8) combine to vanish and $\hat{V}$ is a constant. Since the linearized equations are separable we may decompose the perturbations into normal modes proportional to $\exp(i\omega t - ikx)$ to find the dispersion relation,

$$\Gamma(\omega, k) = 1 + \frac{1}{2[(\omega - kM_1)^2 - k^2]} + \frac{\beta}{2[(\omega + ckM_2)^2 - c^2k^2]}$$

$$\equiv 1 + \Gamma_1 + \Gamma_2 = 0. \tag{9}$$

The quantity Γ, which we call the diagravic function by analogy with the dielectric function of electrodynamics, measures the collective response of the fluid to disturbances in the gravitational field and will serve to indicate the energy signature of any normal mode (Section 4).

For real k the solutions of (9) with complex ω correspond to instability; if ω is real, then solutions of (9) with complex k can give rise to wave amplification instability. Here we analyze only the case of real k. However, we have to deal with both the traditional Jeans instability as well as the two-stream instability, the latter of which involves a sympathetic bunching of particles and is effective for creating instability when the phase speed of the disturbance conspires to create a resonance between different modes.

TABLE 1[a]

Mach Range	Mode Type	k_{crit}^2	$\lim_{M\to 1} k_{crit}^2$	$\lim_{M\to\infty} k_{crit}^2$
$0 \leq M < 1$	Jeans	$\dfrac{1}{1-M^2}$	∞	Not applicable
$1 \leq M$	Two-Stream	$\dfrac{\sqrt{M^2-1}}{4M\left(M^2-1+M\sqrt{M^2-1}\right)}$	$\dfrac{1}{4}$	0
$1 \leq M$	Jeans	$\dfrac{\sqrt{M^2-1}}{4M\left(1-M^2+M\sqrt{M^2-1}\right)}$	$\dfrac{1}{4}$	$\dfrac{1}{2}$

[a]The value of k_{crit}^2 for the two-stream modes with $M \geq 1$ more accurately refers to the k value at which the unstable two-stream and Jeans branches merge.

3.1. Symmetric Case

If both fluids have the same basic properties ($c = 1$ and $\beta = 1$), a frame exists in which $M_1 = M_2 = M$. The dispersion relation then simplifies into a manageable biquadratic with solutions,

$$\omega_\pm^2 = -\frac{1}{2} + k^2(M^2+1) \pm \sqrt{\frac{1}{4} - 2k^2M^2 + 4M^2k^4}. \qquad (10)$$

For $M = 0$, we recover a simple version of the previously studied two-fluid Jeans problem [10], [12], [13]. We find $\omega_+^2 = k^2$, corresponding to sound waves at all k, and $\omega_-^2 = k^2 - 1$, which is the conventional Jeans dispersion relation. The new acoustic modes arise because the aggregate fluid now allows motions unaffected by the gravitational field; for these modes the perturbed gravitational potential is zero.

With relative velocity in the subsonic regime ($0 < M < 1$), there is only a single unstable mode that branches continuously from the Jeans mode at $M = 0$. This mode is unstable for all wavenumbers below a critical value that approaches infinity as M tends to unity from below (see TABLE 1). To study this limit, we let $M^2 = 1 - \alpha/k^2$ with $0 < \alpha < 1$. As $k \to \infty$ we find the approximate solution $\omega_-^2 \sim -\alpha(1-\alpha)/4k^2$, which reveals a weak instability at large k. Thus, weak relative streaming allows gravitational instability at arbitrarily small wavelengths. These large-k instabilities do not arise for Maxwellian velocity distributions within the context of the Vlasov equation [9].

For supersonic motion ($M > 1$) the large-k gravitational instability is no longer present, but a new instability emerges that we call a two-stream instability since it owes its presence to the energy contained in the initial streaming motion. As M ranges from 1 to ∞, the critical wavenumber for instability increases from $k_{crit} < 1/2$ to $k_{crit} < \sqrt{2}/2$.

The upper half of FIGURE 1 shows that near $k = 0$ the two-stream modes are wholly contained within the Jeans band. This fact coupled with the larger growth rates of the Jeans modes has led some to believe that the two-stream instability is swamped by the Jeans instability and is essentially unimportant [11]. As k increases the two-stream and Jeans modes collapse upon each other and together bifurcate into grow-

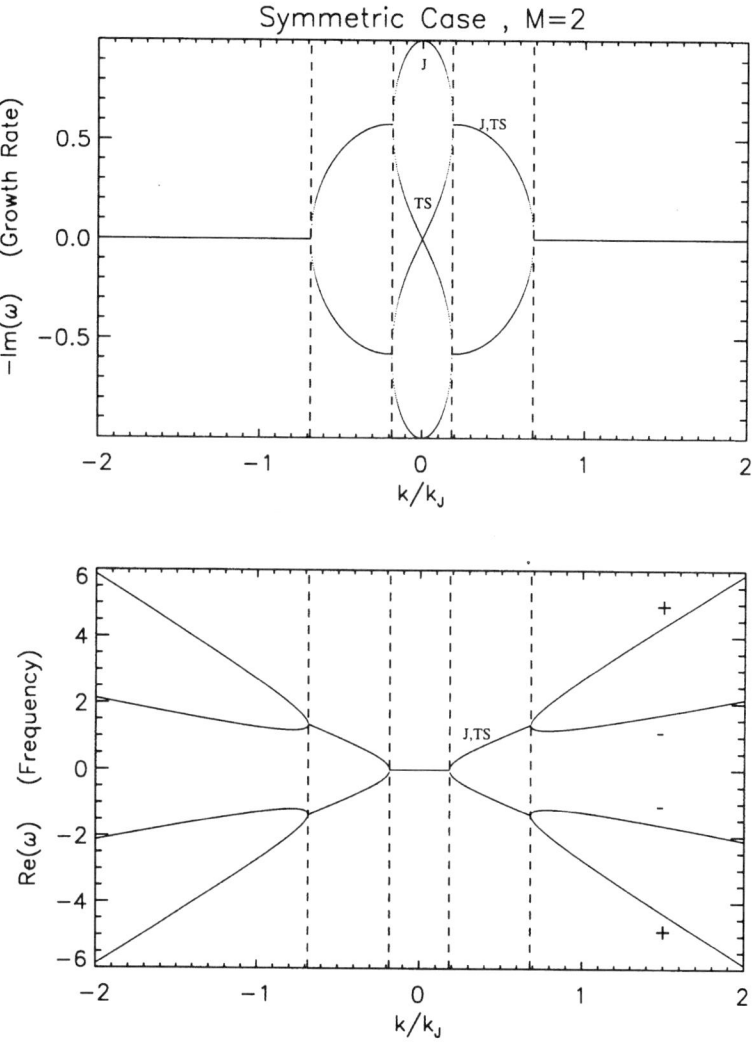

FIGURE 1. Dispersion curves for supersonic motion, $M = 2$. The Jeans modes are indicated with a "J" and the two-stream modes with a "TS." The growth rates of the J and TS modes merge at the onset of oscillations shown in the lower panel. Dashed vertical lines demarcate the critical wavenumbers for growth of the Jeans and two-stream modes. The "+" and "−" signs in the frequency plot denote the energy signature discussed in Section 4.

ing and damped oscillations. At still larger wavenumbers all motions are stable, propagating waves. The critical wavenumbers below which growth is possible at any Mach number are shown in TABLE 1.

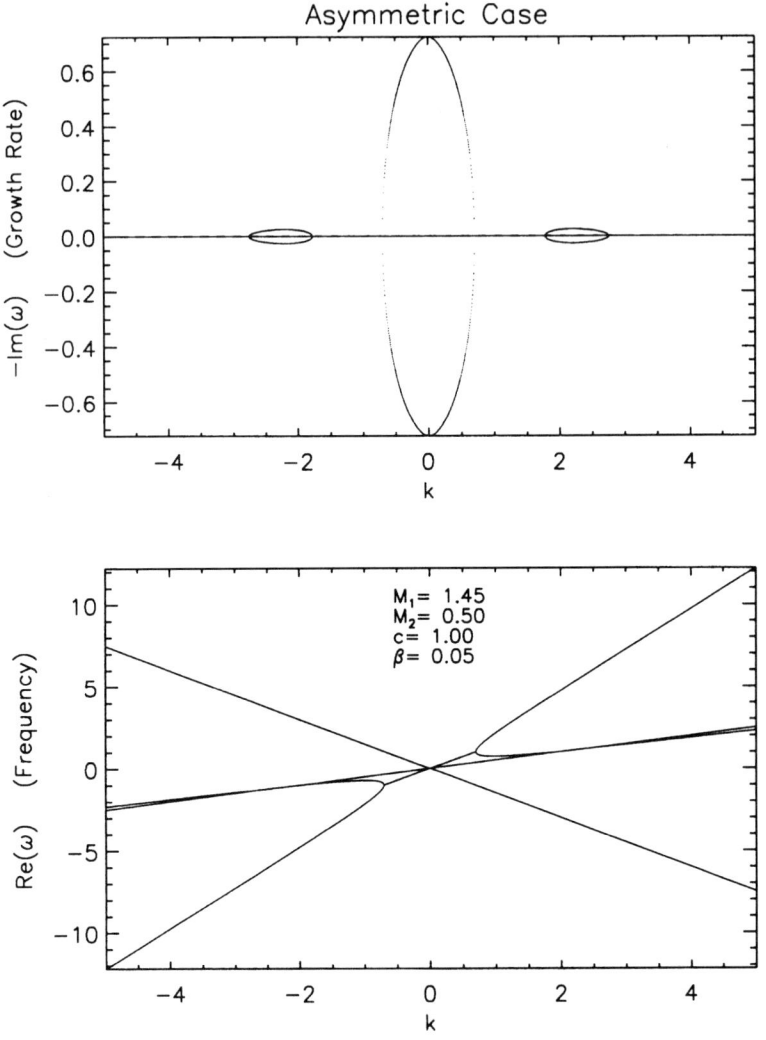

FIGURE 2. Asymmetric case: $\beta = .05$, $c = 1$, $M_1 + M_2 = 1.95$. Illustration of the pinched two-stream bubbles. The frequencies corresponding to the two-stream modes are degenerate along the entire k-band of the bubble. At the edge of the bubble the frequencies once again separate.

3.2. Asymmetric Case

When we relax the constraint of identical conditions in the two fluids, one of the more interesting consequences is the possibility of large wavenumber two-stream instability bands well-separated from the Jeans instabilities clustered at small k. For illustration we consider the effect of changing the initial relative streaming $M_1 + cM_2$ for fixed β and c. In fact, β turns out to be the crucial parameter in achieving the spectral separation; variations in the sound speed ratio c widens both bands together. The distancing of a bubble of two-stream modes from the Jeans band is illustrated in FIGURE 2.

The large-k two-stream modes can be explained qualitatively by examining the separate pieces of the dispersion relation (9). Since the streaming instability is related to resonant motions, it is revealing to examine the solutions to $1 + \Gamma_1(\omega, k) = 1 + \Gamma_2(\omega, k) = 0$ in isolation and see where the curves intersect. The frequencies of these noninteracting modes are given by,

$$\omega_1 = kM_1 \pm \sqrt{k^2 - \frac{1}{2}} \tag{11}$$

$$\omega_2 = -ckM_2 \pm \sqrt{c^2k^2 - \frac{\beta}{2}}. \tag{12}$$

When $\omega_1 = \omega_2$, the assumed independent frequencies match one another for wavenumbers satisfying,

$$k(M_1 + cM_2) = \begin{cases} \pm\left(\sqrt{k^2 - \frac{1}{2}} + \sqrt{c^2k^2 - \frac{\beta}{2}}\right) \\ \pm\left(\sqrt{k^2 - \frac{1}{2}} - \sqrt{c^2k^2 - \frac{\beta}{2}}\right). \end{cases} \tag{13}$$

In the symmetric case where $M_1 = M_2 \equiv M$, $c = \beta = 1$, we see that two of the resonances are lost except in the irrelevant cases $M = 0$ and $k = 0$. The other pair of possibilities, $kM = \pm(\sqrt{k^2 - \frac{1}{2}})$, just restate the critical wavenumber condition for what we know to be the modified Jeans instability when $M < 1$. In the general case, we can expect another pair of intersections that account for the two-stream bubbles of FIGURE 2.

4. NONLINEAR THEORY

4.1. Hamiltonian Formulation and Energy Signature

The dynamical equations (4)–(8) derive from a variational principle and a conserved Hamiltonian functional. The variational formulation has the advantage of shedding light on the relation between the energy content of the disturbances and nonlinear stability. Here we present results for the symmetric case, though the formalism follows through for the asymmetric case as well.

The Hamiltonian and associated equations are,

$$H = \frac{1}{M}\sum_{j=1}^{2}\int_0^L dx\left(\frac{M^2}{2}\rho_j\phi_{jx}^2 + \rho_j U_j - V_x^2\right), \tag{14}$$

$$\partial_t \phi_j = \{\phi_j, H\} = -\frac{\delta H}{\delta \rho_j}, \tag{15}$$

$$\partial_t \rho_j = \{\rho_j, H\} = -\frac{\delta H}{\delta \phi_j}, \tag{16}$$

where $U_j(\rho_j) = \rho_j^{\gamma-1}/\gamma(\gamma-1)$ is the internal energy for the jth fluid and the Poisson bracket is defined by,

$$\{F, G\} = \sum_{j=1}^{2}\int_0^L dx'\left(\frac{\delta G(x)}{\delta\phi_j(x')}\frac{\delta F(x)}{\delta\rho_j(x')} - \frac{\delta F(x)}{\delta\phi_j(x')}\frac{\delta G(x)}{\delta\rho_j(x')}\right), \tag{17}$$

For definiteness we have chosen a box-geometry of length L.

The energy content of a particular mode when the perturbation amplitude is small is given by the second variation of H evaluated at equilibrium (the free energy). After some calculation this is seen to be,

$$\delta^2 H = \frac{1}{2M}\int_0^L dx (M^2[\delta\phi_{1x}^2 + \delta\phi_{2x}^2] + 2M^2[\delta\rho_1\delta\phi_{1x} - \delta\rho_2\delta\phi_{2x}] \\ + \delta\rho_1^2 + \delta\rho_1^2 - 2\delta V_x^2) \quad . \tag{18}$$

It may be verified that this functional is conserved by the equations of motion. Suppose we now insert into (18) an eigenfunction corresponding to a single stable mode with $\text{Im}(\omega) = 0$. Employing overbars to denote eigenvector components and * for complex conjugation, we write,

$$\delta V = \bar{V}e^{i(\omega t - kx)} + \bar{V}_j^* e^{-i(\omega^* t - kx)}, \tag{19}$$

and similarly for the other perturbation variables. Upon effecting the integrations, we can make use of the dispersion relation,

$$\Gamma(\omega, k) = 1 + \frac{1}{2[(\omega - kM)^2 - k^2]} + \frac{1}{2[(\omega + kM)^2 - k^2]} = 0, \tag{20}$$

to express the modal free energy in the compact form,

$$\delta^2 H = -2Lk^2|\bar{V}|^2 \omega \frac{\partial \Gamma}{\partial \omega}. \tag{21}$$

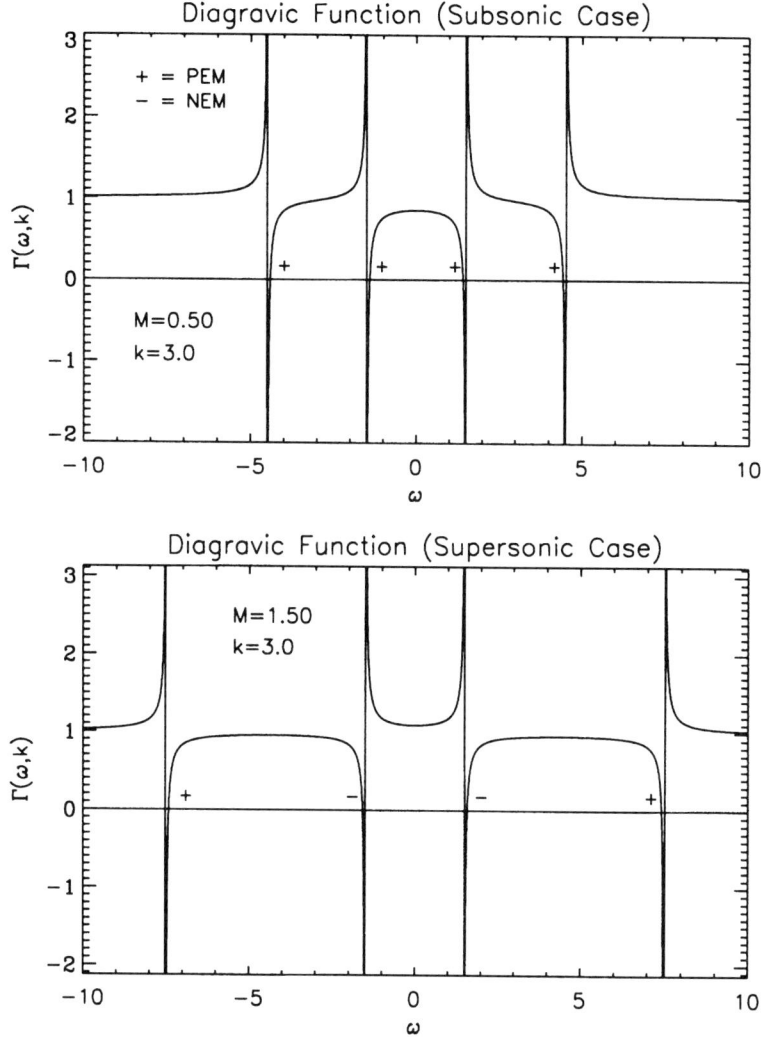

FIGURE 3. Diagravic function in the symmetric case for linearly stable modes with $k = 3$. A stable mode exists wherever Γ crosses the ω-axis. The positive energy modes are indicated with + and the negative energy modes with –.

Wherever $\omega(\partial\Gamma/\partial\omega) < 0$ a positive energy mode (PEM) is implied by (21), while the condition $\omega(\partial\Gamma/\partial\omega) > 0$ defines a negative energy mode (NEM). This possibility of modes of either signature has been elucidated in the plasma physics literature [14] and a gravitational analog was suggested by Lovelace *et al.* [7] in the context of thin, counterrotating stellar disks.

FIGURE 3 shows the diagravic function for both a subsonic and a supersonic case of stable modes with $k = 3$. In the subsonic regime, we find the Hamiltonian to be positive definite near equilibrium since $\omega(\partial\Gamma/\partial\omega) < 0$ at every crossing of Γ on the ω-axis. Right at the border of supersonic streaming ($M = 1$), concomitantly with the appearance of the two-stream instability, the Γ curves undergo a topological transition that allows the coexistence of positive and negative energy modes. The NEMs are *slow modes* in that they have smaller frequencies than their PEM counterparts. This is the typical situation; ω must pass through zero if the energy signature changes [15]. We expect from the precedents of plasma physics [6] that the simultaneous presence of positive and negative energy modes has dramatic consequences on the nonlinear stability of the system.

4.2. Reduction to Action Angle Variables

In the rest of this section we will concentrate on the nonlinear interactions between linearly stable modes in the symmetric problem (see FIGURE 3). From a physical standpoint, attention is focused on situations where the disturbances are of sufficiently small scale so that the Jeans instability can be ignored, though there are no compelling reasons why this ought to be the case. We will further assume supersonic motion in order to examine the interaction of positive and negative energy modes, a situation we expect to be the most interesting. Under these assumptions the equations of motion achieve their simplest form in action-angle coordinates that we now develop.

First we Fourier transform the field variables:

$$\rho_j = \sum_{m=-\infty}^{\infty} \rho_m^{(j)}(t) e^{ik_m x}, \quad \rho_{-m}^{(j)} = \rho_m^{(j)*}, \quad k_m = \frac{2\pi m}{L}. \tag{22}$$

We can then write the free energy in terms of real variables as,

$$\delta^2 H = \frac{1}{2} \sum_{m=1}^{\infty} (\mathbf{q}^T \mathbf{A} \mathbf{q} + \mathbf{p}^T \mathbf{B} \mathbf{p}), \tag{23}$$

where $\mathbf{q} \equiv (q_1, q_2, q_3, q_4)^T$, $\mathbf{p} \equiv (p_1, p_2, p_3, p_4)^T$, are linear combinations of the complex modal amplitudes, $\mathbf{A}$ and $\mathbf{B}$ are symmetric matrices (given in Casti [16]), and the "T" indicates transpose.

Defining the configuration variables $\mathbf{z} = (q_1, \ldots, p_4)^T$, we recast the linearized equations in the form,

$$\frac{d\mathbf{z}}{dt} = \mathbf{J}\nabla_z \delta^2 H \equiv \mathbf{L}\mathbf{z}, \quad \mathbf{L} = \begin{pmatrix} 0 & \mathbf{B} \\ -\mathbf{A} & 0 \end{pmatrix}, \quad \mathbf{J} = \begin{pmatrix} 0 & \mathbf{I} \\ -\mathbf{I} & 0 \end{pmatrix}, \tag{24}$$

where $\mathbf{J}$ is the canonical 8×8 cosymplectic form. The next order of business is to construct a symplectic transformation that puts $\delta^2 H$ in its normal form [17], [18]. This can be achieved by writing $\mathbf{z} = \mathbf{SZ}$, where the matrix $\mathbf{S}$ consists of suitably ordered eigenvectors of $\mathbf{L}$ satisfying the symplectic condition, $\mathbf{S}^T \mathbf{J} \mathbf{S} = \mathbf{J}$. After a final transformation to action-angle coordinates, the free energy expression (23) becomes a superposition of harmonic oscillators,

$$\delta^2 H = \sum_{m=1}^{\infty} (\omega_+ J_1 + \omega_+ J_2 - \omega_- J_3 - \omega_- J_4). \tag{25}$$

The free energy is thus manifestly composed of two pairs each of positive and negative energy modes.

4.3. Three-Wave Resonance and Explosive Growth

Energy conservation forbids nonlinear runaway growth if H is definite (Dirichlet's theorem), as is the case here for subsonic motion. When the relative streaming is supersonic, however, interacting PEMs and NEMs can circumvent this restriction since they contribute energy of opposite sign.

We demonstrate the possibility of explosive growth with a three-wave resonant interaction between two NEMs and one PEM. Since the energy signature of a mode is not Galilean invariant, the existence of a reference frame in which all three modes have the same signature implies nonlinear stability. It may be shown that there is no reference frame in which all three modes have the same energy signature if the highest frequency wave has opposite signature to that of the other two [19]. This provides a criterion for three-wave interaction leading to instability.

The third-order resonance conditions for a triplet of modes are,

$$\begin{aligned} m_1 k_1 + m_2 k_2 + m_3 k_3 &= 0 \\ m_1 \omega_1 \pm m_2 \omega_2 \pm m_3 \omega_3 &= 0 \\ |m_1| + |m_2| + |m_3| &= 3 \quad (m_1, m_2, m_3 \text{ integers}), \end{aligned} \tag{26}$$

which here may be satisfied by $(m_1, m_2, m_3) = (1, 1, -1)$, $(k_1, k_2, k_3) = (k_m, k_m, 2k_m)$, and $(\omega_1, \omega_2, \omega_3) = (\omega_+, \omega_-, \omega_-)$, where the ω_j are taken to be positive. One can see from FIGURE 1 that $\omega_1 > \omega_2, \omega_3$, so the relative signatures of this triplet are immune to a Galilean shift. Note from FIGURE 1 that a resonant triplet involving two PEMs and one NEM would not have robust relative signatures under a frame shift since the PEMs have larger frequencies.

The lowest-order nonlinear terms come from the third variation of H expanded about the dynamical equilibrium,

$$\delta^3 H = \frac{1}{M} \sum_{j=1}^{2} \int_0^L dx \left(\frac{M^2}{2} \delta\rho_j \delta\phi_{jx}^2 + \frac{\gamma}{6} \delta\rho_j^3 \right). \tag{27}$$

In terms of the Fourier amplitudes this expression is,

$$\delta^3 H = \frac{L}{2} \sum_{j=1}^{2} \sum_{m,n=1}^{\infty} \left[-M k_m k_n (\rho_{m+n}^{(j)} \phi_{-m}^{(j)} \phi_{-n}^{(j)} - \rho_{m-n}^{(j)} \phi_{-m}^{(j)} \phi_{-n}^{(j)} + \text{c.c.}) \right. \\ \left. + \frac{\gamma}{3M} (\rho_m^{(j)} \rho_n^{(j)} \rho_{-m-n}^{(j)} - \rho_m^{(j)} \rho_{-n}^{(j)} \rho_{n-m}^{(j)} + \text{c.c.}) \right]. \tag{28}$$

We then effect the same transformations on (28) that led to the diagonalized free energy (25). This spawns a myriad of nonlinear terms, only some of which survive an averaging process that leads to the Birkhoff normal form [20].

For a three-wave resonance, one finds after near-identity transformations that the only higher-order terms contributing to the normal form are of the type [14],

$$O(3) \text{ Terms} \sim J_1^{|l|/2} J_2^{|m|/2} J_3^{|n|/2} \sin(l\theta_1 + m\theta_2 + n\theta_3), \tag{29}$$

with $|l| + |m| + |n| = 3$. The Hamiltonian up to third-order terms for the resonant NEM/PEM triplets can be written,

$$H = \omega_1 J_1 - \omega_2 J_2 - \omega_3 J_3 + \alpha \sqrt{J_1 J_2 J_3} \sin(\theta_1 + \theta_2 + \theta_3), \tag{30}$$

where $\alpha = \alpha(k_m, M, L)$ is a nonlinear coupling constant that is neither especially large or small in the parameter regime of the three-wave resonance considered here.

Since the angles $(\theta_1, \theta_2, \theta_3)$ appear in only one combination in H, further simplification of (30) is possible via the generating function,

$$F_2(\mathbf{I}, \boldsymbol{\theta}) = \frac{1}{2} I_1 (\theta_1 + \theta_2) + I_2 \theta_2 + I_3 \theta_3, \tag{31}$$

with $\psi_j = \partial F_2 / \partial I_j$ and $J_j = \partial F_2 / \partial \theta_j$. This canonical transformation yields,

$$H = \frac{1}{2} \tilde{\omega}_1 I_1 - \omega_3 I_3 + \frac{\alpha}{2} \sqrt{I_1(I_1 + 2I_2)I_3} \sin(2\psi_1 + \psi_3) \tag{32}$$

with $2\tilde{\omega}_1 \equiv \omega_1 - \omega_2$. If one chooses initial conditions satisfying $I_2 \equiv J_2 - J_1 = 0$, then H is identical to the normal form of a two-wave interaction originally presented by Cherry [21], [22]. In terms of the $(\mathbf{q}, \mathbf{p})$ variables, Cherry's Hamiltonian is

$$H = \frac{1}{2} \tilde{\omega}_1 (p_1^2 + q_1^2) - \frac{1}{2} \omega_3 (p_3^2 + q_3^2) + \frac{\epsilon}{2} (2 q_1 p_1 p_3 - q_3 [q_1^2 - p_1^2]), \tag{33}$$

where $\epsilon = \sqrt{2} \alpha / 4$.

The dynamical system generated by the Cherry Hamiltonian is integrable. In the special case of a third-order resonance with $\omega_3 = 2\tilde{\omega}_1$, there exists a family of two-parameter solutions,

$$q_1 = \frac{\sqrt{2}}{\xi - \epsilon t} \sin(\tilde{\omega}_1 t + \eta), \quad p_1 = -\frac{\sqrt{2}}{\xi - \epsilon t} \cos(\tilde{\omega}_1 t + \eta),$$

$$q_3 = -\frac{1}{\xi - \epsilon t} \sin(2\tilde{\omega}_1 t + 2\eta), \quad p_3 = -\frac{1}{\xi - \epsilon t} \cos(2\tilde{\omega}_1 t + 2\eta),$$

$$\tag{34}$$

where ξ and η are constants depending on the initial conditions.

The solutions (34) show the possibility of *finite-time density singularities* when two negative energy modes interact resonantly with a positive energy mode. A system exhibiting this behavior is said to undergo *explosive growth*, and it could be an important mechanism for structure formation in galactic and cosmological settings when relative motion between different fluid species is involved. If the resonance is detuned, separatrices bounding stable orbits emerge in phase space, but the dynamics are still prone to finite-amplitude instability.

5. DISSIPATIVE INSTABILITY

We close our investigation of the consequences of negative energy modes by examining the effects of dissipation on the linear stability of the system. With negative energy modes propagating through a dissipative medium, we may expect new instabilities since the damping can pump more negative energy into the wave. This somewhat counterintuitive effect of frictional forces in other contexts was first pointed out by Kelvin and Tait [23] (see also Zajac [24]).

Suppose that collisions are important at some stage in the development of a gravitationally bound structure. A simple model of this effect incorporates a dynamical friction term $(-1)^j \nu (\mathbf{u}_1 - \mathbf{u}_2)$ on the right side of the momentum equations (2), where ν is a positive damping coefficient.

In the dimensionless symmetric case the dispersion relation becomes,

$$\omega^4 - 2i\nu\omega^3 + [1 - 2k^2(M^2+1)]\omega^2 + 2i\nu[k^2(M^2+1) - 1]\omega + k^2(M^2-1)[k^2(M^2-1)+1] = 0. \tag{35}$$

If we assume the damping is weak, $\nu \ll 1$, we may develop (35) in a regular perturbation series in ν to find the lowest-order corrections to the frequencies (10),

$$\omega_1^\pm = \frac{i}{2}\left(1 \pm \frac{1}{\sqrt{1 - 8k^2M^2 + 16M^2k^4}}\right), \tag{36}$$

where we assume $k^2 \gg (M \pm \sqrt{M^2-1})/(4M)$ to avoid the singularity accompanying the vanishing denominator in (36) (see Casti [16] for the details). If we assume $M > 1$ so that negative energy modes are present, then a close examination of the corrections reveals that the dissipation promotes instability in the wavenumber band $(M \pm \sqrt{M^2-1})/(4M) \ll k < \sqrt{2}/2$ for any $M \geq 1$ *no matter how weak the damping*. Since the instability as $k^2 \to \frac{1}{2}$ is realized only in the $M \to \infty$ limit of the undamped problem, we see that the dissipation indeed has the effect of destabilizing modes that were stable in the conservative case. A numerical investigation revealed that this result holds for any $\nu > 0$.

The modal bands destabilized by the damping become more significant in the asymmetric case. As remarked in Section 3.2, bubbles of unstable two-stream modes can pinch off from the Jeans-unstable bubble and result in well-separated instability bands. The inclusion of dissipation can destabilize the entire band of modes separating the bubbles, as well as some higher-k modes beyond the undamped two-stream bubble. This is illustrated in FIGURE 4.

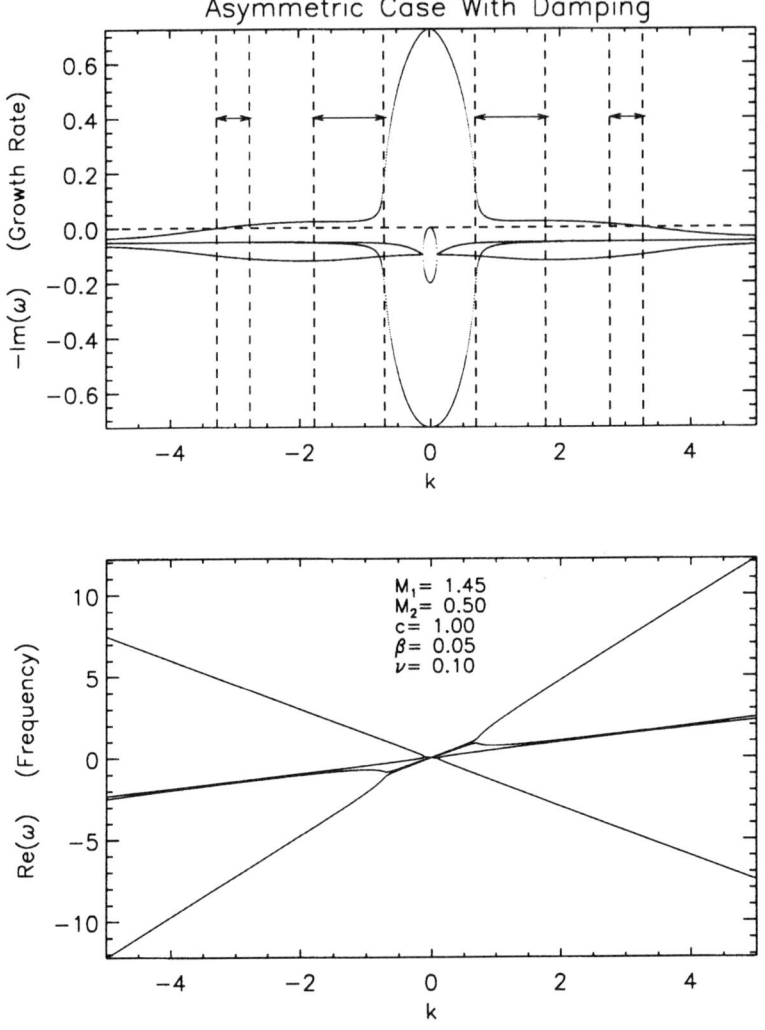

FIGURE 4. Dispersion curves for damped supersonic motion, $\nu = 0.1$, $M_1 = 1.45$, $M_2 = 0.5$, in the asymmetric case with $\beta = .05$ and $c = 1$. The entire k-band separating the Jeans bubble and two-stream bubbles of FIGURE 2 are now unstable. The modal bands destabilized by the dissipation are demarcated by the arrows between the dashed vertical lines.

One should not assume that any form of dissipation will destabilize negative energy modes. For instance, if each fluid feels only a drag proportional to its own velocity, there are no new instabilities even with relative motion. In other words, the dissipation must in some sense project onto the eigenspace spanned by the negative energy modes in a way that decreases their energies. This depends not only on the nature of the dissipation, but also upon the initial equilibrium about which one perturbs.

The effect of damping can be understood by examining the time evolution of the free energy. If the dissipation acts to increase the energy of a positive energy mode or decrease the energy of a negative energy mode, then one can show that pure imaginary eigenvalues take on a positive real part [25]. To see that this is possible here, consider the temporal change in the free energy, which in the symmetric case can be written,

$$\delta^2 \dot{H} = -\nu M \int_0^L dx (\delta\phi_{1x} - \delta\phi_{2x})^2 - \nu M \int_0^L dx (\delta\phi_1 - \delta\phi_2)(\delta\rho_{1x} + \delta\rho_{2x}) . \qquad (37)$$

Since the first term of $\delta^2 \dot{H}$ is negative definite, the conditions for which the free energy decays or grows in time are determined by the relative phasings of the velocity and density perturbations comprising the second term. One may deduce the effect of the dissipation on any particular mode of the conservative problem by inserting the undamped modes into (37), which yields a formula for $\delta^2 \dot{H}$ valid up to $\mathcal{O}(\nu^2)$. For subsonic relative motion, $M < 1$, the expression (37) is negative definite and the damping lives up to its name and causes the PEMs to decay in time. When $M > 1$, $\delta^2 \dot{H}$ can be either positive or negative for an NEM depending on the value of k, which explains why some NEMs are destabilized and others are damped in the usual sense.

6. DISCUSSION

The formation of structures through the action of gravity is much analyzed in cosmology, galactic structure and cosmogony. Most of this analysis is centered on the operation of gravitational instability, though streaming fluids can resonantly interact via the gravitational field to cause linear instability in spectral ranges inaccessible to the traditional Jeans instability. As we have brought out here, the distinction between the two types of unstable modes, the Jeans and the two-stream, becomes sharper when one constituent is far denser than the other. Much of the previous work on the subject failed to take advantage of this crucial feature by focusing attention on situations where each component exists in equal abundance. Even when the two-stream instability does not occur, if the total energy of the gravitational two-stream interaction is indefinite, the positive and negative energy modes that are *stable* in the linear theory can interact to produce explosive development of disturbances of arbitrarily small amplitude. This can be a significant aspect of the theory of structure formation.

There are many clear instances where the dynamics of interpenetrating fluids may play a role in developing structures, but we close here by suggesting that even when the streaming is not apparent, two-stream dynamics may be relevant. An interesting example is provided by the coexistence of dark matter and luminous (baryonic) matter that is generally believed to occur throughout the cosmos. The locations of the two kinds of matter seem to be well correlated, which would not be the case if they were now streaming through each other. On the other hand, it might be reasonable to ask why there is this apparent correlation (or anticorrelation in the case of negative gravitational density) of the two kinds of material. Even if they had once been in relative motion, this situation would not long persist, as we have seen. But the outcome, as far as large-scale structure is concerned, could be quite different if the kinematic history of the interaction of the two matters had been richer than has been supposed hitherto. Given the indefiniteness of the free energy if the initial streaming is large enough, waves of short-length scale could have interacted in an explosive manner to quickly produce highly nonlinear density fluctuations. This is a feature of gravitational structure development that could be profitably studied, particularly in situations where the background Hubble expansion cannot be ignored. The dynamics of a two-fluid system with initially *time-dependent* relative motion is currently under investigation.

REFERENCES

1. THEYS, J.C. & E.A. SPIEGEL. 1977. Ap. J. **212:** 616.
2. BARNES, J. E. 1992. Ap. J. **393:** 484.
3. BUNEMAN, O. 1959. Phys. Rev. **115:** 503.
4. SWEET, P.A. 1963. Monthly Notices Roy. Astron. Soc. **125:** 285.
5. LYNDEN-BELL, D. 1967, in Relativity Theory and Astrophysics 2. Galactic Structure. J. Ehlers, Ed. American Math. Soc. Providence, RI.
6. DAVIDSON, R.C. 1972. Methods in Nonlinear Plasma Theory. Academic Press, New York.
7. LOVELACE, R.V.E., K.P. JORE & M.P. HAYNES. 1997. Ap. J. **475:** 83.
8. JEANS, J. 1902. Phil. Trans. R. Soc. **199A:** 49.
9. IKEUCHI, S., T. NAKAMURA & F. TAKAHARA. 1974. Prog. Theor. Phys. **52:** 1807.
10. FRIDMAN, A.M. & V.L. POLYACHENKO. 1984. Physics of Gravitating Systems II. Springer-Verlag. New York.
11. ARAKI, S. 1987. Astron. J. **94:** 99.
12. SPIEGEL, E.A. 1972, in Symposium on the Origin of the Solar System: 165. H. Reeves, Ed. Editions du Centre National de la Recherche Scientifique, Paris.
13. DE CARVALHO, J.P.M. & P.G. MACEDO. 1995. Astron. Astrophys. **299:** 326.
14. KUENY, C.S. & P.J. MORRISON. 1995. Phys. Plasmas. **2:** 1926.
15. MORRISON, P.J. 1998. Rev. Mod. Phys. **70:** 467.
16. CASTI, A.R R. 1998. Ph. D. Thesis. Columbia University.
17. MOSER, J.K. 1958. Comm. Pure Appl. Math. **11:** 81.
18. MEYER, K.R. & G.R. HALL. 1991. Introduction to Hamiltonian Dynamical Systems and the N-Body Problem. Springer-Verlag. New York..
19. WEILAND, J. & H. WILHELMSSON. 1977. Coherent Nonlinear Interaction of Waves in Plasmas. Pergamon. Oxford.
20. OZORIO DE ALMEIDA, A.M. 1988. Hamiltonian Systems: Chaos and Quantization. Cambridge University Press. Cambridge.
21. CHERRY, T.M. 1925. Trans. Cambridge Philos. Soc. **23:** 199.
22. WHITTAKER, E.T. 1937. Analytical Dynamics: 142. Cambridge, London.
23. THOMPSON, W. (Lord Kelvin) & P.G. TAIT. 1921. Treatise on Natural Philosophy, Part I: 388. Cambridge University Press. Cambridge.
24. ZAJAC, E.E. 1964. Journal of the Astronautical Sciences **11:** 46.
25. MACKAY, R.S. 1991. Phys. Lett. A. **155:** 266.

Invariants and Labels in Lie–Poisson Systems

JEAN-LUC THIFFEAULT[a] AND P.J. MORRISON[b]

Institute for Fusion Studies and Department of Physics, University of Texas at Austin, Austin, Texas, 78712-1060

> ABSTRACT: Reduction is a process that uses symmetry to lower the order of a Hamiltonian system. The new variables in the reduced picture are often not canonical: there are no clear variables representing positions and momenta, and the Poisson bracket obtained is not of the canonical type. Specifically, we give two examples that give rise to brackets of the noncanonical Lie–Poisson form: the rigid body and the two-dimensional ideal fluid. From these simple cases, we then use the *semidirect product* extension of algebras to describe more complex physical systems. The Casimir invariants in these systems are examined, and some are shown to be linked to the recovery of information about the configuration of the system. We discuss a case in which the extension is not a semidirect product, namely compressible reduced MHD, and find for this case that the Casimir invariants lend partial information about the configuration of the system.

INTRODUCTION

This paper explores the Casimir invariants of Lie–Poisson brackets, which generate the dynamics of some discrete and continuous Hamiltonian systems. Lie–Poisson brackets are a type of noncanonical Poisson bracket and are ubiquitous in the reduction of canonical Hamiltonian systems with symmetry. Casimir invariants are constants of motion for all Hamiltonians; they are associated with the degeneracy of noncanonical Poisson brackets. Finite-dimensional examples of systems described by Lie–Poisson brackets include the heavy top and the moment reduction of the Kida vortex, while infinite-dimensional examples include the 2D ideal fluid, reduced magnetohydrodynamics (MHD), and the 1D Vlasov equation. (See Ref. [1] and references therein for a full review.) The Casimir invariants determine the manifold on which the system is kinematically constrained to evolve. Understanding the nature of these constraints is thus of paramount importance.

In Section I we examine specific Lie–Poisson brackets, namely, those that arise from the reduction to Eulerian variables of a Lagrangian system with relabeling symmetry. We make use of two prototypical examples, the rigid body (finite-dimensional) and the 2D ideal fluid (infinite-dimensional), and we interpret their Casimir invariants. This is done to motivate the introduction of such brackets and to show their physical relevance. In Section II we turn to building Lie–Poisson brackets directly from Lie algebras by the procedure of extension. We introduce the semidirect product extension and illustrate it with two physical examples: the heavy top and

---

[a]Address for telecommunication: phone, 512-471-6121; fax, 512-471-6715; e-mail: jeanluc@physics.utexas.edu
[b]E-mail: morrison@hagar.ph.utexas.edu

low-beta reduced MHD. Finally, in Section III we look at a nonsemidirect example, compressible reduced MHD, and discuss work in progress.

I. LIE–POISSON BRACKETS AND REDUCTION

For our purposes, a reduction is a mapping of the dynamical variables of a system to a smaller set of variables, such that the transformed Hamiltonian and bracket depend only on the smaller set of variables. (See, for example, Ref. [2] for a detailed treatment.) The simplest example of a reduction is the case in which a cyclic variable is eliminated, but more generally a reduction exists as a consequence of an underlying symmetry of the system. We present two examples of reduction.

A. Reduction of the Free Rigid Body

The Hamiltonian for the free rigid body is an unwieldy function of three Euler angles ϕ, ψ, θ and their conjugate momenta p_ϕ, p_ψ, p_θ. The motion is described by Hamilton's equations using the canonical bracket

$$\{f, g\}_{can} = \frac{\partial f}{\partial \phi}\frac{\partial g}{\partial p_\phi} - \frac{\partial g}{\partial \phi}\frac{\partial f}{\partial p_\phi} + \frac{\partial f}{\partial \psi}\frac{\partial g}{\partial p_\psi} - \frac{\partial g}{\partial \psi}\frac{\partial f}{\partial p_\psi} + \frac{\partial f}{\partial \theta}\frac{\partial g}{\partial p_\theta} - \frac{\partial g}{\partial \theta}\frac{\partial f}{\partial p_\theta}.$$

Here we have 3 degrees of freedom (6 coordinates), the configuration space is the rotation group $SO(3)$, and the phase space is its contangent bundle $T^*SO(3)$.

A reduction is possible for this system. In terms of angular momenta ℓ_i about the principal axes, we have

$$H(\phi, \psi, \theta, p_\phi, p_\psi, p_\theta) \longrightarrow H(\ell_1, \ell_2, \ell_3) = \sum_{i=1}^{3} \frac{\ell_i^2}{2I_i}.$$

Under this (noncanonical) mapping, the bracket obtains the Lie–Poisson form

$$\{f, g\} = -\ell \cdot \frac{\partial f}{\partial \ell} \times \frac{\partial g}{\partial \ell}.$$

The equations of motion generated by this bracket and $H(\ell_1, \ell_2, \ell_3)$ are Euler's equations for the rigid body,

$$\dot{\ell}_i = \frac{I_j - I_k}{I_j I_k} \ell_j \ell_k,$$

where i,j,k are cyclic permutations of $1,2,3$. The energy is conserved, and so is the quantity

$$C(\ell) = \sum_{i=1}^{3} \ell_i^2,$$

which commutes with any function of ℓ. Such functions are called Casimir invariants (or Casimirs for short). Casimirs are conserved quantities for any Hamiltonian, so they tell us about the topology of the manifold on which the motion takes place. For the simple case of the rigid body, the motion takes place on the two-sphere, S^2 (not in physical space, but in angular momentum space). The symmetry that permits the reduction is the invariance of the equations of motion for $(\phi, \psi, \theta, p_\phi, p_\psi, p_\theta)$ under rotations (elements of $SO(3)$). This symmetry amounts to the freedom of choosing axes from which the Euler angles are measured. In that sense it is a relabeling symmetry, since the choice of axes amounts to making "marks," or labels, on the rigid body.

We shall say that the original system is a Lagrangian description (by analogy with the fluid case below) because at any time the exact configuration of the system (including orientation) is known, whereas the reduced system is Eulerian because only the angular momentum of the body is known.

B. Reduction of the 2D Ideal Fluid

As our prototype for reduction in an infinite-dimensional system we take an ideal 2D fluid confined to some domain, D. The Lagrangian description involves fluid elements labeled by some coordinate a, which is usually taken to be the initial position of the fluid elements. These labels are analogous to the choice of axes in the rigid body example above (the "marks" on the rigid body). The Hamiltonian functional is

$$H[q;\pi] = \int_D \left[\frac{\pi^2}{2\rho_0} - p(a,t)\left(\left|\frac{\partial q}{\partial a}\right| - 1\right) \right] d^2 a,$$

where $q(a,t)$ is the position of the fluid element labeled by a and $\pi(a,t)$ is its momentum. The Jacobian of the transformation from the labels a to the position of the fluid elements at a later time is $|\partial q/\partial a|$. The density ρ_0 is taken to be constant, and the pressure $p(a,t)$ appears here as a Lagrange multiplier that enforces the incompressibility condition, $|\partial q/\partial a| = 1$ (see Ref. [3]). This Hamiltonian together with the canonical bracket

$$\{F, G\}_{can} = \int_D \left(\frac{\delta F}{\delta q}\frac{\delta G}{\delta \pi} - \frac{\delta G}{\delta q}\frac{\delta F}{\delta \pi} \right) d^2 a$$

generates the equations of motion for a fluid in Lagrangian variables. The information about the position of every fluid element at any time is contained in the model. The dynamical evolution of the system is independent of the particular choice of labels for the fluid elements. This *relabeling* symmetry of the initial condition labels, a, suggests a reduction.

We introduce the streamfunction ϕ defined by $v(\mathbf{x},t) = (-\partial_y \phi, \partial_x \phi)$, so that $\nabla \cdot v = 0$ is automatically satisfied, and the vorticity $\omega(\mathbf{x},t) = \nabla^2 \phi$. The mapping from the Lagrangian momentum to the Eulerian velocity field is

$$\rho_0 v(\mathbf{x},t) = \int_D \pi(a, t)\delta(\mathbf{x} - q(a, t))\, d^2 a.$$

We take $\rho_0 = 1$ for simplicity. Taking the curl of v and dotting with $\hat{z}$ gives

$$\omega(\mathbf{x},t) = \hat{z} \cdot \nabla \times v(\mathbf{x},t) = \int_D \epsilon_{ij}\pi_i(a,t)\partial_j \delta(\mathbf{x}-q(a,t))\, d^2a$$

where repeated indices are summed, $\partial_i := \partial/\partial x_i$, and the antisymmetric symbol ϵ_{ij} is defined by $\epsilon_{12} = 1$. The variation of ω is

$$\delta\omega(\mathbf{x},t) = \int_D (\epsilon_{ij}\delta\pi_i(a,t)\partial_j\delta(\mathbf{x}-q(a,t)) - \epsilon_{ij}\pi_i(a,t)\partial_j\partial_k\delta(\mathbf{x}-q(a,t))\delta q_k(a,t))d^2a$$

which we can insert into

$$\delta F = \int_D \frac{\delta F}{\delta \omega}\delta\omega\, d^2x = \int_D \left(\frac{\delta F}{\delta \pi}\delta\pi - \frac{\delta F}{\delta q}\delta q\right) d^2a$$

to find

$$\frac{\delta F}{\delta \pi} = \hat{z}\times \nabla\left(\frac{\delta F}{\delta \omega}\right)\bigg|_{\mathbf{x}=q(a,t)},$$

$$\frac{\delta F}{\delta q} = -\int_D \epsilon_{ij}\pi_i(a,t)\partial_j \nabla\delta(\mathbf{x}-q(a,t))\frac{\delta F}{\delta \omega}\, d^2x.$$

We then insert these two expressions into the canonical bracket. After some manipulation involving integration by parts (we assume boundary terms vanish) we obtain the Lie–Poisson bracket

$$\{F,G\} = \int_D \omega\left[\frac{\delta F}{\delta \omega},\frac{\delta G}{\delta \omega}\right] d^2x,$$

where

$$[f,g] := \frac{\partial f}{\partial x}\frac{\partial g}{\partial y} - \frac{\partial f}{\partial y}\frac{\partial g}{\partial x}.$$

Note that the incompressibility of the fluid (the fact that the Jacobian $|\partial q/\partial a|$ is unity) was not used in the derivation of this bracket. However, in order to write the Hamiltonian in terms of ω one must introduce the streamfunction ϕ, which is possible only if $\nabla \cdot v = 0$. The equation of motion generated by the bracket and the transformed Hamiltonian

$$H[\omega] = -\frac{1}{2}\int_D \phi\omega\, d^2x = \frac{1}{2}\int_D |\nabla\phi|^2\, d^2x$$

is just Euler's equation for the ideal fluid

$$\dot{\omega}(\mathbf{x}) = -[\phi,\omega].$$

This has a Casimir given by

$$C[\omega] = \int_D f(\omega(\mathbf{x})) d^2x,$$

where f is an arbitrary function. The interpretation of this invariant is given in detail in Ref. [4]. It implies the preservation of contours of ω, so that the value ω_0 on a contour labels that contour for all times. This is a consequence of the dissipationless and divergence-free nature of the system. Substituting $f(\omega) = \omega^n$ we also see that all the moments of vorticity are conserved. By choosing $f(\omega) = \theta(\omega(\mathbf{x}) - \omega_0)$, a heavyside function, it follows that the area inside of any ω-contour is conserved.

II. EXTENSIONS AND THE SEMIDIRECT PRODUCT

We now investigate systems involving Lie algebras by *extension*, a procedure for combining two or more Lie algebras to make a new Lie algebra. There are a myriad of ways to extend algebras, and we will only touch on a few here. All the extensions discussed here have their equivalent for Lie groups, but we choose the algebra approach here because it leads more directly to a Lie–Poisson bracket. In this section we let $\mathfrak{h}$ be an extension of the Lie algebra $\mathfrak{g}$ by the algebra $\mathfrak{v}$. The elements of $\mathfrak{h}$ are written as 2-tuples, (ξ, η), where $\xi \in \mathfrak{g}$ and $\eta \in \mathfrak{v}$.

The simplest extension is the *direct product* (or *direct sum*) of Lie algebras. Let ξ and ξ' be elements of a Lie algebra $\mathfrak{g}$ and η and η' be elements of a vector space $\mathfrak{v}$ (which is an Abelian Lie algebra under addition). The direct product of these two algebras is an algebra $\mathfrak{h}$ with bracket

$$[(\xi, \eta), (\xi', \eta')] := ([\xi, \xi'], [\eta, \eta']).$$

Given the same $\mathfrak{g}$ and $\mathfrak{v}$ as above there is a less trivial way to make a new Lie algebra called the *semidirect product* with an operation defined by

$$[(\xi, \eta), (\xi', \eta')] := ([\xi, \xi'], [\xi, \eta'] + [\eta, \xi']).$$

A simple example of a semidirect product structure is when $\mathfrak{g}$ is the Lie algebra $so(3)$ associated with the rotation group $SO(3)$ and $\mathfrak{v}$ is $\mathbb{R}^3$. Their semidirect product is the algebra of the 6-parameter Euclidean group of rotations and translations. Both the elements of $\mathfrak{g}$ and $\mathfrak{v}$ can then be represented by vectors in $\mathbb{R}^3$, with bracket $[\xi, \eta] = \xi \times \eta$, the cross product of vectors. Since $\mathfrak{h}$ is itself a Lie algebra, it can be extended again as needed to make an n-fold extension (an algebra of n-tuples).

We can build Lie–Poisson brackets from these algebras by extension [6]. For an n-fold extension $\mathfrak{h}$ of the Lie algebra $\mathfrak{g}$, we define

$$\{F, G\} := \pm \left\langle \mu, \left[\frac{\delta F}{\delta \mu}, \frac{\delta G}{\delta \mu} \right] \right\rangle$$

where $\mu \in \mathfrak{h}^*$, the dual of $\mathfrak{h}$ under the pairing $\langle\,,\,\rangle : \mathfrak{h}^* \times \mathfrak{h} \to \mathbb{R}$. The dynamical variables of the system are the elements of the n-tuple $\mu = \mu(t)$. These elements may be fields or variables, so the Lie–Poisson bracket derived from an algebra by extension generates the dynamics for a system involving several dynamical quantities. The functions (or functionals) F and G are maps from $\mathfrak{h}^*$ to $\mathbb{R}$. Here $\delta/\delta\mu$ is a derivative or a functional derivative, depending on the dimensionality of the algebra (finite or infinite). For $n = 1$, $\mathfrak{h} = so(3)$, we have $\mu := \ell$ and we recover the bracket for the free rigid body (Section IA). The overall sign of the Lie–Poisson bracket has to do with left- or right-invariance of vector fields and will not be discussed here (see Ref. [2]).

Using this procedure to make a Lie–Poisson bracket from a direct product of algebras leads to a sum of n independent brackets. This will not interest us further, since the coupling between dynamical variables can only come from the Hamiltonian. However, there are interesting physical examples of a direct product structure. (This is the case for the model in Ref. [5], although a coordinate transformation is needed to exhibit the structure.)

We illustrate the process of building a Lie–Poisson bracket from a semidirect product of algebras by two examples, which are extensions of the rigid body and ideal fluid examples of Section I.

A. The Heavy Top

The Lie–Poisson bracket for the semidirect product of the rotation group $SO(3)$ and the vector space $\mathbb{R}^3$ is

$$\{f, g\} = -\ell \cdot \left(\frac{\partial f}{\partial \ell} \times \frac{\partial g}{\partial \ell}\right) - \alpha \cdot \left(\frac{\partial f}{\partial \ell} \times \frac{\partial g}{\partial \alpha} + \frac{\partial f}{\partial \alpha} \times \frac{\partial g}{\partial \ell}\right)$$

where α denotes a 3-vector. By using

$$H(\ell, \alpha) = \sum_{i=1}^{3} \frac{\ell_i^2}{2I_i} + \alpha \cdot \mathbf{c}$$

where $\mathbf{c}$ is a vector representing the position of the center-of-mass, we get the prototypical example of a semidirect product system, the heavy rigid body (in the body frame):

$$\dot{\ell}_i = \frac{I_j - I_k}{I_j I_k} \ell_j \ell_k + \alpha_j c_k - \alpha_k c_j, \quad \dot{\alpha}_i = \frac{\ell_k \alpha_j}{I_k} - \frac{\ell_j \alpha_k}{I_j},$$

where i, j, k are cyclic permutations of $1, 2, 3$. The vector α rotates rigidly with the body, which is always true for a Hamiltonian quadratic in ℓ. The Casimirs for this bracket are

$$C_1 = \alpha^2, \quad C_2 = \ell \cdot \alpha.$$

Looking at the bracket as derived by reduction of the heavy top in Euler angles (as we did in Section IA, but here with gravity), the Casimir C_2 expresses conservation of p_ϕ, since ϕ is cyclic. Knowing α does not lead to a determination of the orientation of the rigid body: there is still a symmetry of rotation about α. Taking the semidirect product has led to the recovery of some of the Lagrangian (configuration) information.

B. Low-Beta Reduced MHD

The semidirect product bracket for two fields ω and ψ is

$$\{F,G\} = \int_D \left(\omega\left[\frac{\delta F}{\delta \omega}, \frac{\delta G}{\delta \omega}\right] + \psi\left(\left[\frac{\delta F}{\delta \omega}, \frac{\delta G}{\delta \psi}\right] + \left[\frac{\delta F}{\delta \psi}, \frac{\delta G}{\delta \omega}\right]\right) \right) d^2x.$$

If $\omega = \nabla^2 \phi$, where ϕ is the electric potential, ψ is the magnetic flux, and $J = \nabla^2 \psi$ is the current, then the Hamiltonian

$$H[\omega;\psi] = \frac{1}{2}\int_D (|\nabla\phi|^2 + |\nabla\psi|^2)\, d^2x$$

with the above bracket gives us

$$\dot{\omega} = [\omega,\phi] + [\psi,J], \quad \dot{\psi} = [\psi,\phi],$$

a model for low-beta reduced MHD [7]. (Reference [8] contains a system with a similar structure, but for waves in a density-stratified fluid.)

The bracket has two Casimir invariants,

$$C_1[\psi] = \int_D f(\psi)\, d^2x, \quad C_2[\omega;\psi] = \int_D \omega g(\psi)\, d^2x.$$

The first has the form of the Casimir for 2D Euler of Section IB and has the same interpretation. To understand the second one we let $g(\psi) = \theta(\psi - \psi_0)$, a heavyside function. In this case we have

$$C_2[\omega;\psi] = \int_{\Psi_0} \omega\, d^2x,$$

where Ψ_0 represents the (not necessarily connected) region of D enclosed by the contour $\psi = \psi_0$, and $\partial \Psi_0$ is its boundary. The contour $\partial \Psi_0$ moves with the fluid, so this just expresses Kelvin's circulation theorem: the circulation around a closed material loop is conserved.

This theorem is true for any ψ-contour, therefore it holds in the region between two contours $\psi = \psi_0$ and $\psi = \psi_0 + \delta$. Letting $\delta \to 0$ we see that the two contours delineate a "line" of fluid elements with value ψ_0 of the magnetic flux. Knowledge of the value of ψ on a fluid element thus only determines which contour it is on, but not its location on the contour. Therefore, there is still a relabeling symmetry: the fluid elements can be shifted around the contour without changing the Casimirs C_1 and C_2. As with the heavy top, the semidirect product has led to the recovery of some, but not all, of the Lagrangian information.

C. Putting Labels on a Rigid Body

Remember that taking a semidirect product restricted the symmetry group of the body to rotations about α. If we take another semidirect product to get

$$\{f,g\} = -\ell \cdot \left(\frac{\partial f}{\partial \ell} \times \frac{\partial g}{\partial \ell}\right) - \alpha \cdot \left(\frac{\partial f}{\partial \ell} \times \frac{\partial g}{\partial \alpha} + \frac{\partial f}{\partial \alpha} \times \frac{\partial g}{\partial \ell}\right) - \beta \cdot \left(\frac{\partial f}{\partial \ell} \times \frac{\partial g}{\partial \beta} + \frac{\partial f}{\partial \beta} \times \frac{\partial g}{\partial \ell}\right)$$

where β is a 3-vector, we have a bracket that can model a rigid body with two forces acting on it, for example, a charged, rigid insulator in an electric field. The new bracket has Casimirs

$$C_1 = \alpha^2, \quad C_2 = \beta^2, \quad C_3 = \alpha \cdot \beta.$$

The angular momentum ℓ has disappeared from the Casimirs. This is because knowing α and β completely specifies the orientation of the rigid body (unless the two are colinear). In other words, by taking semidirect products we have reintroduced the Lagrangian information into the bracket. Note that taking more than two semidirect products is redundant as far as the Lagrangian information is concerned: knowing the orientation of more than two vectors does not add new information. This is reflected by the fact that the number of variables minus the number of Casimirs is six for two or more "advected" quantities. This is the dimension of $T^*SO(3)$, the original phase space (before reduction).

D. Advection in an Ideal Fluid

We now take a second semidirect product for the ideal fluid, say low-beta MHD with a second advected quantity, the pressure p. In that case we get a model for high-beta reduced MHD [9]. The Casimir is

$$C[\psi;p] = \int_D f(\psi, p)\, d^2x, \quad f \text{ arbitrary}.$$

This Casimir amounts to being able to label two contours. Locally, this permits a unique labeling of the fluid elements as long as p and ψ are not constant in some region. However, globally there is some ambiguity, because contours can cross in several places. Thus, in the infinite-dimensional case the semidirect product is not equivalent to recovering the full Lagrangian information, unless the contours do not close, are monotonic, and nonparallel ($\nabla \psi \times \nabla p$ does not vanish). A third advected quantity will in general break this degeneracy. Note that if the advected quantities label the fluid elements unambiguously at $t = 0$ then they will do so for all times.

III. BEYOND THE SEMIDIRECT PRODUCT: COCYCLES

In general there are other ways to extend Lie algebras besides the semidirect product. One example is the model derived in references [10] and [11] for 2D compressible reduced MHD. The model has four fields, and is obtained from an expansion in the inverse aspect ratio of the tokamak. The Hamiltonian is

$$H[\omega, v, p, \psi] = \frac{1}{2}\int_D \left(|\nabla \phi|^2 + v^2 + \frac{(p - 2\beta x)^2}{\beta} + |\nabla \psi|^2\right) d^2x,$$

where v is the parallel velocity, p is the pressure, and β is a parameter that measures compressibility. The bracket is

$$\{A,B\} = \int_D d^2x \left(\omega\left[\frac{\delta A}{\delta \omega}, \frac{\delta B}{\delta \omega}\right] + v\left(\left[\frac{\delta A}{\delta \omega}, \frac{\delta B}{\delta v}\right] + \left[\frac{\delta A}{\delta v}, \frac{\delta B}{\delta \omega}\right]\right) + p\left(\left[\frac{\delta A}{\delta \omega}, \frac{\delta B}{\delta p}\right] + \left[\frac{\delta A}{\delta p}, \frac{\delta B}{\delta \omega}\right]\right) + \right.$$
$$\left. \psi\left(\left[\frac{\delta A}{\delta \omega}, \frac{\delta B}{\delta \psi}\right] + \left[\frac{\delta A}{\delta \psi}, \frac{\delta B}{\delta \omega}\right]\right) - \beta\left[\frac{\delta A}{\delta p}, \frac{\delta B}{\delta v}\right] - \beta\left[\frac{\delta A}{\delta v}, \frac{\delta B}{\delta p}\right] \right).$$

The term proportional to β is an obstruction to the semidirect product structure, and it cannot be removed by a coordinate transformation.

The theory that deals with the classification of extensions is Lie algebra cohomology. In general the way to extend a bracket is by adding a nontrivial *cocycle*. Though *a priori* there are an infinite number of ways to make an extension, for low dimensions, after allowing for coordinate transformations, very few possibilities remain; we have classified these in Ref. [12]. We have also found all the Casimir invariants for the low-dimensional brackets (five fields or less).

The Casimirs of the above bracket are

$$C_1[\psi] = \int_D f(\psi)\, d^2x, \quad C_2[p;\psi] = \int_D p\, g(\psi)\, d^2x,$$

$$C_3[v;\psi] = \int_D v h(\psi)\, d^2x, \quad C_4[\omega,v,p,\psi] = \int_D \left(\omega k(\psi) - \frac{vp}{\beta} k'(\psi)\right) d^2x.$$

Finding the invariant C_4 directly from the equations of motion would be tedious, but is straightforward from the bracket. These Casimirs do not allow a labeling of the fluid elements. The meaning of invariants of the form of C_1, C_2, and C_3 was discussed in Sections IB and IIB: the total magnetic flux, pressure, and parallel velocity inside of any ψ-contour are preserved. To understand C_4 we use the fact that $\omega = \nabla^2 \phi$ and then integrate by parts to obtain

$$C_4[\omega,v,p,\psi] = -\int_D \left(\nabla \phi \cdot \nabla \psi + \frac{vp}{\beta}\right) k'(\psi)\, d^2x.$$

The quantity in parentheses is thus invariant inside of any ψ-contour. It can be shown that this is a remnant of the conservation in the full MHD model of the cross helicity,

$$V = \int_D \mathbf{v} \cdot \mathbf{B}\, d^2x,$$

at second order in the inverse aspect ratio, while C_3 is a consequence of preservation of this quantity at first order. Here **B** is the magnetic field. As for C_1 and C_2 they are, respectively, the first- and second-order remnants of the preservation of helicity,

$$W = \int_D \mathbf{A} \cdot \mathbf{B}\, d^2 x,$$

where **A** is the magnetic vector potential.

IV. CONCLUSIONS

We gave an introduction to the reduction of physical systems based on their symmetries. The prototypical examples were shown, the rigid body and the 2D ideal fluid. For these two cases some information about the configuration of the system was lost after reduction, correponding to the symmetry used to reduce the system.

The semidirect product allowed us to build larger brackets from a "base" algebra in a systematic manner. We were thus able to describe the heavy top and low-beta reduced MHD. Examining the invariants, we concluded that the semidirect product had recovered some or all of the Lagrangian information.

For general extensions (not necessarily semidirect) things are different: the Lagrangian information is not necessarily a consequence of the Casimirs. However, for compressible reduced MHD the Casimirs represent constraints that are remnants of invariants of the full MHD equations from which the model is derived asymptotically.

As mentioned in Section III, when considering a general extension, all brackets can be reduced to a small number of normal forms, at least for low-dimensional extensions. It will be interesting to see if physical systems can be found that are realized by these brackets, both in the finite and infinite degree-of-freedom cases. We are currently investigating a toy model that we call the Leibniz top, which is one of the simplest non-semidirect system one can build. It is a straightforward generalization of the Lagrange top (a heavy top with $I_1 = I_2$). The Lagrange top is integrable, and we have found that so is the Leibniz top.

ACKNOWLEDGMENTS

This work was supported by the U.S. Department of Energy Contract No. DE-FG03-96ER-54346. J.-L.T. also acknowledges support from the Fonds pour la Formation de Chercheurs et l'Aide à la Recherche du Canada.

REFERENCES

1. MORRISON, P.J. 1998. Hamiltonian description of the ideal fluid. Rev. Modern Phys. **70:** 467.
2. MARSDEN, J.E. & T.S. RATIU. 1994. Introduction to Mechanics and Symmetry. Springer-Verlag. Berlin.
3. NEWCOMB, W.A. 1962. Lagrangian and Hamiltonian methods in magnetohydrodynamics. Nuclear Fusion: Supplement, Part 2.
4. MORRISON, P.J. 1987. Variational principle and stability of nonmonotonic Vlasov–Poisson equilibria. Z. Naturforsch **42a:** 1115.
5. HAZELTINE, R.D., D.D. HOLM & P.J. MORRISON. 1985. Electromagnetic solitary waves in magnetized plasmas. J. Plasma Physics **34:** 103.

6. MARSDEN, J.E. & P.J. MORRISON. 1984. Noncanonical Hamiltonian field theory and reduced MHD. Contemp. Math. **28**:133.
7. MORRISON, P.J. & R.D. HAZELTINE. 1984. Hamiltonian formulation of reduced magnetohydrodynamics. Phys. Fluids **27**: 886.
8. BENJAMIN, T.B. 1984. Impulse, flow force, and variational principles. IMA J. Appl. Math **32**: 3.
9. STRAUSS, H.R. 1977. Dynamics of high β tokamaks. Phys. Fluids **20**: 1354.
10. HAZELTINE, R.D., M. KOTSCHENREUTHER & P.J. MORRISON. 1985. A four-field model for tokamak plasma dynamics. Phys. Fluids **28**: 2466.
11. HAZELTINE, R.D., C.T. HSU & P.J. MORRISON. 1987. Hamiltonian four-field model for nonlinear tokamak dynamics. Phys. Fluids **30**: 3204.
12. THIFFEAULT, J.-L. & P.J. MORRISON. 1998. Classification and Casimir invariants of Lie–Poisson brackets. Institute for Fusion Studies Report 837.

From Jupiter's Great Red Spot to the Structure of Galaxies: Statistical Mechanics of Two-Dimensional Vortices and Stellar Systems

PIERRE-HENRI CHAVANIS

Ecole Normale Supérieure de Lyon, 46 Allée d'Italie 69364 Lyon, France and Istituto di Cosmogeofisica, Corso Fiume 4, 10133 Torino, Italia

ABSTRACT: The statistical mechanics of two-dimensional vortices and stellar systems both at equilibrium and out of equilibrium are discussed, with emphasis on the analogies (and on the differences) between these two systems. Limitations of statistical theory and problems posed by the long-range nature of the interactions are described in detail. Special attention is devoted to the problem of "incomplete relaxation" and, in the case of stellar systems, to the "gravothermal catastrophe." The relaxation toward equilibrium, possibly restricted to a "maximum entropy bubble," is described with the aid of a maximum entropy production principle (MEPP). The relation with Fokker-Planck equations is made explicit and the structure of the diffusion current analyzed in terms of a pure diffusion compensated by an appropriate friction or a drift.

1. INTRODUCTION

Two-dimensional flows with high Reynolds numbers, described by Euler-like equations, have the striking property of organizing spontaneously into coherent structures [26]. The robustness of Jupiter's Great Red Spot, a huge vortex persisting for more than two centuries in a turbulent shear between two zonal jets, is probably related to this general phenomenon [41]. Some other coherent structures like dipoles (a pair of cyclone/anticyclone) and sometimes tripoles have been found in atmospheric or oceanic systems and can persist during several days or weeks responsible for atmospheric blocking. Some astrophysicists invoke the existence of organized vortices in the gaseous component of disk galaxies [29] in relation with the emission of spiral density waves. It has also been proposed that planet formation began inside persistant gaseous vortices born out of the protoplanetary nebula [2], [48]. As a result, hydrodynamical vortices occur in a wide variety of geophysical or astrophysical phenomena and their robustness demands a general understanding.

Similarly, it is striking to observe that galaxies themselves follow a kind of organization revealed in the Hubble classification or in de Vaucouleur's $R^{1/4}$ law [3]. Now, the dynamics of galaxies is dominated by stars under collective gravitational interaction rather than gas or hydrodynamical processes. In particular, for most stellar systems the collisions (i.e., close encounters) between stars are quite negligible (the corresponding relaxation time exceeds the age of the Universe), and the galaxy dynamics can be modeled by the Vlasov equation.

In this paper, I stress the deep analogy between these two systems [9], [10] and show that their structure and organization can be understood from relatively similar statistical mechanics (Section 2). In Section 3, I discuss the existence of equilibrium states for stellar systems with particular attention devoted to the "gravothermal instability." In Section 4, I present a classification of the "zoology" of vortices usually met in two-dimensional flows. In Section 5, I discuss the relaxation toward equilibrium, using a maximum entropy production principle. I show how we can account for "incomplete relaxation" thanks to a variable diffusion coefficient and interpret the evolution equations in terms of generalized Fokker-Planck equations. In Section 6, I show that a point vortex (or a piece of vorticity) in 2D turbulence experiences a "systematic drift" in the same way that a star in a galaxy experiences a "dynamical friction." I conclude on unresolved problems posed by the use of the Euler and the Vlasov equations in the statistical description.

2. ANALOGY BETWEEN TWO-DIMENSIONAL VORTICES AND STELLAR SYSTEMS

At a first level, the analogy between 2D vortices and galaxies resides in the similar morphology of the Euler:

$$\frac{\partial \omega}{\partial t} + \mathbf{u}\nabla\omega = 0, \qquad (2.1)$$

$$\nabla\psi = -\omega; \qquad (2.2)$$

and the Vlasov equation:

$$\frac{\partial f}{\partial t} + \mathbf{v}\frac{\partial f}{\partial \mathbf{r}} + \mathbf{F}\frac{\partial f}{\partial \mathbf{v}} = 0 \qquad (2.3)$$

$$\Delta\Phi = 4\pi G \int f d^3\mathbf{v}, \qquad (2.4)$$

which apply, respectively, to two-dimensional incompressible perfect flows and collisionless stellar systems. These two equations describe the advection of a density (the vorticity $\omega = \hat{\mathbf{z}}(\nabla \wedge \mathbf{u})$ or the distribution function f, i.e., the average mass of stars in $\mathbf{r}, \mathbf{v}$ at time t) by an incompressible velocity field $\mathbf{u} = -\hat{\mathbf{z}} \wedge \nabla\psi$ or $\mathbf{U}_6 \equiv (\mathbf{v}, \mathbf{F})$ $= (\mathbf{v}, -\nabla\Phi)$. However, these quantities are not simply advected by the flow (as passive scalars) but are coupled to its motion via a Poisson equation (2.2) or (2.4) relating the streamfunction ψ to the vorticity ω or the gravitational potential Φ to the distribution function f. This interaction produces strong fluctuations and generates a "mixing process" in the plane or in phase space (see FIGURE 1). There is therefore a clear analogy between these two systems: the vorticity plays the role of the distribution function and the streamfunction the role of the gravitational potential.

In general, the Euler and the Vlasov equations never achieve equilibrium, but develop a very intricate filamentation at smaller and smaller scales. However, if we average locally over these filaments, defining a "coarse-grained" vorticity $\overline{\omega}$ or a "coarse-grained" distribution function $\overline{f}$, these locally averaged fields are expected to reach a kind of equilibrium. This was first shown by Lynden-Bell [23] in the case

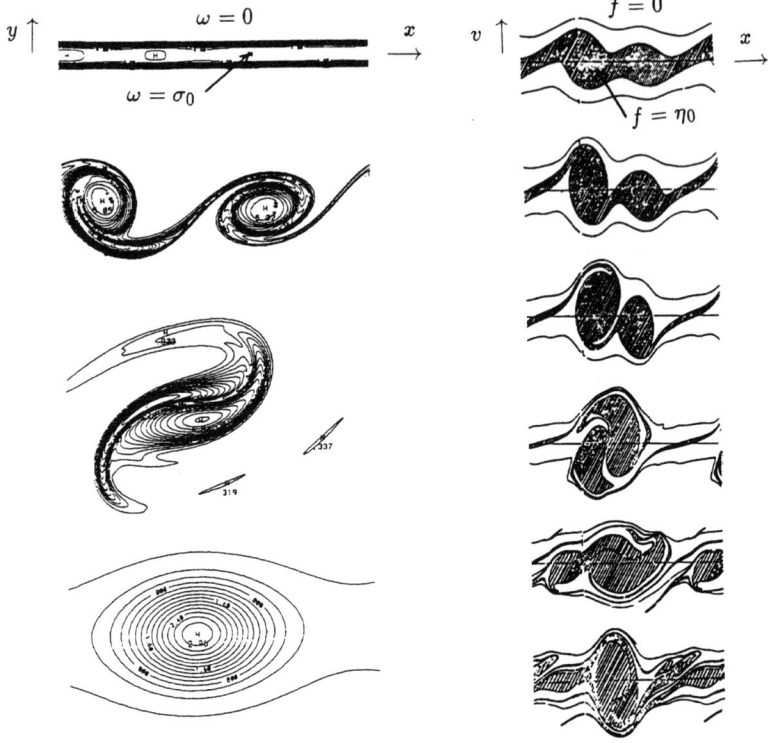

FIGURE 1.

of stellar systems, and rediscovered more recently by Kuzmin [22], Miller [28], and Robert and Sommeria [36] in the context of 2D turbulence. Robert [33] has provided a firm mathematical justification of these ideas in terms of Young measures. Physically, this equilibrium state is the outcome of a mixing process restricted by the conservation of energy. Due to the strong fluctuations of the gravitational potential or of the stream function at the early stage of the evolution, this form of relaxation is extremely "violent" (a few dynamical times t_D). Of course, the validity of the theory is conditioned by a hypothesis of ergodicity, which may not be completely fulfilled.

If the initial condition consists of a patch of uniform distribution function $f = \eta_0$ surrounded by vacuum $f = 0$ or a patch of uniform vorticity σ_0 surrounded by irrotational flow $\omega = 0$, the equilibrium state belongs to the Fermi-Dirac statistics [23], [22], [28], [36]:

$$\bar{f} = \frac{\eta_0}{1 + \lambda e^{\beta \eta_0 \left(\frac{v^2}{2} + \Phi\right)}} ; \tag{2.5}$$

$$\varpi = \frac{\sigma_0}{1 + \lambda e^{\beta \sigma_0 \psi}}. \tag{2.6}$$

There is an effective "exclusion principle" as in quantum mechanics but arising here for a different reason. Due to the averaging procedure, the "coarse-grained" distribution function or the "coarse-grained" vorticity can increase or decrease by internal mixing (as vacuum or irrotational flow are being incorporated into the initial patch) but, because of the incompressibility of the Vlasov or the Euler equations, they can never exceed the extremum values of the initial condition. Therefore, the constraints $0 \leq \bar{f} \leq \eta_0$ and $0 \leq \varpi \leq \sigma_0$ must be satisfied everywhere. This explains qualitatively why the equilibrium state is described by a Fermi-Dirac (not a Maxwell-Boltzmann) statistics.

More precisely, the equilibrium state (2.5) is obtained by maximizing the mixing entropy

$$S = -\int \left\{ \frac{\bar{f}}{\eta_0} \ln \frac{\bar{f}}{\eta_0} + \left(1 - \frac{\bar{f}}{\eta_0}\right) \ln \left(1 - \frac{\bar{f}}{\eta_0}\right) \right\} d^3\mathbf{r} d^3\mathbf{v} \tag{2.7}$$

at fixed energy and mass, and the equilibrium state (2.6) by maximizing the mixing entropy:

$$S = -\int \left\{ \frac{\varpi}{\sigma_0} \ln \frac{\varpi}{\sigma_0} + \left(1 - \frac{\varpi}{\sigma_0}\right) \ln \left(1 - \frac{\varpi}{\sigma_0}\right) \right\} d^2\mathbf{r} \tag{2.8}$$

at fixed energy and circulation (the integral of the vorticity). These results can be generalized to any initial condition involving a continuous spectrum of phase levels $\{\eta\}$ or vorticity levels $\{\sigma\}$.

We can also consider a collection of N-point stars (of mass m) or N-point vortices (of circulation γ). These N-body systems have a Hamiltonian structure and their statistical mechanics is more conventional. In the case of point-mass stars, the Hamiltonian has a kinetic and a potential part:

$$H = \sum_{i=1}^{N} \frac{1}{2} m v_i^2 - \sum_{i<j} \frac{G m^2}{|\mathbf{r}_i - \mathbf{r}_j|}, \tag{2.9}$$

whereas for point vortices, the Hamiltonian is purely "potential":

$$H = -\frac{1}{2\pi} \sum_{i<j} \gamma^2 \ln |\mathbf{r}_i - \mathbf{r}_j|. \tag{2.10}$$

This has strong physical consequences. In particular, in the case of stars, the temperature is necessarily positive whereas for point vortices *negative temperatures* are possible. As shown by Onsager [31], this is precisely at negative temperatures that point vortices cluster into "supervortices" (the interaction becomes "attractive").[1] On the other hand, stellar systems can undergo a "gravothermal catastrophe": a stellar system can always achieve higher and higher concentrations by increasing its temperature (kinetic term) in order to compensate for the loss of potential energy (maintaining its total energy unchanged). There is no such collapse instability for point vortices: the energy wouldn't be conserved.

Due to the development of correlations between stars or point vortices (caused by discrete interactions), these Hamiltonian systems are expected to relax toward a Maxwell-Boltzmann equilibrium state (see, e.g, [30], [18]):

$$f = Ae^{-\beta m\left(\frac{v^2}{2} + \Phi\right)}, \tag{2.11}$$

$$\omega = Ae^{-\beta\gamma\psi}. \tag{2.12}$$

As before, these equilibrium states can be obtained from a variational principle by maximizing the Boltzmann entropy:

$$S = -\int f \ln f \, d^3\mathbf{r} d^3\mathbf{v} \tag{2.13}$$

$$S = -\int \omega \ln \omega \, d^2\mathbf{r} \tag{2.14}$$

at fixed energy and mass or circulation.

3. GRAVITATIONAL COLLAPSE OF STELLAR SYSTEMS

Because of the long-range nature, and the singularity when $r \to 0$, of the gravitational potential, the statistical mechanics of stellar systems creates problems and the notion of equilibrium is not always well-defined. I will recall some standard results about the existence (or nonexistence!) of equilibrium states, first in the case of collisional stellar systems (such as globular clusters) described by the Maxwell-Boltzmann statistics (2.11).

THEOREM 1: There is no global entropy maximum in an unbounded domain. A self-gravitating system can always increase its entropy by spreading its density.

Proof. This is clearly seen by considering the family of distribution functions [34]:

$$f_\lambda(\mathbf{r}, \mathbf{v}) = \left(\frac{\beta_\lambda}{2\pi}\right)^{\frac{3}{2}} n_\lambda(\mathbf{r}) e^{-\beta_\lambda \frac{v^2}{2}}, \tag{3.1}$$

where

$$n_\lambda(\mathbf{r}) = \frac{1}{\lambda^3} n\left(\frac{\mathbf{r}}{\lambda}\right)$$

---

[1]Whereas the organization of stars is relatively clear because of the attractive nature of gravity, the organization of point vortices at negative temperatures is much less intuitive. I will give a physical interpretation of this phenomenon in Section 6 in terms of a "systematic drift."

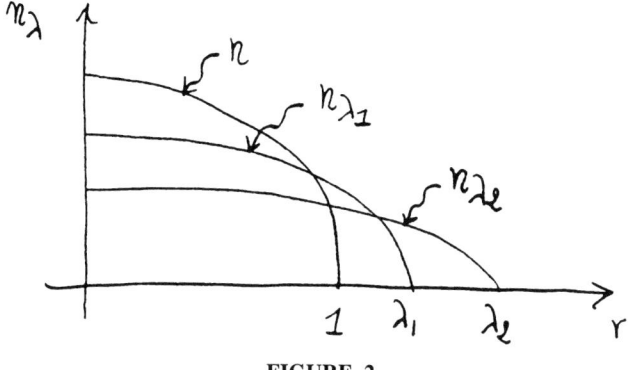

FIGURE 2.

and $n(\mathbf{r})$ is a density profile defined on the compact $[0,1]$ and satisfying $M = \int n d^3\mathbf{r}$. The temperature β_λ is determined by the conservation of energy and the entropy can be calculated explicitely for any value of λ. Clearly, increasing λ amounts to spreading the density profile n_λ (see FIGURE 2). When $\lambda \to +\infty$, the entropy behaves like $S_\lambda \sim 3M \ln \lambda$ and diverges.

THEOREM 2: There is *not even a local* entropy maximum in an unbounded domain.

Proof. The solution of the Poisson equation (2.4) coupled to the Maxwell-Boltzmann statistics (2.11) has an infinite mass ($n \sim 1/r^2$, at large distances) [3].

It is therefore necessary to confine the system (in a box, or by using truncated models like the Michie-King model) if we want to make the statistical mechanics of self-gravitating systems. In fact this confinement is justified physically by the realization that the relaxation is necessarily *incomplete* (see Section 5) so the entropy must be maximized only in a *subdomain*. However, even when confined into a box, self-gravitating systems exhibit a peculiar behavior.

THEOREM 3: There is *no global* entropy maximum in a box. A self-gravitating system can always increase its entropy by making its core denser and denser (and hotter and hotter).

Proof. This is clearly seen by considering the family of distribution functions [34]:

$$f_\epsilon(\mathbf{r}, \mathbf{v}) = \left(\frac{\beta_\epsilon}{2\pi}\right)^{\frac{3}{2}} n_\epsilon(\mathbf{r}) e^{-\beta_\epsilon \frac{v^2}{2}}, \qquad (3.2)$$

where $n_\epsilon(r) = 0$ if $r < \epsilon$ and $n_\epsilon(r) = C_\epsilon/r^\alpha$ ($2.5 < \alpha < 3$) if $\epsilon \le r \le R$ (R is the box radius). Clearly, decreasing ϵ amounts to increasing the central density (see FIGURE 3). The constant C_ϵ is determined by the conservation of mass and the inverse temperature β_ϵ by the conservation of energy. When $\epsilon \to 0$, the potential energy diverges to $-\infty$, but the temperature rises to $+\infty$ in order to maintain the total energy fixed. On the

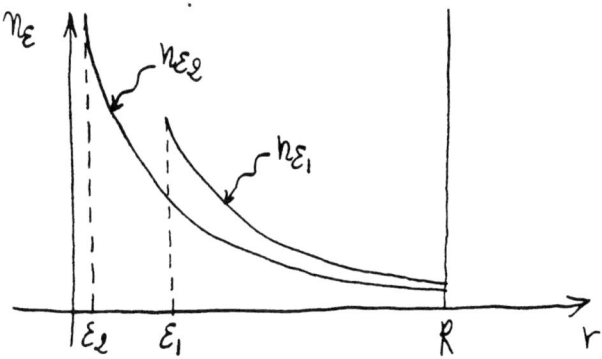

FIGURE 3.

other hand, the entropy behaves like $S_\epsilon \sim -\frac{3}{2}\ln\beta_\epsilon$ and diverges. This shows the general tendency of stellar systems to develop a dense and hot "core" surrounded by a "halo" of stars.

THEOREM 4 [1]: There exists some local entropy maxima for a self-gravitating system confined in a box *only if* $\Lambda \equiv -ER/GM^2 < 0.335$. They have a density contrast $\mathcal{R} \equiv n(0)/n(R) < 709$ (see FIGURES 4 and 5, full curve).

For smaller energies (or higher box radius R) the stable branch disappears and the system undergoes a "gravothermal catastrophe" [24]: it takes a "core-halo" structure and can always create entropy by increasing its central density and temperature. *A priori*, this process continues up to the formation of a Black Hole singularity. Lynden-Bell and Lynden-Bell [25] have related this instability to the *negative heat capacity* of self-gravitating systems: by loosing heat, they grow hotter and evolve away from equilibrium.

The case of collisionless stellar systems (like elliptical galaxies) described by the Fermi-Dirac statistics (2.5) is somewhat different [8], [34].

THEOREM 5: There is still *no global* or local entropy maximum in an unbounded domain. However, when the system is confined in a box, there now exists a global entropy maximum for all the accessible values of the energy.

Therefore, degeneracy has a stabilizing role and is able to stop the "gravothermal catastrophe." When the degeneracy parameter $\mu = \eta_0/\langle f \rangle \equiv \eta_0\sqrt{512\pi^4 G^3 MR^3}$ is small, the Fermi-Dirac spheres depart relatively rapidly from the nondegenerate spiral; the equilibrium diagram is represented on FIGURE 4 (dashed line). A global entropy maximum exists for any accessible energy and it is now possible to overcome the critical density contrast of 709. For high energies the system is nondegenerate. For lower energies, the equilibrium state has a "core-halo" structure with a degenerate nucleus surrounded by a Maxwellian atmosphere. When we decrease the energy, more and more mass is concentrated into the nucleus until a minimum accessible en-

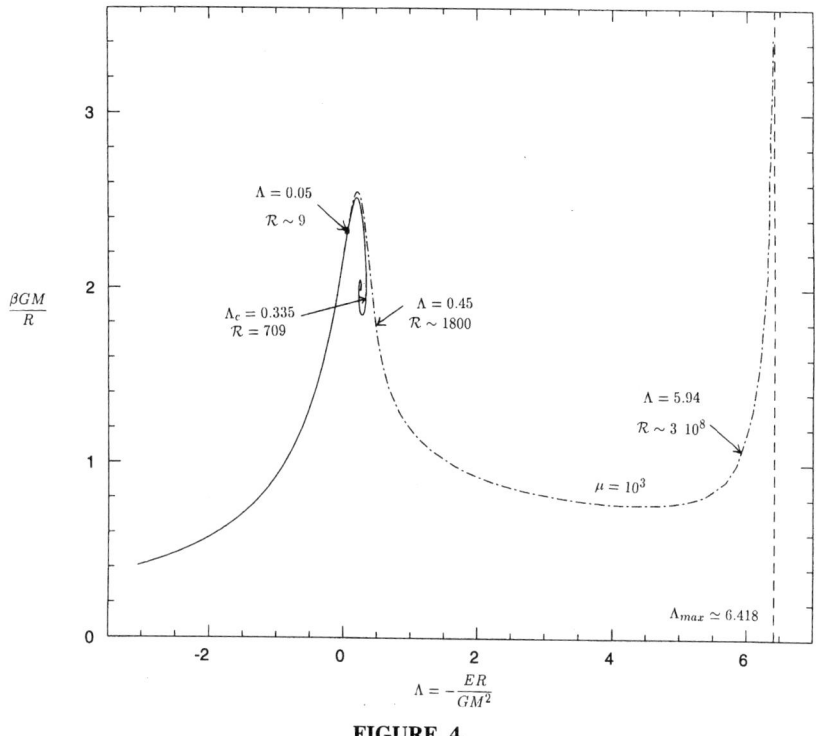

FIGURE 4.

ergy (corresponding to $\Lambda_{max}(\mu)$) at which the nucleus contains all the mass. In that case, the atmosphere has been swallowed and the system has the same structure as a cold White Dwarf Star.

When the degeneracy parameter μ is higher, the Fermi-Dirac spheres follow the nondegenerate spiral longer, then proceed backward along the upper branch until they finally turn over at $\Lambda^*(\mu)$ and come back to the right toward $\Lambda_{max}(\mu)$ (see FIGURE 5, dashed line). We now have several solutions for each single value of the energy in the range $\Lambda_*(\mu) < \Lambda < \Lambda_c = 0.335$: the solution on the upper branch (point A) is nondegenerate and is a local entropy maximum. By contrast, the solution on the lower branch (point C) has a "core-halo" structure, with a degenerate nucleus, and is a global entropy maximum (the intermediate solution, point B, is unstable). The "bassin of attraction" of the nondegenerate solution is expected to be wider so that the system will prefer in general the local entropy maximum. However, when we approach the critical Antonov value Λ_c, the upper branch disappears: in that case, the system inexorably undergoes gravitational collapse, but, for collisionless stellar systems, this collapse stops when the center becomes degenerate. In that case the system falls onto a global entropy maximum with a "core-halo" structure (point D).

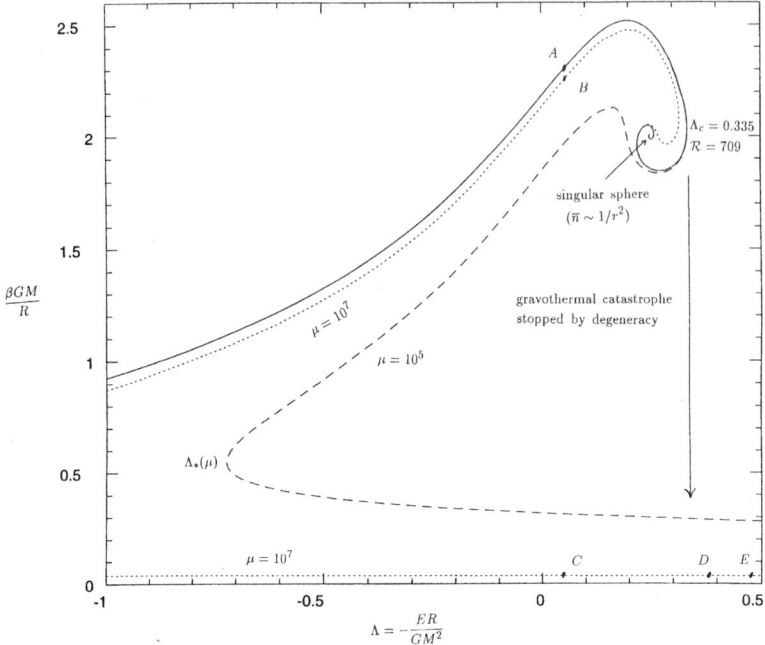

FIGURE 5.

4. CLASSIFICATION OF TWO-DIMENSIONAL VORTICES

In 2D turbulence, the vorticity can be either positive or negative whereas the distribution function of stars is restricted to positive values. This yields a wider variety of structures such as monopoles, rotating or translating dipoles, and sometimes tripoles. The interesting problem in that context is to obtain a classification of this "zoology" of vortices. The statistical theory of the Euler equation provides a general framework to tackle this problem. However, the prediction is not straightforward because, in the general case, we have to take into account an infinite set of constraints in addition to energy E, namely, the total area $\gamma(\sigma)$ of each vorticity level σ (or equivalently all the vorticity moments $\Gamma_n = \int \overline{\omega^n} \, d^2\mathbf{r}$). The relationship predicted by the statistical theory is now a superpostion of Fermi-Dirac states and the mean-field equation to be solved writes:

$$\overline{\omega} = \frac{\int \sigma e^{-\alpha(\sigma) - \beta\sigma\psi} d\sigma}{\int e^{-\alpha(\sigma) - \beta\sigma\psi} d\sigma} = -\Delta\psi, \qquad (4.1)$$

where β and $\alpha(\sigma)$ are the Lagrange multipliers associated with the conservation of E and $\gamma(\sigma)$.

This problem is highly nonlinear and a lot of bifurcations appear (in particular because of the energy constraint). A numerical algorithm has been developed by Turkington and Whitaker [43] and several calculations have been performed in rectangular or circular domains for a finite number of vorticity levels and for particular values of the integral constraints. However, many structures are found and it is difficult to have a clear picture of the bifurcation diagram in parameter space. For that reason, we have considered a limit of "strong mixing" where an analytical study is possible and where a nice classification of the coherent structures can be obtained in terms of a single control parameter [8], [12]. This "strong mixing" limit is valid when $\beta\sigma\psi \ll 1$. This condition is fulfilled when β is small, which corresponds to a weak energy constraint, or when ψ is small which corresponds to a weak value of energy (normalized by enstrophy). In that case, it is possible to expand the equations of the problem in terms of the small parameter $\beta\sigma\psi$. To lowest order in the expansion, we obtain a linear relationship between vorticity and streamfunction and we can justify an inviscid minimum enstrophy principle [8], [12], [16]. Moreover, this limit makes a hierarchy between the constraints: to lowest order, only circulation and enstrophy are relevant to characterize the equilibrium state (in addition, of course, to energy) but when we go to higher orders in the expansion, more and more vorticity moments are necessary to describe the structure of equilibrium. Because of this hierarchy, we can make some relevant predictions without the detailed knowledge of the initial condition.

As in the case of stellar systems, the relaxation of 2D vortices is in general incomplete. In fact, when the vorticity is everywhere positive or everywhere negative, unconfined monopoles exist mathematically as solutions of (4.1) in the whole plane, but numerical simulations and laboratory experiments reveal that these complete equilibria are not reached by the system. On the other hand, unconfined dipoles or tripoles do not exist at all. It is therefore necessary to consider *restricted equilibrium states*. Because of kinetic effects (discussed in Section 5), vorticity organizes spontaneously in a *subdomain* of space. The vortex structure therefore resembles a kind of "bubble" isolated from the rest of the flow. The boundary of this bubble is free to deform but its area remains constant as kinetic effects restrict mixing with the surrounding. These considerations led us to propose that the system organizes inside a "maximum entropy bubble" with the usual constraints of the Euler equation and the additional (kinetic) constraint of a given area [12].

The determination of this restricted equilibrium state is a well-posed problem: the shape and the position of the bubble, and the vorticity field inside, result from the maximization of entropy and the matching conditions between the bubble and the irrotational surrounding flow. Nevertheless there remains a free parameter in the problem, namely, the bubble area characterizing kinetic restriction to mixing. In fact, it turns out that the vortex size is determined by stability conditions: the bubble must be an entropy maximum with respect to small changes in the probability field but also when we deform its boundary. These stability conditions are very stringent and constrain terribly the possible sizes of the vortex. If the vortex has not the right area, it will deform irreversibly incorporating irrotational flow (or ejecting vorticity) so as to become stable (if this is not possible it should break up in different pieces, as observed in some experiments).

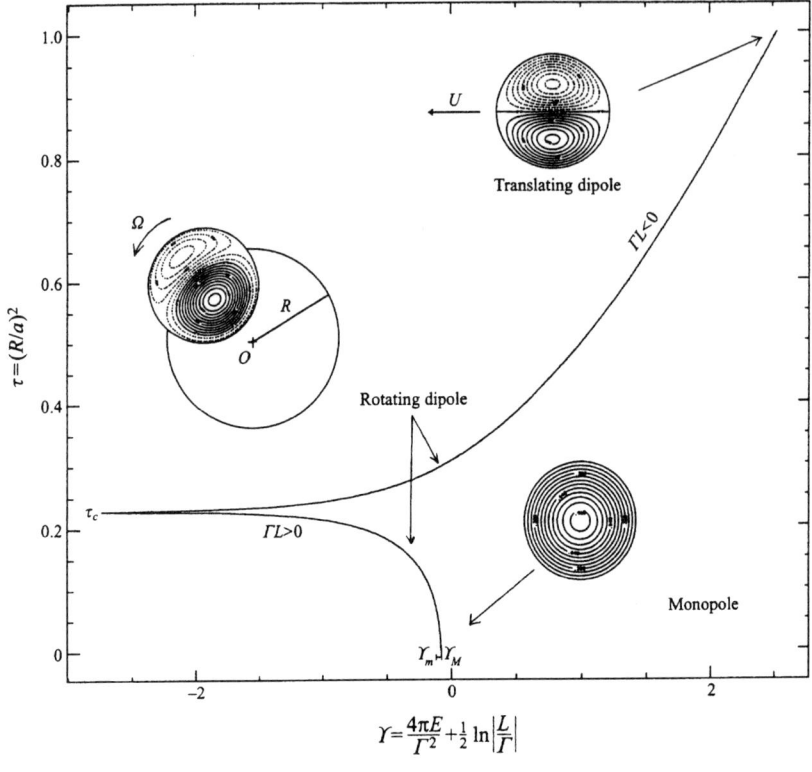

FIGURE 6.

It is possible to calculate "maximum entropy bubbles" analytically in the limit of "strong mixing" and obtain a complete prediction and classification of the equilibrium states in terms of a single control parameter

$$\Upsilon = \frac{4\pi E}{\Gamma^2} + \frac{1}{2}\ln\left|\frac{L}{\Gamma}\right|,$$

depending on energy E, circulation Γ, and angular momentum L (see FIGURE 6). Stable monopoles exist only in a very narrow range of parameters, between $\Upsilon_m \simeq -0.09$ and $\Upsilon_M \simeq -0.08$. When $\Upsilon > \Upsilon_M$, the vorticity profile does not decrease monotonously and the monopoles are unstable. However, they can gain stability by expulsing a tiny amount of vorticity far away, changing their angular momentum but not the other constraints. By this process, the parameter Υ can be reduced to the range $[\Upsilon_m, \Upsilon_M]$ [5]. This is not possible, however, when $\Gamma = 0$; monopoles with zero circulation generally evolve toward nonlinear tripoles (linear tripoles are believed to be unstable) or break down into two dipoles translating in opposite directions (to conserve the total impulse). Stable dipoles exist for all the values of Υ when $\Gamma L < 0$ and for $\Upsilon < \Upsilon_M$ when $\Gamma L > 0$. When $\Gamma \neq 0$, the dipoles are asymmetric and rotate around the center of vorticity. When $\Gamma L < 0$ and $\Gamma \to 0$ (corresponding to $\Upsilon \to \infty$), the two lobes become more and more symmetrical, the radius of rotation increases and a translating dipole is obtained asymptotically. On the contrary, when $\Gamma L > 0$ and $\Upsilon \to \Upsilon_M$ one of the lobes swells and swallows the other: the radius of rotation is reduced and for $\Upsilon = \Upsilon_M$, the branch of rotating dipoles connects the branch of monopoles.

5. RELAXATION TOWARD EQUILIBRIUM

The previous discussion has revealed that complete equilibrium does not exist in systems with long-range interactions such as self-gravitating systems or 2D vortices. It is therefore necessary to consider nonequilibrium statistical mechanics and try to derive relaxation equations for the "coarse-grained" distribution function or vorticity. This will provide a precise framework to understand what limits relaxation and why the complete equilibrium is not reached in general. At first this problem seems extremely difficult since "violent relaxation" is a very nonlinear and chaotic process, ruling out any attempt to implement *perturbative* methods. For that reason, there is a strong incentive to explore *variational* methods which are considerably simpler and give a more intuitive understanding of the problem.

In the context of 2D turbulence, Robert and Sommeria [36] have suggested that "the flow could evolve out of equilibrium so as to maximize its rate of entropy production $\dot{S}$ while accounting for all the constraints imposed by the dynamics (in particular the conservation of energy $\dot{E} = 0$)." This is the maximum entropy production principle (MEPP). This principle takes the best advantage of one's ignorance and may be a useful guide in very nonlinear problems when we have to face a complete lack of information at small scales. The MEPP is also the most natural extension, out of equilibrium, of the well-known principle of thermodynamics according to which: "at equilibrium, the system is in a maximum entropy state with appropriate constraints." For a single level of vorticity, the maximum entropy production principle provides the following equation [36]:

$$\frac{\partial \varpi}{\partial t} + \bar{\mathbf{u}} \nabla \varpi = \nabla [D(\nabla \varpi + \beta(t) \varpi (\sigma_0 - \varpi) \nabla \psi)] \qquad (5.1)$$

for the evolution of the "coarse-grained" vorticity. This equation conserves energy at any time (thanks to a Lagrange multiplier $\beta(t)$ which evolves accordingly), increases the entropy with an optimal rate ($\dot{S} \geq 0$) and drives the system toward the Fermi-Dirac equilibrium state (2.6).

This relatively elegant and simple variational formulation shows that the global structure of the relaxation equation is determined by purely thermodynamical arguments. All the explicit reference to the "subdynamics" is rejected in the diffusion coefficient D whose expression cannot be captured by the MEPP (it appears as an ill-defined Lagrange multiplier). The precise value of the diffusion coefficient is nevertheless capital as it determines the characteristic time scale of the relaxation process and, being related to the microscopic fluctuations, it is able to account for "incomplete relaxation." It is well-known indeed that the presence of fluctuations is necessary to drive the system toward equilibrium. Here, the fluctuations are represented by the microscopic oscillations of the vorticity field and may vanish before

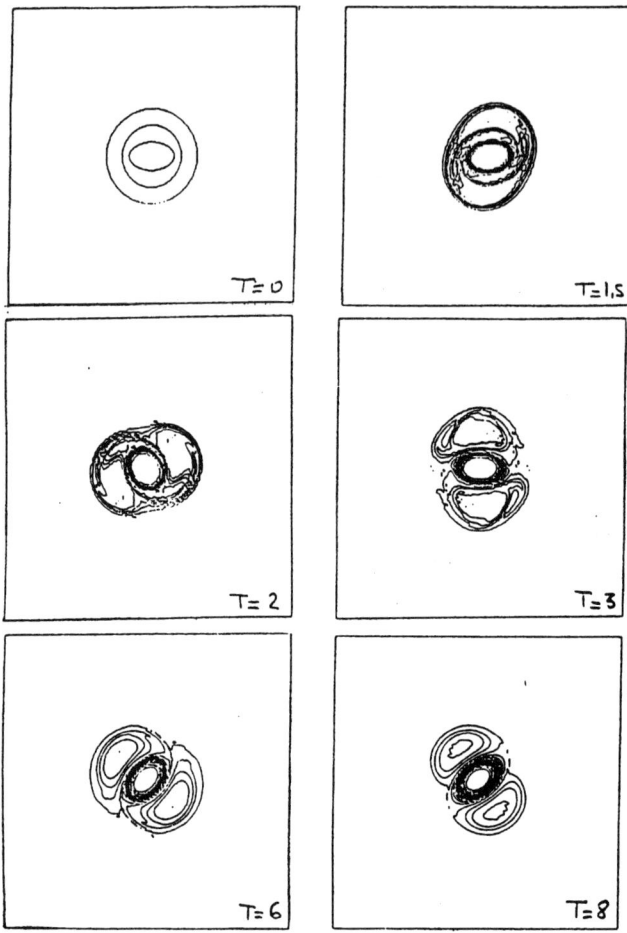

FIGURE 7a.

the system had time to reach complete equilibrium. In that case, the relaxation is strongly slowed down and the system remains "frozen" in a subdomain of space. Developing this heuristic picture and introducing a simple stochastic model, Robert and Rosier [37] have obtained an explicit expression for the diffusion coefficient. For a single level of vorticity, it writes [37] (see also [9]):

$$D(\mathbf{r}, t) = \frac{\tau \epsilon^2}{8\pi} \ln\left(\frac{L}{\epsilon}\right) \varpi (\sigma_0 - \varpi), \tag{5.2}$$

where ϵ is the spatial scale at which the oscillations of the vorticity occur, τ the decorrelation time, and L the typical size of the system. The important point is that the diffusion coefficient (5.2) is not constant *in space* but vanishes where $\varpi = 0$, i.e.,

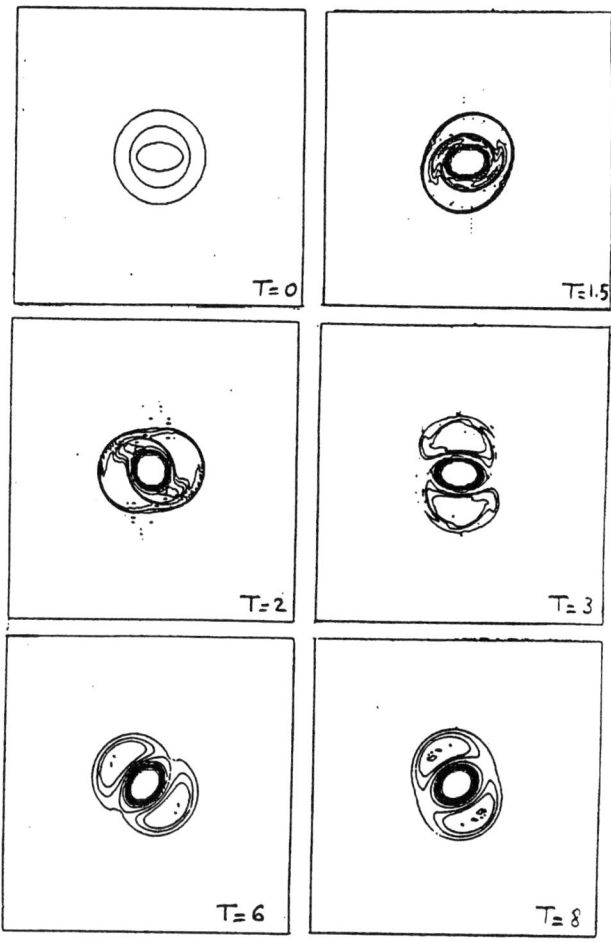

FIGURE 7b.

where there is no mixing of the vorticity at small scales. At the contact with the unmixed flow, the eddy flux also vanishes and this results in a confinement of the structure. This confinement is amplified by the *temporal* decay of the fluctuations as the system develops finer and finer filaments. These kinetic considerations are able to justify restricted equilibrium states and give a more solid basis to the concept of "maximum entropy bubble" presented in Section 4.

Robert and Rosier [37] have studied numerically the formation of a tripole from an initial condition consisting of an ellipse of uniform (negative) vorticity surrounded by an annulus (of positive vorticity). A direct Navier-Stokes simulation with high Reynolds number, represented on FIGURE 7a, is compared with the result of the relaxation equation (5.1) (generalized to two levels). FIGURE 7c is obtained with a constant diffusion coefficient: in that case the vorticity diffuses in the whole domain and

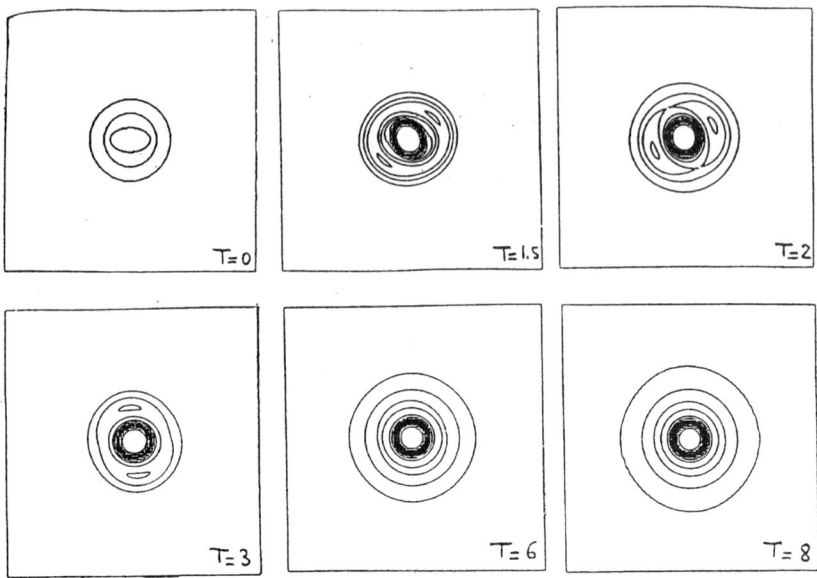

FIGURE 7c.

the relaxation equation converges toward a monopole which is the complete maximum entropy state but which does not correspond to reality. By contrast, when the space dependant diffusion coefficient (5.2) is introduced, the diffusion is slowed down and the agreement with the DNS simulation is excellent (FIGURE 7b). This clearly shows that the tripole can be interpreted as a restricted maximum entropy state confined in a "bubble."

The previous relaxation equations are useful to understand the formation and the structure of a *single* vortex. By contrast, since the inverse temperature β is uniform (as a result of the *global* conservation of energy), they clearly cannot describe the organization of the flow into *several* independent vortices (which would have *a priori* different temperatures). This physical problem is closely related to the fact that (5.1) does not respect the *invariance properties* of the Euler equation: invariance by translation or rotation of the coordinates, Galilean invariance and invariance by rotation of the referential. However, it is possible to reformulate the MEPP under a local form so as to take these properties into account. More general (but also more complicated) equations are then obtained [11] and should be able to describe the organization of large-scale turbulent flows into several independent structures ("multibubble" flows).

Since these general thermodynamical considerations give relatively good results in 2D turbulence, we can also try to apply them in the similar context of stellar systems. The outcome of the MEPP is now an equation of the form [9], [15]:

$$\frac{\partial \bar{f}}{\partial t} + \mathbf{v}\frac{\partial \bar{f}}{\partial \mathbf{r}} + \bar{\mathbf{F}}\frac{\partial \bar{f}}{\partial \mathbf{v}} = \frac{\partial}{\partial \mathbf{v}}\left[D\left(\frac{\partial \bar{f}}{\partial \mathbf{v}} + \beta\bar{f}(\eta_0 - \bar{f})\mathbf{v}\right)\right] \quad (5.3)$$

for the evolution of the "coarse-grained" distribution function. This equation bears strong resemblances with the conventional Fokker-Planck equation of collisional stellar systems (see, e.g, [7]). Indeed, the diffusion current consists of two parts: a pure diffusion term in velocity space which arises naturally from the variations of entropy (fluctuation) and a friction term which is necessary to conserve energy (dissipation). Therefore, a kind of "fluctuation-dissipation" theorem is buried in the MEPP and the Einstein relation $\xi = D\eta_0\beta$ (where ξ is the friction coefficient) is automatically satisfied by the variational principle. It must be noted however that Eq. (5.3) is *not* a Fokker-Planck equation as the friction term is nonlinear in $\bar{f}$. This nonlinearity is the mark of the degeneracy discovered by Lynden-Bell [23] at equilibrium and accounts for the incompressibility of the Vlasov equation in phase space.

The diffusion coefficient is not determined by the MEPP but can be calculated (asymptotically) with a quasilinear theory [19], [39], [15] strictly valid at the late stages of the relaxation when nonlinearities have weaken ("gentle" relaxation). The result is an expression of the form [15]:

$$ D = \frac{16\sqrt{2}\pi^2 G^2 \epsilon_r^3 \epsilon_v^3}{\eta_0^{1/2}\beta^{5/2}v^3} \ln\left(\frac{L}{\epsilon_r}\right) \int_0^{\beta\eta_0\frac{v^2}{2}} \frac{\sqrt{x}}{1+\lambda e^x}dx, \qquad (5.4) $$

where ϵ_r, ϵ_v are the resolution scales in position and velocity and L the typical size of the system. Using (5.4), we find that the time of relaxation, estimated by $t_r \sim \xi^{-1} = 1/D\eta_0\beta$ is of order t_D, the dynamical time. Therefore, the relaxation by "phase mixing" is extremely "violent" as stressed originally by Lynden-Bell [23]. Therefore, it can explain the apparent regularity of elliptical galaxies without recourse to collisions, which operate on a much longer time $t_{coll} \sim (N/lnN)t_D$.

However, we know that the relaxation is incomplete. Many different physical processes can limit the progression toward equilibrium, and it is not easy to determine which one is the most relevant. Like in 2D turbulence, incomplete relaxation can be justified by the decay of the fluctuations as time goes on or when we depart from the relatively well-mixed central region of the galaxy. In addition, the evaporation of stars may play a certain role. During violent relaxation, the stars extract their energy from the rapid fluctuations of the gravitational field. By this process, some stars may acquire very high energies and escape from the system (being ultimately captured by the gravity of other systems). This leads to a depletion of high energy states in Lynden-Bell's statistics. More precisely, we can obtain the following truncated model [15]:

$$ \bar{f} = \eta_0 \frac{e^{-\beta\eta_0\epsilon} - e^{-\beta\eta_0\epsilon_m}}{\lambda + e^{-\beta\eta_0\epsilon_m}}, \qquad (5.5) $$

where $\epsilon = (v^2/2) + \Phi$, as a particular solution of (5.3) and (5.4). Evaporation is justified as long as fluctuations are relatively strong, but when they die away the distribution function (5.5) is maintained as a stationary solution of the Vlasov equation. This distribution function has a finite mass and is a generalization of the Michie-King model [27], [21] to the case of (possibly degenerate) collisionless systems. Indeed, when $\epsilon \to \epsilon_m$ we recover the usual Michie-King model of collisional stellar systems (without the bar on f!) and when $\epsilon \to -\infty$, Eq. (5.5) reduces to the Fermi-Dirac distribution function (2.5). Therefore, the distribution function (5.5) connects continuously the two limits considered by Lynden-Bell [23] in his basic paper. As mentioned previously, other processes may account for incomplete relaxation and lead to other truncated models (see, e.g., [44], [17]).

6. SYSTEMATIC DRIFT EXPERIENCED BY A POINT VORTEX IN 2D TURBULENCE

Equations (5.1) and (5.3) have a very similar structure as they incorporate a term in $\beta \bar{f}(\eta_0 - \bar{f})\mathbf{v}$ or $\beta \varpi(\sigma_0 - \varpi)\nabla \psi$ which compensates the diffusion in velocity or position space. In the framework of the MEPP, these additional terms arise naturally as a consequence of the conservation of energy (β is the corresponding Lagrange multiplier). However, we would like to find a more physical interpretation of these terms. Owing to the analogy between (5.3) and the Fokker-Planck equation (to which it reduces in the nondegenerate limit) we can interpret the term $\beta \eta_0 \bar{f} \mathbf{v}$ as a kind of friction. Now, for a system of point mass stars, Kandrup [20] has shown that it was possible to understand the "dynamical friction" as a result of a polarization process. It is therefore tempting to apply the same line of thought to the case of 2D vortices as well. This leads to the finding that a point vortex in 2D turbulence experiences a "systematic drift" (the counterpart of the "dynamical friction") normal to its mean-field velocity [14]. Indeed, as it travels in the field of other vortices, it alters their distribution and in response the system exerts a back reaction which modifies its initial trajectory. This is the physical reason for the drift. It can be calculated precisely with a linear response theory yielding the expression [14]:

$$\langle \mathbf{V} \rangle_{\text{drift}} = -\beta \gamma D \nabla \psi. \quad (6.1)$$

The drift coefficient $\xi = \beta \gamma D$ is an amusing generalization of the Einstein relation to the case of point vortices. The diffusion coefficient is given by a Kubo formula which can be made explicit, in a mean-field approximation, as:

$$D = \frac{\gamma \tau}{8\pi} \ln \Lambda \langle \omega \rangle, \quad (6.2)$$

where $\tau \sim 2\pi/\langle \Sigma \rangle$ is related to the local shear Σ of the flow and $\ln \Lambda = \int_0^{+\infty} \frac{dY}{Y}$ is the ubiquitous Coulomb factor. There is a divergence at both large and small scales, like for stellar systems, due to the long-range nature (and the singularity for $r \to 0$) of a potential in $\ln r$. The divergence at small Y accounts for the failure of the mean-field approximation on scale $\delta \sim L/N$ when the velocity fluctuations become comparable to the mean-field velocity. The divergence at large scales is solved by the finite extent of the system; in plasma physics we would stop the integration at λ_D (the Debye length) but, in our case, the interactions are unshielded (except in the geophysical context where the Rossby radius plays the same role as the Debye length). It is therefore natural to cut the integral at L, the system size. Therefore, we shall take $\ln \Lambda = \ln(L/\delta) \sim \ln N$. Since the divergence is weak (logarithmic) the result does not depend too much on the precise value of the cut-offs. Comparing the diffusion coefficients

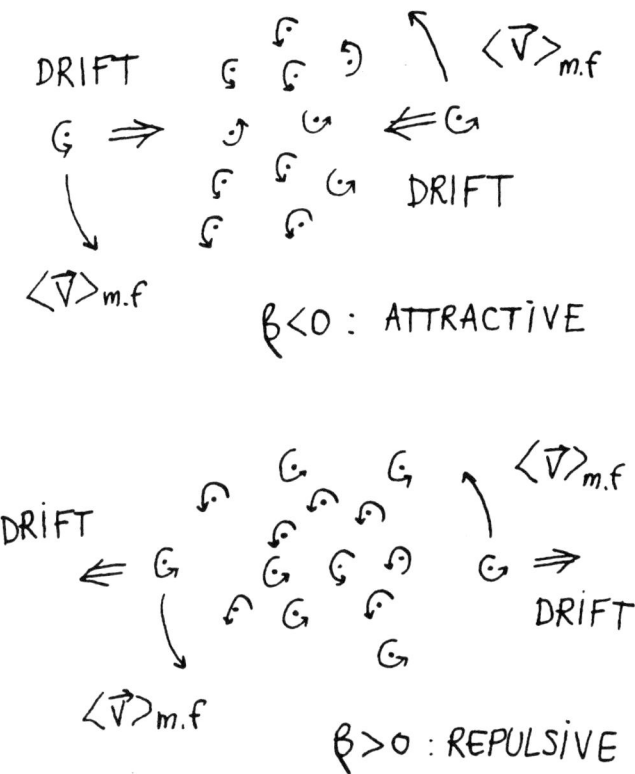

FIGURE 8.

(6.2) and (5.2), we see that the relaxation of point vortices (of order $(N/\ln N)t_D$) is much slower than the (violent) relaxation of continuous vorticity fields (of order t_D).

The direction of the drift has important physical implications. Consider a point vortex moving at the periphery of the system (FIGURE 8). Its motion is anticlockwise if we assume positive circulation. For negative temperatures, the drift is directed to its left: the vortex is attracted to the center of the domain. On the contrary, for positive temperatures, the drift is directed to its right: the vortex is rejected against the boundary. This reflects the general structure of the equilibrium state [31] and provides a physical mechanism for the organization of point vortices at negative temperatures.

Taking into account this "systematic drift," we can write down a Langevin equation for the motion of a point vortex in 2D turbulence:

$$\Delta \mathbf{r} = \langle \mathbf{V} \rangle \Delta t - \xi \nabla \psi \Delta t + \mathbf{B}(\Delta t). \tag{6.3}$$

The first term corresponds to the mean-field velocity, the second to the drift, and the third to fluctuations arising from the difference between the exact distribution of the vortices and their "smoothed out" distribution. We can now apply the standard technics of Brownian theory (see, e.g, [6]). Assuming that the motion of the point vortex is Markovian and that the fluctuations $\mathbf{B}(\Delta t)$ can be described by a gaussian stochastic process, we can derive the following Fokker-Planck equation:

$$\frac{\partial \langle \omega \rangle}{\partial t} + \langle \mathbf{V} \rangle \nabla \langle \omega \rangle = \nabla [D(\nabla \langle \omega \rangle + \beta \gamma \langle \omega \rangle \nabla \psi)] \qquad (6.4)$$

for the average vorticity. This equation is similar to the equation of the MEPP (5.1). This correspondence shows clearly that the term in $\nabla \psi$ can be interpreted as a "systematic drift" of the vorticity. However, Eq. (6.4) applies to a collection of point vortices instead of continuous vorticity fields. Therefore, the nonlinear term $\overline{\omega}(\sigma_0 - \overline{\omega})$ is replaced by $\gamma \langle \omega \rangle$ since there is not the constraint of incompressibility. Except for this (important) difference, Eqs. (5.1) and (6.4) are similar in structure. This may enlarge the interest of the point vortex model.

7. CONCLUSION

This paper stresses the analogy between 2D vortices and stellar systems and shows that statistical mechanics can give some insight on their structure and organization. In this description, the vorticity and the streamfunction in two-dimensional turbulence play the same role as the distribution function and the gravitational field in galaxies. Moreover, vortices experience a "systematic drift" in the same way that stars are subjected to a "dynamical friction." However, the drift has a very different physical effect as it accounts for the *organization* of vorticity at negative temperatures. In this respect, it plays the role of a "binding force." The statistical mechanics of long-range systems faces some difficulties due in particular to the absence of equilibrium in the whole space and the need for self-confinement. These problems can be resolved by considering nonequilibrium thermodynamics and invoking "incomplete relaxation." Moreover, in the case of gravity, global equilibrium does not exist and the system shows the tendency to collapse. It can settle, however, on a local equilibrium state until "gravothermal catastrophe" finally comes into play and creates a central condensation (Black Hole?). There are additional, conceptual, problems posed by the use of the Euler and the Vlasov equations in the statistical description. The Euler equation applies to inviscid flows and it is not clear if the limit of small viscosity is equivalent to zero viscosity. In particular a small but finite viscosity alters the "inviscid invariants" (in particular the high-order moments of the vorticity distribution) and, in case of a long evolution, the prediction of the statistical theory from the initial vorticity field can lead to strong discrepancies [38], [4]. On the other hand, the Vlasov equation is an approximation of a discrete system of stars when encounters can be neglected. Here again, it is not clear to what extent "discreteness" effects can be ignored even in the collisionless regime (see, e.g., [40], [32]). The predictive power of the statistical theories is then limited as long as a correct treatment of the viscous effects and of the granularities is not given.

ACKNOWLEDGMENTS

I would like to express my gratitude to J. Sommeria and R. Robert. I am also grateful to A. Provenzale and the hospitality of the Istituto di Cosmogeofisica. I acknowledge very stimulating discussions with D. Lynden-Bell, H. Kandrup, P. Morrisson, B. Miller, and C.C. Lin during the workshop.

REFERENCES

1. ANTONOV, V.A. 1962. Vest. Leningr. Gos. Univ. **7:** 135. (English version available in Dynamics of Stars Clusters, P. Hut, Ed. IAU Symp. 113, 1985. Reidel, Dordrecht.).
2. BARGE, P. & J. SOMMERIA. 1995. Did planet formation begin inside persistant gaseous vortices? Astron. Astrophys. **295:** L1–L4.
3. BINNEY, J. & S. TREMAINE. 1987. Galactic Dynamics. Princeton Series in Astrophysics. Princeton, NJ.
4. BRANDS, H., J. STULEMEYER, R.A. PASMANTER & T.J. SCHEP. 1997. A mean field prediction of the asymptotic state of decaying 2D turbulence. Phys. Fluids **9**(10): 2815–2817.
5. BRANDS H., P.H. CHAVANIS, R. PASMANTER & J. SOMMERIA. 1998. Maximum entropy vs minimum enstrophy vortices. Phys. Fluids. Submitted.
6. CHANDRASEKHAR. 1943. Stochastic problems in physics and astronomy. Rev. Mod. Phys. **15:** 1.
7. CHANDRASEKHAR. 1949. Brownian motion, dynamical friction, and stellar dynamics. Rev. Mod. Phys. **21**(3): 383–388.
8. CHAVANIS, P.H & J. SOMMERIA. 1996. Classification of self-organized vortices in two-dimensional turbulence: The case of a bounded domain. J. Fluid. Mech. **314:** 267–297.
9. CHAVANIS P.H, J. SOMMERIA & R. ROBERT. 1996. Statistical mechanics of 2D vortices and collisionless stellar systems. Astrophys. J. **471:** 385–399.
10. CHAVANIS, P.H. 1996. Contribution à la mécanique statistique des tourbillons bidimensionnels. Analogie avec la relaxation violente des systèmes stellaires. Thèse de Doctorat, Ecole Normale Supérieure de Lyon.
11. CHAVANIS, P.H. & J. SOMMERIA.1997. Thermodynamical approach for small-scale parametrization in 2D turbulence. Phys. Rev. Lett. **78**(17): 3302–3305.
12. CHAVANIS, P.H. & J. SOMMERIA. 1998. Classification of robust isolated vortices in two-dimensional hydrodynamics. J. Fluid Mech. **356:** 259–296.
13. CHAVANIS, P.H. & J. SOMMERIA. 1998. Degenerate equilibrium states of collisionless stellar systems. Mon. Not. R. Astr. Soc. **296**(3): 569–578.
14. CHAVANIS, P.H. 1998. Systematic drift experienced by a point vortex in 2D turbulence. Phys. Rev. E **58**(2): R1199–R1202.
15. CHAVANIS, P.H. 1998. On the coarse-grained evolution of collisionless stellar systems. Mon. Not. R. Astr. Soc. To appear.
16. CHAVANIS, P.H. & J. SOMMERIA. 1998. Justification of an inviscid minimum enstrophy principle. In preparation.
17. HJORTH J. & J. MADSEN. 1991. Violent relaxation and the $R^{1/4}$ law. Mon. Not. R. Astr. Soc. **253:** 703–709.
18. JOYCE, G. & D. MONTGOMERY. 1973. Negative temperature states for the two-dimensional guiding-center plasma. J. Plasma Phys. **10**(1): 107–121.
19. KADOMTSEV, B.B. & O.P. POGUTSE. 1970. Collisionless relaxation in systems with Coulomb interactions. Phys. Rev. Lett. **25**(17): 1155–1157.
20. KANDRUP, H. 1983. Dynamical friction in a mean field approximation. Astrophys. & Space Sci. **97:** 435–452.

21. KING, I.R. 1966. Astr. J. N.Y. **71:** 64.
22. KUZMIN. 1982. Statistical mechanics of the organization into two-dimensional coherent structures. *In* Structural Turbulence, M.A. Goldshtik, Ed.: 103–114. Acad. Nauk CCCP Novosibirsk, Institute of Thermophysics.
23. LYNDEN-BELL, D. 1967. Statistical mechanics of violent relaxation in stellar systems. Mon. Not. R. Astr. Soc. **136:** 101–121.
24. LYNDEN-BELL, D. & R. WOOD. 1968. The gravothermal catastrophe in isothermal spheres and the onset of red-giants structure for stellar systems. Mon. Not. R. Astr. Soc. **138:** 495–525.
25. LYNDEN-BELL, D. & R. LYNDEN-BELL. 1977. On the negative specific heat paradox. Mon. Not. R. Astr. Soc. **181:** 405.
26. MCWILLIAMS, J. 1984. The emergence of isolated coherent vortices in turbulent flows. J. Fluid. Mech. **146:** 21–43.
27. MICHIE, R.W. 1963. On the distribution of high energy stars in spherical stellar systems. Mon. Not. R. Astr. Soc. **125:** 127.
28. MILLER, J. 1990. Statistical mechanics of the Euler equations in two dimensions. Phys. Rev. Lett. **65**(171)**:** 2137–2140.
29. NEZLIN, M.V. & E.N. SNEZHKIN. 1993. Rossby Vortices, Spiral Structure, Solitons. Springer-Verlag. New York.
30. OGORODNIKOV, K.F. 1963. Dynamics of Stellar Systems. Pergamon.
31. ONSAGER, L. 1949. Statistical hydrodynamics. Nuovo Cimento Suppl. **6:** 279–287.
32. PADMANABHAN, T. 1990. Statistical mechanics of gravitating systems. Phys. Rep. **188**(5)**:** 285–362.
33. ROBERT, R. 1998. On the statistical mechanics of 2D Euler and 3D Vlasov-Poisson equations. Comm. Math. Phys. Submitted.
34. ROBERT, R. 1998. On the gravitational collapse of stellar systems. In preparation.
35. ROBERT, R. & J. SOMMERIA. 1991. Statistical equilibrium states for two-dimensional flows. J. Fluid. Mech. **229:** 291–310.
36. ROBERT, R. & J. SOMMERIA. 1992. Relaxation towards a statistical equilibrium state in two-dimensional perfect fluid dynamics. Phys. Rev. Lett. **69:** 2776–2779.
37. ROBERT, R. & C. ROSIER. 1997. The modelling of small scales in 2D turbulent flows: A statistical mechanics approach. J. Stat. Phys. **86:** 481–515.
38. SEGRE, E. & S. KIDA S. 1996. Final states of incompressible two dimensional decaying vorticity fields. *In* Advances in Turbulence, VI: 137. Kluwer Academic Publishers.
39. SEVERNE, G. & M. LUWEL. 1980. Dynamical theory of collisionless relaxation. Astrophys. & Space Sci. **72:** 293–313.
40. SHU, F.H. 1978. On the statistical mechanics of violent relaxation. Astrophys. J. **225:** 83–94.
41. SOMMERIA, J., C. NORE, T. DUMONT & R. ROBERT. 1991. Théorie statistique de la tache rouge de Jupiter. C.R. Acad. Sci. Paris., Ser. II. **312:** 999–1005.
42. TANGA, P., A. BABIANO, B. DUBRULLE & A. PROVENZALE. 1996. Forming planetesimals in vortices. Icarus **121:** 158–170.
43. TURKINGTON B. & N. WHITAKER. 1996. Statistical equilibrium computations of coherent structures in turbulent shear layers. SIAM J. Sci. Comput. **17**(16)**:** 1414.
44. TREMAINE, S. 1987. *In* Structure and Dynamics of Elliptical Galaxies. T. de Zeeuw, Ed.: 367–374. IAU Symp. No. 127. Reidel, Dordrecht.

N-Body Simulations of Galaxies and Groups of Galaxies with the Marseille GRAPE Systems

E. ATHANASSOULA[a]

Observatoire de Marseille, 2 place le Verrier, 13248 Marseille cedex 04, France

ABSTRACT: I review the Marseille GRAPE systems and the *N*-body simulations done with them. First I briefly describe the available hardware and software, their possibilities and their limitations. I then describe work done on interacting galaxies and groups of galaxies. This includes simulations of the formation of ring galaxies, simulations of bar destruction by a massive compact satellite, of merging in compact groups and of the formation of brightest members in clusters of galaxies.

GRAPE HARDWARE AT MARSEILLE OBSERVATORY

The idea behind GRAPE systems is at the same time very simple and very efficient. It stems from the realization that most of the CPU time in *N*-body simulations is spent calculating the forces, with only a small percentage devoted to the remaining parts, like moving the particles. Thus the group around D. Sugimoto and J. Makino realized GRAPE (from GRAvity piPE), a card which performs the force calculation on custom-made chips and which can be put in a standard workstation, allowing one to achieve at relatively low cost an excellent CPU performance. A series of such GRAPE boards have been built by the group in Tokyo University, starting with GRAPE-1 and evolving steadily to GRAPE-4, while new members of this family, like GRAPE-5 and GRAPE-6, are in progress. For a brief history of this project and descriptions of the various GRAPE systems see Makino and Taiji [1] and references therein. Boards with even numbers have high accuracy arithmetic and can be used for collisional simulations, where close encounters play an important role in the evolution of the system, as for globular clusters and planetesimals. Boards with odd numbers have lower precision arithmetic and can only be used for simulating collisionless systems.

Two main GRAPE systems are presently working in Marseille Observatory. A 5-board GRAPE-3AF system, coupled via an Sbus/VMEbus converter to an Ultra 2/200 front end, and a GRAPE-4 system coupled via a PCI interface to an Alpha 500/500 workstation. Since the latter system was only made operational a few months before this conference, most of this paper will be devoted to the GRAPE-3AF system and the results obtained with it.

Our GRAPE-3AF system has 40 chips in total and gives us a peak speed equivalent of more than 20 Gflops. The boards are hardware limited to 131,072 particles, but it is possible to use them for a much larger number by splitting the particles into

---

[a]E-mail: lia@paxi,cnrs-mrs.fr, lia@obmara.cnrs-mrs.fr

groups of 131,072 particles or less, presenting the groups successively to the board and then adding the forces from all the groups on the front-end machine.

Since GRAPE-3 boards are meant only for collisionless simulations they use low accuracy arithmetic (14 bits for the masses, 20 bits for the positions and 56 bits for the forces). As discussed by Athanassoula et al. [2], this accuracy is sufficient for collisionless simulations.

Doing on-line analysis of the simulation on the same processor as that used to pilot the GRAPE boards would slow down the simulations in an unacceptable way. Thus a second processor is necessary. Such on-line analysis can include calculation of pattern speeds, amplitudes and shapes of different structures, energy and angular momentum exchange between different components, etc. The processor piloting the boards spawns a task starting the analysis scripts at regular intervals. Several tasks, such as calculating the mass still bound to a given galaxy, can be carried out much faster on a GRAPE board. In such cases the analysis scripts are executed on a smaller GRAPE-3 system, piloted by another workstation.

A good description of the GRAPE-4 boards and of their performances has been given by Makino et al. [3]. A description more specific to the Marseille system will be given in a forthcoming paper. Since the front end of our GRAPE-4 system has no second processor, the on-line analysis is spawned to another workstation, acting as a slave to the Alpha 500/500 driving the GRAPE-4 system.

GRAPE SOFTWARE AT MARSEILLE OBSERVATORY

Two codes are routinely used on the Marseille GRAPE-3 systems: a direct summation code and a tree code [2], [4]. The latter follows the vectorization scheme proposed by Barnes [5]. Thus the particles are first divided into blocks and then the tree traversal is executed for a block of particles at a time, rather than for each particle separately, as in the standard tree code. The optimum number of particles per block depends of course on the number of boards, the power of the front end and, to a lesser extent, on the number of particles and the tolerance parameter. We find that, for our 5-board GRAPE-3 configuration and the type of simulations we run, 7,000 to 15,000 particles per block are a good choice. The CPU time necessary for one time step increases roughly linearly with the number of particles N. It also increases with decreasing tolerance (or opening angle), but the dependence is less strong than for the standard tree code, being particularly small for values of the tolerance larger than 0.8. Because of the increased role of the direct summation in the force calculation, this tree code is much more accurate than the standard one.

The accuracy of the force calculations by GRAPE-3 was tested in [2] with the help of the MISE/MASE formalism introduced by Merritt [6]. It was found that the forces are calculated as accurately as when full precision is used on the front end. The reason is that the errors on GRAPE-3 are due to round-off and thus can be considered as random. They thus cancel out when the force contributions of a large number of particles are added. MISE/MASE tests also showed that the accuracy of the tree code is comparable to that of the direct summation. That can be easily understood since, in the version of the tree code in operation on our GRAPE systems, the force from nearby particles is calculated by direct summation.

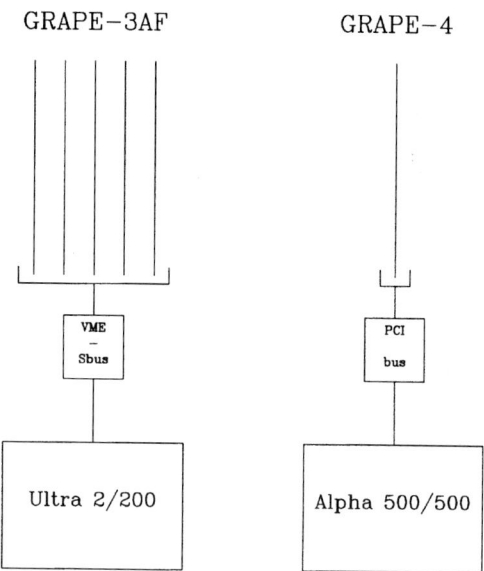

FIGURE 1. Schematic representation of the two main GRAPE systems in Marseille Observatory. To the left GRAPE-3 and to the right GRAPE-4.

Further tests include energy conservation during the simulations and the comparison of results of runs with different number of particles. Finally, a few simulations were performed both on GRAPE-3 and GRAPE-4 and the comparison shows very good agreement. Thus it can be concluded that GRAPE-3 is well suited for N-body simulations of collisionless systems, both because of its accuracy and because of its high speed.

It is at present possible to execute three codes on our GRAPE-4 system, a direct summation and a tree code, both with a constant time step, and a direct N-body code with a variable time step. The latter uses an implementation of the Ahmad-Cohen scheme based on on a fourth-order Hermite integrator [7]. A description of their accuracy and performance, as well as a comparison with those of GRAPE-3, will be given elsewhere.

MAIN RESEARCH AREAS

Our GRAPE-3 and GRAPE-4 systems are used for N-body simulations in many different areas of astronomical research, ranging from dynamics and evolution of clusters of galaxies, to the dynamical evolution of planetesimals. Most of it, however, centers around galaxies and groups of galaxies. Some of the latest results are briefly discussed below. To this list should be added the study of cusps (in collaboration with Ch. Siopis and H. Kandrup), the study of the effect of black holes in the central parts of elliptical galaxies (in collaboration with F. Leeuwin), the study of the

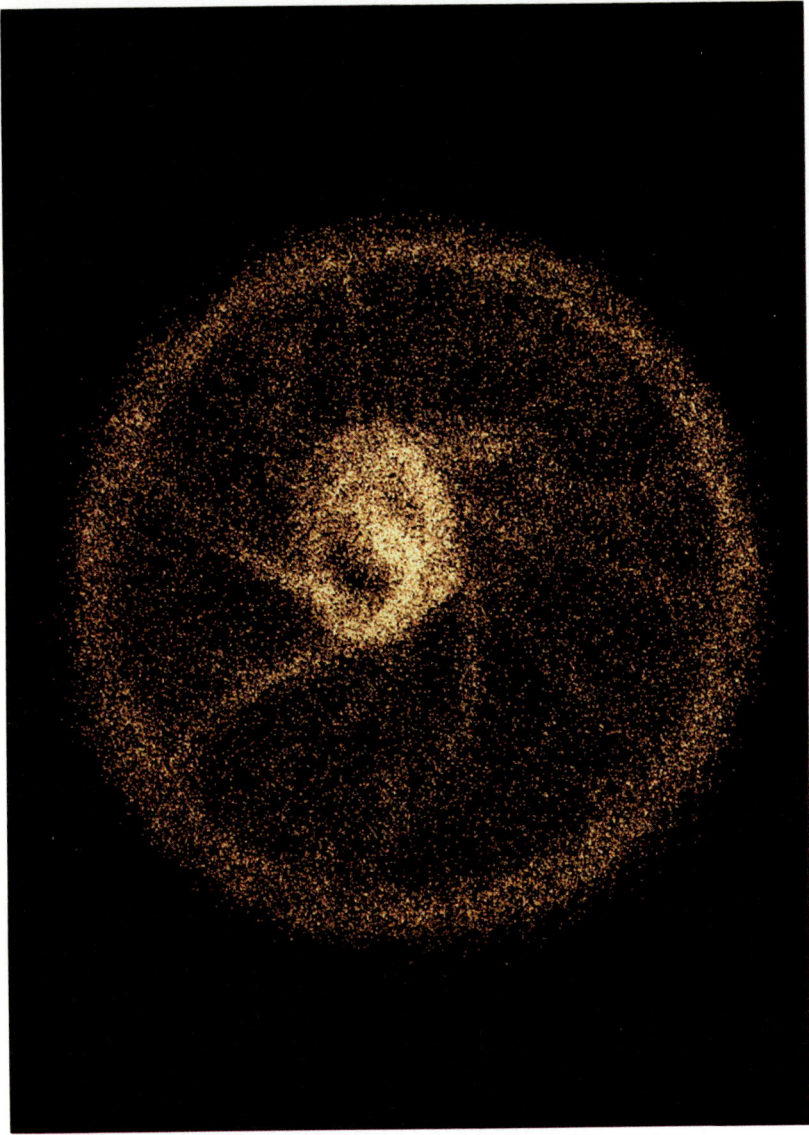

FIGURE 2. Snapshot from an *N*-body simulation of the formation of a ring galaxy. Only particles initially in the target disc are plotted. Both rings, as well as the spokes in the region between them, are clearly visible.

dynamical evolution of planetesimals (in collaboration with P. Barge) and that of the gas flows in bars (in collaboration with I. Berentzen and C. Heller).

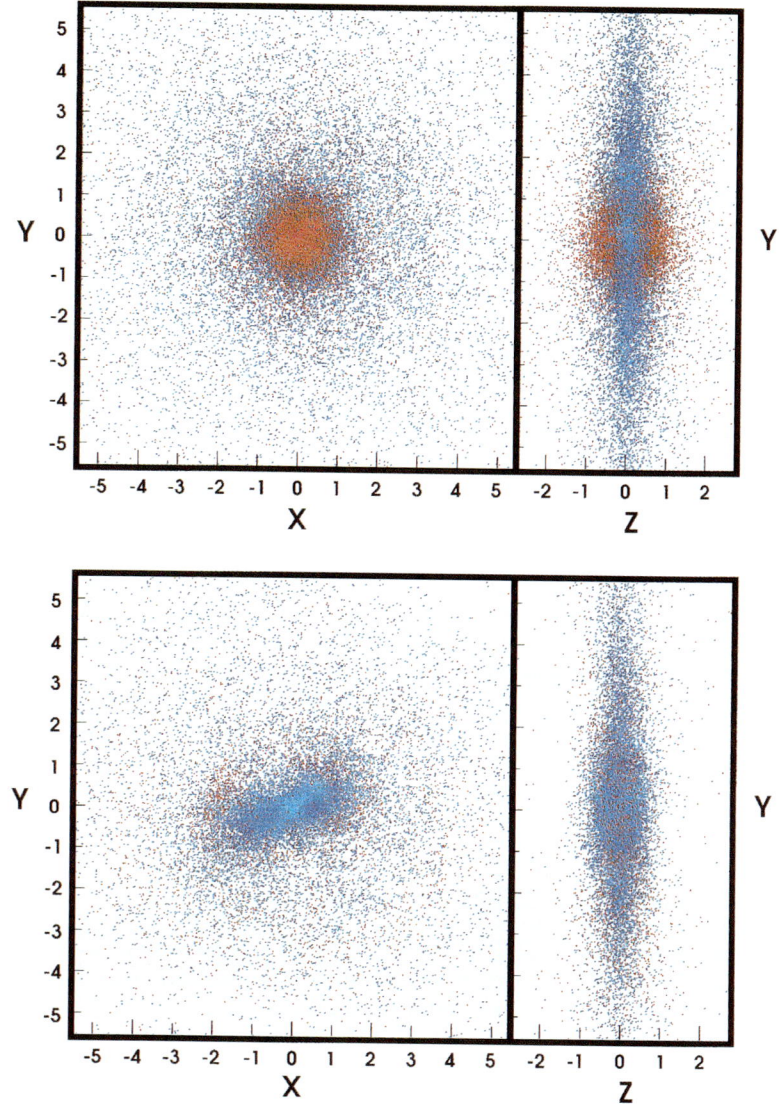

FIGURE 3. One of the final instants from an *N*-body simulation of a target disc galaxy and a satellite, after the merging has been completed. The particles of the target disc are shown in blue and those of the satellite in red, while particles in the halo of the target are not plotted. Face-on views are given by the left panels and edge-on views by the right panels. In the simulation shown in the top two panels the companion was initially massive and compact. After merging, it has lost little of its mass and forms a bulge in the center of the target disc. In the simulation shown in the bottom two panels the companion was initially fluffy and less massive and has been shredded to pieces while spiraling in the target's disc.

RING GALAXIES

When a small compact galaxy hits the disc of a target disc galaxy head-on and near-vertically, then one or more expanding rings can be generated [8]–[12]. Indeed as the companion approaches the target it exerts an extra inward gravitational force on the disc particles, which causes their orbits to contract. This is followed by a rebound, which, because of the decrease of the epicyclic frequency with radius, will result in a crowding of the orbits and the generation of a high amplitude, transient density wave, propagating outwards. We performed a number of fully self-consistent N-body simulations of such encounters, both on barred and on nonbarred target disc galaxies [13]. One or two transient and short-lived rings form, the second considerably after the first one. The expansion velocity of the first ring is bigger than that of the second, and both decrease with time. The amplitude, width, lifetime, and expansion velocity of the first ring are considerably higher for impacts of large mass companions, than for lower mass ones. After the second ring has formed several simulations showed spokes in the region between the two rings. They are trailing, nearly straight and short-lived. An example is shown in FIGURE 2. Rings formed from such head-on encounters need not be mistaken with those observed at the resonances of disc galaxies. Indeed, even when they are symmetric and have no spokes, they have considerable expansion velocities, which should be detectable spectroscopically.

The Cartwheel is probably the best studied ring galaxy [14]–[19]. It has two clear rings and several spokes in the region between them. Three small galaxies can be found in its neighborhood and one of them should be responsible for its structure. Although Higdon [15] proposed that it is the one farthest from the Cartwheel that is the culprit, a careful comparison of simulations and observations (Bosma, Puerari & Athanassoula, in preparation), taking into account both the morphological and the kinematical data, argues that it is G2 (in Higdon's notation) which is to blame.

IS IT POSSIBLE TO DESTROY A BAR WITHOUT DESTROYING THE DISC IT RESIDES IN?

An interesting question that can be asked in this context is whether it is possible for a companion to destroy a bar in a disc galaxy, while not destroying the disc. To answer it I first tried trajectories where the companion, initially on a rectilinear orbit, hit the central part of the disc either vertically, or at a skew angle [20]. Such trajectories can bring substantial changes to the pattern speed of the bar, as well as to the amplitude of its $m = 2$ component. The lowering of the $m = 2$ component is in many cases very spectacular, so one could easily talk of a bar dissolution. In all these cases, however, the disc thickens too much. I was unable to find a case where the disc stayed thin and at the same time the bar was destroyed, although I must admit that my search was not exhaustive.

I then tried a different kind of trajectory [21]. Now the companion starts off in a quasi-circular orbit outside the halo, and spirals, via dynamical friction, to the central part of the galaxy. If it is sufficiently massive and compact, it will lose only a small fraction of its mass by the time it has reached the center. It will then contribute to the bulge population, or, if there was no bulge present before the companion fell in, it will form it. Thus the target galaxy will evolve along the Hubble sequence, from a late to an early type disc galaxy. While the companion spirals through the target disc it heats it up and makes it thicker (cf. [22]–[25]). On the other hand the target also expands, because the system has to conserve angular momentum, so, comparing its axial ratio before and after the merging, we see that the disc has been somewhat, but not much, thickened. The $m = 2$ amplitude of the disc decreases very abruptly when the companion reaches the center. At that time it increases considerably the central concentration of the galaxy, and, by so doing, increases the fraction of irregular orbits present in the disc to the detriment of the x_1 stable periodic orbits and the regular orbits trapped around them [26]–[29]. Thus the bar, deprived of its most ardent supporters, will be destroyed. The final stage of such a simulation is shown in the two upper panels of FIGURE 3, the particles initially in the target disc shown in blue and the particles initially in the companion in red. FIGURE 4 shows separately the particles initially in the target disc (upper two panels) and the satellite (lower two panels). Note that the satellite, which was initially spherical, has, after the merging, become oblate.

The situation is totally different in the case of a small mass, fluffy companion. In this case the companion loses a lot of its mass while spiraling through the disc and no substantial fraction of it will reach the center. Thus the bar will not be destroyed. Material stripped from the companion will form a thick disc, thicker than that of the original target. The final stage of such a simulation is shown in the lower two panels of FIGURE 3. Again the particles initially in the target disc are shown in blue and the particles initially in the companion in red. FIGURE 5 shows separately the particles initially in the target disc (upper two panels) and the satellite (lower two panels). Note that a substantial fraction of the companion mass is concentrated along the ends of the bar.

In both the above examples the plane of the companion's orbit is the same as that of the target's disc. Let us now consider cases where the two planes are initially at an angle. Then during the simulation the plane of the target disc will tilt [21], [30], so that the angular momentum of the system be conserved. In the case shown in [21], and other unpublished simulations, the mass of the companion is equal to that of the target disc, and the angle by which the disc tilts is comparable to, although smaller than, the angle between the orbit of the companion and the plane of the target disc at the beginning of the simulation. The remaining results are as in the case where the companion orbits initially in the plane of the target's disc, except for an increased thickening of both the target disc and the disc made by the material shredded from the companion.

MERGING RATES IN COMPACT GROUPS

Compact groups are groups of a few galaxies, close together in the sky, and far from other galaxies or groups of galaxies. Hickson, using a precise definition along these lines, catalogued 100 such groups from the Palomar sky survey [31]. He has also given a review of the relevant observational data about such groups [32]. One of the main questions that they raise is that of their life time. Indeed, if one simply

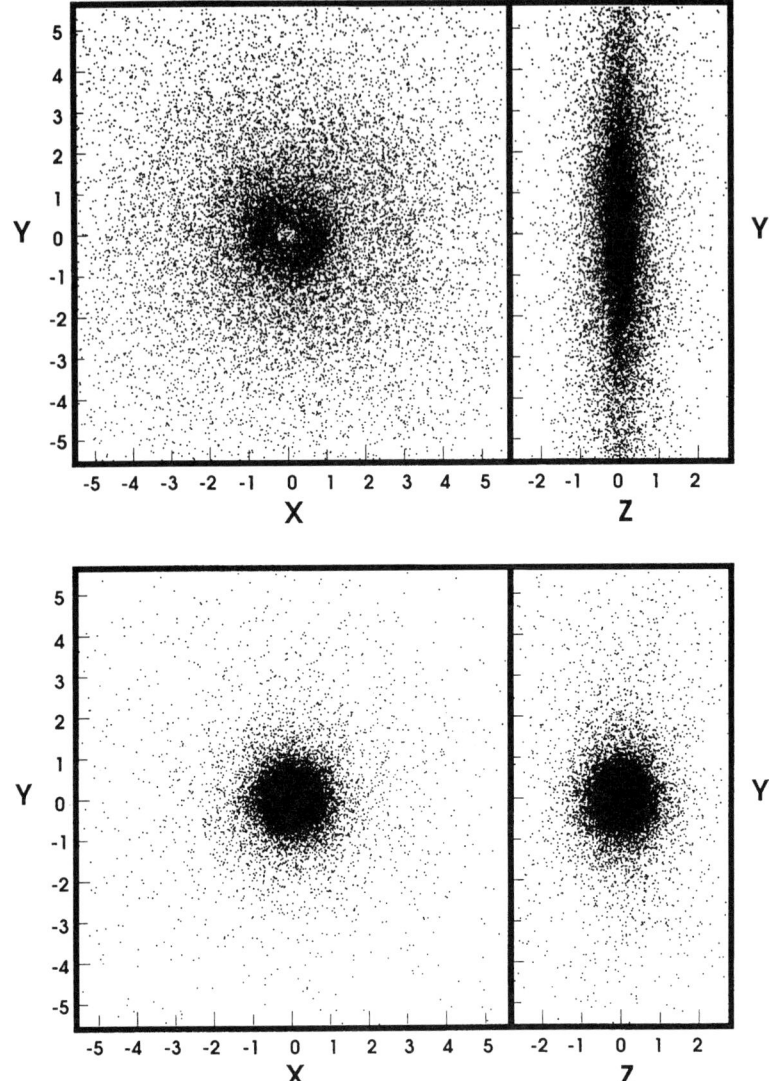

FIGURE 4. Same simulation as the top two panels of FIGURE 3, but now the particles from the target disc and companion are shown separately, the target disc in the top two panels and the companion in the two bottom ones.

calculates the crossing time in such groups from their size and their velocity dispersion, one finds very low values, from which it has often been inferred that mergings should occur quite frequently. Thus the question of why so many such groups are still observed is raised. One possibility is to generate such groups continuously from

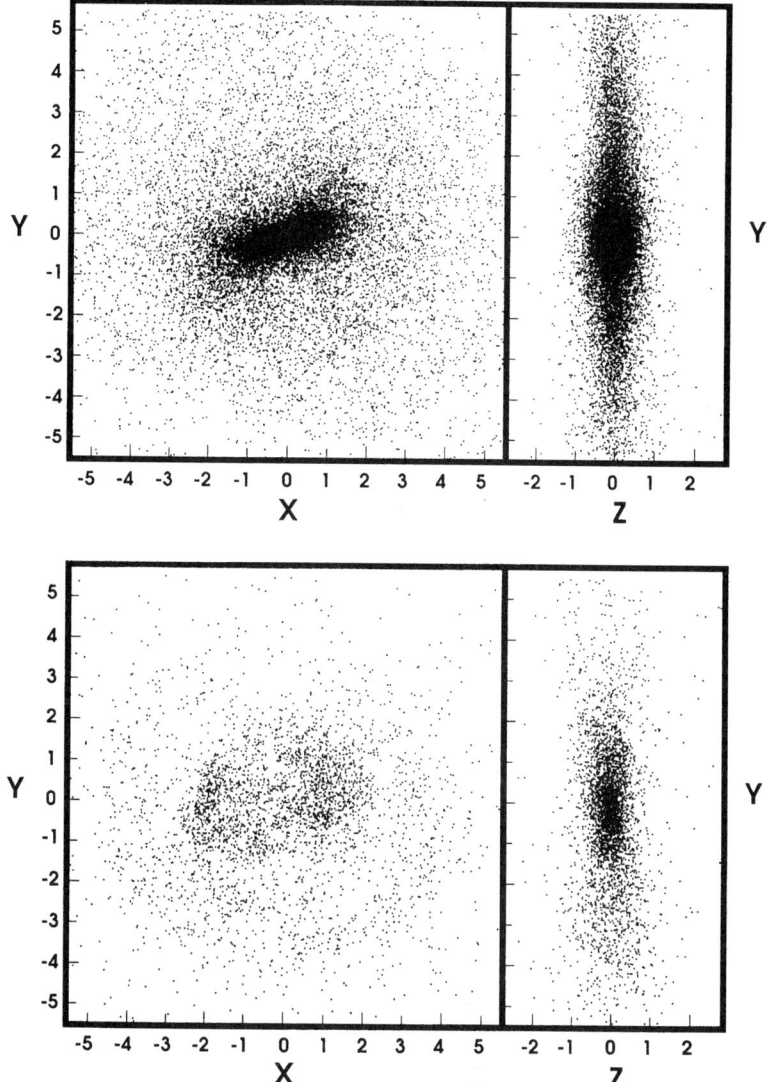

FIGURE 5. Same simulation as the bottom two panels of FIGURE 3, but now the particles from the target disc and companion are shown separately, the target disc in the top two panels and the companion in the two bottom ones.

loose groups [33], [34], or add new galaxies from infalling new material [35]. The first of these alternatives introduces the problem of the merger remnants, both because the integrated luminosity of a compact group is three to four times higher than that of an average isolated elliptical galaxy, and because the number of such remnants could be quite high. For the second alternative, one or more of the galaxies in each compact group would have to be the result of previous mergings, also introducing problems concerning the fraction of ellipticals in compact groups, as well as their total luminosity. Together with Makino and Bosma [36] I have followed another alternative, namely, we studied what parameters of the group affect the merging rate and thus obtained clues about how to form long-lived compact groups.

Our simulations start with five identical spherical galaxies disposed in a compact group which has, in all cases, the same mass and binding energy. Two different types of halos have been considered: Either halos around individual galaxies (hereafter individual halos, or IH), or halos encompassing the whole group and centered at its center (hereafter common halos, or CH). We have also considered different luminous-to-dark mass ratios, individual halos of different spatial extents, as well as different density distributions and different kinematics, both for the common halos and the distributions of the centers of the galaxies. Once these parameters were fixed, we made five different realizations, with different random number seeds, in order to allow some, albeit small, statistics, and be able to make averages over different realizations. We thus ran over 200 simulations, but even so, we are far from covering all possible cases. For all simulations we counted the number of galaxies still present in the group as a function of time and thus were able to measure merging rates.

The first question we addressed is whether groups with individual halos merge faster or slower than corresponding groups with common halos, since two contradictory results had been previously reported in the literature [37]–[39]. Indeed, there are two different effects influencing the outcome in an opposite sense. On the one hand dynamical friction is more important in the case of denser halos, and this should lead to shorter merging times. On the other hand more massive halos would entail less massive galaxies (since the total mass of the group is the same in all simulations) and therefore less mutual attractions between them, which would lead to longer merging times. Which of the two effects is dominant depends on the configurations. Thus groups with individual halos merge faster than groups with common halos if the configuration is centrally concentrated. For less centrally concentrated groups the merging is initially faster for IH cases, and slower after part of the group has merged.

In the case of common halos we find that the more massive the common halo, the longer it will take the group to merge. This can be easily understood, since the mutual attractions between galaxies are smaller for relatively more massive common halos, and, in the extreme case where the masses of the individual galaxies are zero, then they could be considered as test particles and encounters would happen only accidentally, determined by their trajectories and cross sections.

Another factor influencing the merging rate is the central concentration of the configuration. In particular, for common halos and a high halo-to-total mass ratio, groups in centrally concentrated halos merge considerably faster. This can be understood because of the important dynamical friction that galaxies will feel in the central regions of such centrally concentrated configurations. As far as the initial kinematics of the group are concerned, groups with initially cylindrical rotation merge slower.

Taking into account all the effects enumerated above, it is possible to construct long-lived compact groups. Thus Athanassoula, Makino, and Bosma [36] followed the evolution of a group with a common halo, a high halo-to-total mass ratio and a density distribution with little central concentration and found that the merging occurred only after a large number of crossing times, corresponding to a time larger than a Hubble time. This provides a solution to the longevity problem of compact groups, and could explain why we observe so many of them.

FORMATION OF BRIGHTEST CLUSTER MEMBERS AND cD GALAXIES

We have performed a number of simulations to follow the dynamical evolution of groups of 50 to 100 identical spherical galaxies (cf. [40], [41] and in preparation). They can be thought of as simulating groups, poor clusters, or subcondensations within rich clusters, provided that the dynamical influence of the remaining part of the cluster can be neglected. A large variety of initial conditions have been considered. This includes the case of individual halos (where each halo is centered around a galaxy), or a common halo encompassing the whole group or cluster. We have also considered different density distributions of the halo material and of the galaxies in the group or cluster, different ratios of halo-to-total mass and different initial kinematics.

The standard evolution shows important merging in the central regions and the formation of a massive central object. The only way to avoid this is to consider initial conditions such that the central parts do not contain any galaxies. This is obviously artificial, but has the advantage of stressing the role of the initial seed in the formation of the massive central object. It also predicts that initial configurations with low central concentration should form the central massive object slower than configurations with high initial central concentrations, as we were able to confirm with further simulations. Let me also note that it is in good agreement with our results on compact groups described above.

Two mechanisms contribute to the formation of the massive central objects. One is cannibalism of the small galaxies by the big central object (e.g., [42], [43]), and the other is accretion onto the central object of material that has been stripped off individual small galaxies [44]–[46]. Both are present in all simulations, but to a varying degree, depending on the initial conditions.

We compared the properties of the central massive object with those of observed brighter cluster members and found fairly good agreement, concerning the morphology, the surface photometry, and the kinematics. For example we find that, as is observed [47], [48], the triaxiality is stronger in the outer than in the inner parts. Also we found that, in the case of nonspherical anisotropic initial conditions, the central massive objects "remember" the orientation of the initial configuration. This should be linked to the fact that the orientations of brighter cluster members are not random, but correlate with that of the cluster in which they reside [49]–[56].

Schombert [57] did photometry of a large sample of brightest cluster members and showed that they fall in three classes. In the first class we find objects whose projected light profile follows a $r^{1/4}$ over most of the galaxy. In the second, the projected light distribution falls, in the outer parts, somewhat below the $r^{1/4}$ that fits the main body of the galaxy, while in the third class it is higher than that of the $r^{1/4}$ law. Brighter cluster members falling in that third category are called cD galaxies, are rather frequent, and have a light "halo." The term "halo" is perhaps rather unfortunate, since it could lead to confusion with the dark halos around galaxies, and it must be stressed that the "halo" of cD galaxies is luminous material in the outer part of the galaxy, in excess of the $r^{1/4}$ law.

Assuming that the M/L ratio does not depend on radius, we also calculated projected density profiles of the central massive objects formed in our simulations and found that they fall in the same three classes as those outlined by Schombert for his observed sample. Thus, we have set out to determine which properties in the initial conditions determine in which of the three classes a central massive object will fall. Although our results for this are still preliminary, they nevertheless allow us to make a few tentative conclusions. Some of the objects we have found so far in the second class originated from rather nonspherical initial conditions. More interesting are the objects falling in the third class. Our simulations suggest a link between the percentage of the mass of the central massive object that came via accretion and the excess matter in the outer parts, in the sense that we find in the third class objects for which a high fraction of their mass is due to accretion. This of course shifts the question to what types of initial conditions result in a considerable accretion, a question which we are currently investigating. In order to have considerable accretion we have to have a considerable amount of material which was stripped from the initial galaxies. One way of achieving this is to have a quite centrally concentrated common dark matter halo. In such a case, the galaxies passing near the center of the group or cluster are torn to pieces, thus creating the material necessary for the accretion. An alternative way, which we have not yet verified by numerical simulations, would be to have important interactions between the individual galaxies before they merge to form the central massive object. Such interactions could tear material off the small galaxies by tidal forces, and this material could be at later times accreted by the central object. The amount of material thus stripped should depend on the initial distribution of matter in the individual galaxies, e.g., on whether they are disc or elliptical galaxies. Thus more elaborate N-body simulations are necessary to verify this possibility.

ACKNOWLEDGMENTS

I am grateful to Albert Bosma and Jean-Charles Lambert, since without their help and encouragement this work would not have been possible. It is also a pleasure to thank all my collaborators in the projects mentioned in this paper, and particularly Jun Makino, Carlos Garcia-Gomez, Tony Garijo, and Ivanio Puerari. I would also like to thank Philippe Balard for producing FIGURES 2 to 5. I also thank the INSU/CNRS, the University of Aix-Marseille I, and the Institut Gassendi (IGRAP) for funds to develop the necessary computing facilities. The final draft of this manuscript was written at the Newton Institute for Mathematical Sciences, whose support I gratefully acknowledge.

REFERENCES

1. MAKINO, J. & M. TAIJI. 1998. Scientific simulations with special purpose computers: The GRAPE. Wiley . Chichester.

2. ATHANASSOULA, E., A. BOSMA, J.-C. LAMBERT & J. MAKINO. 1998. Performance and accuracy of a GRAPE-3 system for collisionless N-body simulations. Mon. Not. R. Astron. Soc. **293**: 369–380.
3. MAKINO, J., M. TAIJI, T. EBISUZAKI & D. SUGIMOTO. 1997. GRAPE-4: A Massively Parallel Special-Purpose Computer for Collisional N-body Simulations. Astrophys. J. **480**: 432–446.
4. MAKINO, J. 1991. Treecode with a Special-Purpose Processor. Publ. Astron. Soc. Japan **43**: 621–638.
5. BARNES, J. 1990. A modified tree code: Don't Laugh; It Runs. J. Comp. Phys. **87**: 161–170.
6. MERRITT, D. 1996. Optimal smoothing for N-body codes. Astron. J. **111**: 2462–2464.
7. MAKINO, J. & S. AARSETH. 1992. On a Hermite Integrator with Ahmad-Cohen Scheme for Gravitational Many-Body Problems. Publ. Astron. Soc. Japan **44**: 141–151.
8. LYNDS, R. & A. TOOMRE. 1976. On the interpretation of ring galaxies: The binary ring system II Hz 4. Astrophys. J. **209**: 382–388
9. THEYS, J.C. & E.A. SPIEGEL. 1976. Ring galaxies I. Astrophys. J. **208**: 650–661.
10. THEYS, J.C. & E.A. SPIEGEL. 1977. Ring galaxies II. Astrophys. J. **212**: 616–633.
11. TOOMRE, A. 1978. Interacting systems. *In* The large scale structure of the Universe. Eds. M.S. Longair & J. Einasto, I.A.U. Symp. **79**: 109–116.
12. APPLETON, P.N. & C. STRUCK-MARCELL. 1996. Collisional ring galaxies. Fundamentals of Cosmic Phys. **16**: 111–220.
13. ATHANASSOULA, E., I. PUERARI & A. BOSMA. 1997. Formation of rings by infall of a small companion galaxy. Mon. Not. R. Astron. Soc. **286**: 284–302.
14. HIGDON, J. 1995. Wheels of Fire I. Massive Star Formation in the Cartwheel Ring Galaxy. Astrophys. J. **455**: 524–535.
15. HIGDON, J. 1996. Wheels of Fire II. Neutral Hydrogen in the Cartwheel Ring Galaxy. Astron. J. **467**: 241–260.
16. AMRAM, P., C. MENDES DE OLIVEIRA, J. BOULESTEIX & C. BALKOWSKI. 1998. The Hα Kinematics of the Cartwheel Galaxy. Astron. Astrophys. **330**: 881–893.
17. STRUCK, C., P.N. APPLETON, K.D. BORNE, & R.A. LUCAS. 1996. Hubble Space Telescope Imaging of Dust Lanes and Cometary Structures in the Inner Disk of the Cartwheel Ring Galaxy. Astron. J. **112**: 1868–1876.
18. HERNQUIST, L. & M.L. WEIL. 1993. Spokes in ring galaxies. Mon. Not. R. Astron. Soc. **261**: 804–818.
19. STRUCK-MARCEL, C. & J. HIGDON. 1993. Hydrodynamic Models of the Cartwheel Ring Galaxy. Astrophys. J. **411**: 108–124.
20. ATHANASSOULA, E. 1996. Evolution of bars in isolated and interacting disc galaxies *In* Barred Galaxies, R. Buta, D.A. Crocker & B.G. Elmegreen, Eds. Astron. Soc. Pac. Conference Series. **91**: 309–321.
21. ATHANASSOULA, E. 1996. The fate of barred galaxies in interacting and merging systems. *In* Barred Galaxies and Circumnuclear Activity. Nobel Symposium No. 89, Aa. Sandqvist & P.O. Lindblad, Eds. **474**: 59–66. Lecture Notes in Physics, Springer Verlag. Berlin.
22. QUINN, P.J. & J. GOODMAN. 1986. Sinking satellites of Spiral Systems. Astrophys. J. **309**: 472–495.
23. TÓTH, G. & J.P. OSTRIKER. 1992. Galactic disks, infall and the global value of Ω. Astrophys. J. **389**: 5–26.
24. QUINN, P.J., L. HERNQUIST & D. FULLAGAR. 1993. Heating of galactic Disks by Mergers. Astrophys. J. **403**: 74–93.
25. WALKER, I.R., J.C. MIHOS & L. HERNQUIST. 1996. Quantifying the fragility of galactic discs in minor mergers. Astrophys. J. **460**: 121–135.
26. HASAN, H. & C.A. NORMAN. 1990. Chaotic orbits in barred galaxies with central mass concentration. Astrophys. J. **361**: 69–77.

27. HASAN, H., D. PFENNIGER & C.A. NORMAN. 1993. Galactic bars with central mass concentrations. Three-dimensional dynamics. Astrophys. J. **409:** 91–109.
28. NORMAN, C.A., J.A. SELLWOOD & H. HASAN. 1996. Bar dissolution and bulge formation: An example of secular dynamical evolution in galaxies. Astrophys. J. **462:** 114–124.
29. FRIEDLI, D. & W. BENZ. 1993. Secular evolution of isolated barred galaxies. I Gravitational coupling between stellar bars and interstellar matter. Astron. Astrophys. **268:** 65–85.
30. HUANG, S. & R.G. CARLBERG. 1997. Sinking Satellites and Tilting Disk Galaxies. Astrophys. J. **480:** 503–523.
31. HICKSON, P. 1982. Systematic properties of compact groups of galaxies. Astrophys. J. **255:** 382–391
32. HICKSON, P. 1997. Compact Groups of Galaxies. Annu. Rev. Astron. Astrophys. **35:** 357–388
33. DIAFERIO, A., M.J. GELLER & M. RAMELLA. 1994. The formation of compact groups of galaxies. I. Optical properties. Astron. J. **107:** 868–879
34. DIAFERIO, A., M.J. GELLER & M. RAMELLA. 1994. The formation of compact groups of galaxies. II. X-Ray properties. Astron. J. **109:** 2293–2304.
35. GOVERNATO, F., P.A. TOZZI & CAVALIERE. 1996. Small groups of galaxies: A clue to a critical Universe. Astrophys. J. **458:** 18–26.
36. ATHANASSOULA, E., J. MAKINO & A. BOSMA. 1997. Evolution of compact groups of galaxies I. Merging rates. Mon. Not. R. Astron. Soc. **286:** 825–838.
37. BARNES, J.E. 1985. The dynamical state of groups of galaxies. Mon. Not. R. Astron. Soc. **215:** 517–536.
38. BODE, P.W., H.N. COHN & P.M. LUGGER. 1992. Simulations of Compact Groups of Galaxies: The effect of the Dark Matter Distribution. Astrophys. J. **416:** 17–25.
39. ATHANASSOULA, E. & J. MAKINO. 1995. Simulations of compact groups of galaxies: Some preliminary results. *In* Compact Groups of Galaxies, O. Richter & K. Borne, Eds. A.S.P. Conf. Series **70:** 143–149.
40. GARCÍA GÓMEZ, C., E. ATHANASSOULA & A. GARIJO. 1996. Dynamical evolution of galaxy groups: A comparison of two approaches. Astron. Astrophys. **313:** 363–376.
41. GARIJO, A., E. ATHANASSOULA & C. GARCÍA GÓMEZ. 1997. The formation of cD galaxies. Astron. Astrophys. **327:** 930–946.
42. OSTRIKER, J.P. & S.D. TREMAINE. 1975. Another evolutionary correction to the luminosity of giant galaxies. Astrophys. J. **202:** L113–L117.
43. OSTRIKER, J.P. & M.A. HAUSMAN. 1977. Cannibalism among the galaxies — Dynamically produced evolution of cluster luminosity functions. Astrophys. J. **217:** L125–L129.
44. GALLAGHER, J.S. & J.P. OSTRIKER. 1972. A note on mass loss during collisions between galaxies and the formation of giant systems. Astron. J. **77:** 288–291.
45. RICHSTONE, D.O. 1975. Collisions of galaxies in dense clusters. I. Dynamics of collisions of two galaxies. Astrophys. J. **200:** 535–547.
46. RICHSTONE, D.O. 1976. Collisions of galaxies in dense clusters. II. Dynamical evolution of cluster galaxies. Astrophys. J. **204:** 642–648.
47. PORTER, A.C., D.P. SCHNEIDER & J.C. HOESSEL. 1991. CCD observations of Abell clusters. V - Isophotometry. Astron. J. **101:** 1561–1594.
48. MACKIE, G., N. VISVANATHAN & D. CARTER. 1990. The stellar content of central dominant galaxies. I - CCD surface photometry. Astrophys. J. Suppl. **73:** 637–660.
49. SASTRY, G.N. 1968. Clusters associated with supergiant galaxies. Publ. Astron. Soc. Pac. **80:** 252–262.

50. Rood, H.J. & G.N. Sastry. 1972. Static properties of galaxies in the cluster Abell 2199. Astron. J. **77:** 451–458.
51. Austin, T.B. & J.V. Peach. 1974. Studies of rich clusters II. The structure and luminosity function of the cluster A1413. Mon. Not. R. Astron. Soc. **168:** 591–602.
52. Carter, D. & N. Metcalfe. 1980. The morphology of clusters of galaxies. Mon. Not. R. Astron. Soc. **191:** 325–337.
53. Bingelli, B. 1982. The shape and orientation of clusters of galaxies. Astron. Astrophys. **107:** 338–349.
54. Struble, M.F. & P.J.E. Peebles. 1985. A new application of Binggeli's test for large-scale alignment of clusters of galaxies. Astron. J. **90:** 582–589.
55. Rhee, G. & P. Katgert. 1987. A study of the elongation of Abell clusters I. A sample of 37 clusters studied earlier by Binggeli and Struble & Peebles. Astron. Astrophys. **183:** 217–227.
56. Lambas, D.G., E.J. Groth & P.J.E. Peebles. 1988. Alignments of brightest cluster galaxies with large-scale structures. Astron. J. **95:** 996–998.
57. Schombert, J.M. 1986. The structure of brightest cluster members I. Surface photometry. Astrophys. J. Sup. **60:** 603–693.

On Nonlinear Dynamics of Three-Dimensional Astrophysical Disks[a]

A.M. FRIDMAN[b,c] AND O.V. KHORUZHII[b,d]

[b]*Institute of Astronomy of Russian Academy of Sciences, Pyatnitskaya Str. 48, Moscow, 109017, Russia*
[c]*2 Sternberg Astronomical Institute, Moscow State University, University Prospect, 13, Moscow, 119899, Russia*
[d]*National Research Center, Roitsk Institute for Innovation and Fusion Research, Troitsk, Moscow Region, 142092, Russia*

ABSTRACT: Nonlinear processes concerned with different aspects of nonlinear dynamics of astrophysical disks — structures, flows, turbulence — are reviewed. Special attention is paid to the influence of the three dimentionality of disks on their nonlinear behavior.

INTRODUCTION

Compared to other astrophysical objects, disks have the most varied dynamical structures and types of turbulence. So far the origin of many observed structures in disks is a mystery, as are the turbulence mechanisms of different kinds of disks. The problem of galactic spiral structure has remained unsolved for more than 150 years. The origin of narrow Uranian rings and the Cassini division with its complex inner structure, the cause of a turbulent viscosity many orders greater than the molecular one in accretion disks, the non-Kolmogorov turbulence spectrum of the Milky Way, and the problem of double galactic nuclei not connected with merging are still unsolved problems.

TABLE 1. Brief summary of the main topics of the paper

Nonlinear Dynamics of Astrophysical Disks	Stationary Structures	Turbulence
Wave dynamics	Solitons	Turbulence of Rossby waves
Application	Spiral arms	Observed spectrum of turbulence
Vortex dynamics	Stationary vortices: solitary and double	Turbulence of Rossby vortices
Application	Single vortices, double objects	Enhanced star formation in areas with solid-body rotation
Nonlinear wave-medium interaction	Dissipativeless	Large-scale convection
	Dissipative	Inflow-outflow processes

[a]This work was performed under the partial support of RFBR grant N 96-02-17792, grant "Leading Scientific Schools" N 96-15-96648, and grant "Fundamental Space Researches, Astronomy" for the 1998 year N 1.2.3.1, N 1.7.4.3.

In the present paper we review several aspects of nonlinear dynamics of astrophysical disks with special emphasis on the crucial role of three-dimensionality of disks. We consider in great detail two cases when small but finite disk thickness can be never neglected. It is convenient to split the nonlinear dynamics of astrophysical disks into wave and vortex dynamics, which in turn could be subdivided into structures and turbulence. The main topics of the paper are summarized in the TABLE 1.

NONLINEAR DYNAMICS OF A MARGINALLY UNSTABLE SELF-GRAVITATING DISK

Special investigations [1] have shown that gaseous disks of galaxies are near the boundary of self-gravitational instability. This fact is an expected one: as the instability increases, the velocity dispersion grows and the disk approach the boundary of instability.

If the rotation velocity curve for a gaseous disk has a jump or kink, a hydrodynamical instability can develop [2], [3]. In this case, the disk also lies near the boundary of self-hydrodynamical instability [4]. The reason is similar. As a result of the instability, the smearing of the jump begins to grow until the disk reaches marginal stability.

The nonlinear dynamics of a marginally unstable self-gravitating disk was analyzed by Mikhailovskii, Petviashvili, and Fridman [5]–[7] (see also Fridman and Polyachenko [8]). As the eigenfrequency of a marginally unstable disk in a corotating system of reference, $\hat{\omega}$, is small ($\hat{\omega} \ll \Omega$), the consideration of the problem in a 2D approximation is valid (see conditions (10) and discussion therein). When only a small region of wavevectors is unstable, $\Delta k \ll k_0$ where k_0 is the wavevector of the most unstable perturbations, this nonlinear dynamical equation can be derived:

$$\frac{\partial^2 \varepsilon}{\partial \tau^2} = -v_0^2 \varepsilon + \frac{3}{2}(2 - \gamma s)\left(\gamma s - \frac{5}{3}\right)|\varepsilon|^2 \varepsilon. \tag{1}$$

Here ε is the dimensionless amplitude of the wave, τ is nondimensional time,

$$v_0^2 = \frac{(\pi G \sigma_0/c)^2 - \kappa^2}{\Omega_0^2} \ll 1 \tag{2}$$

determines the dimensionless growth rate of the most unstable perturbations, G is the gravitational constant, σ_0 is unperturbed surface density of the disk, κ is epicyclic frequency, and Ω_0 is unperturbed angular velocity of the disk. Equation (1) leads to a nonlinear dispertion relation in the form

$$v^2 = v_0^2 + \frac{3}{2}(2 - \gamma_S)\left(\gamma_S - \frac{5}{3}\right)|\varepsilon|^2. \tag{3}$$

It is easily seen that this relation describes either a soliton propagation or an explosive instability depending on the value of the "surface" polytropic index γ_S.

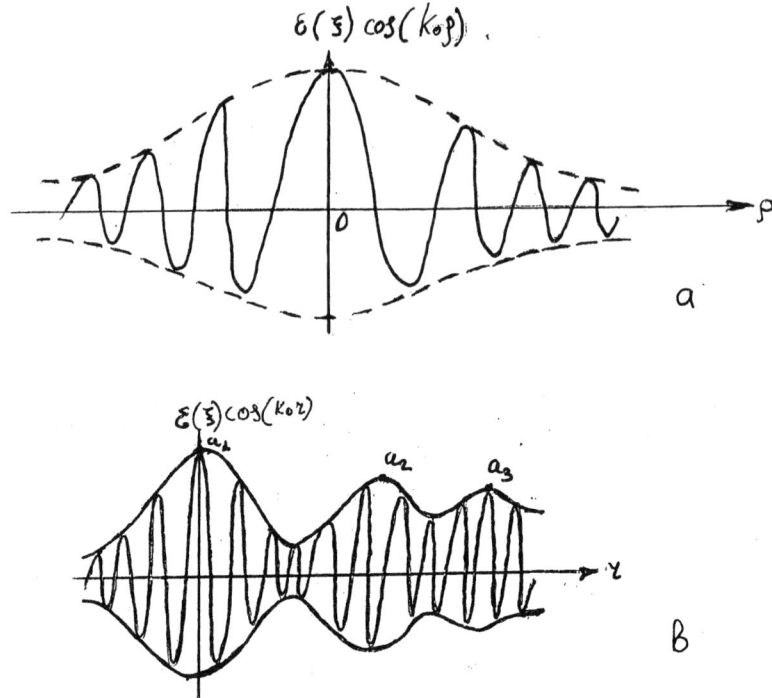

FIGURE 1. Schematic view of the envelope solitones generating in marginally unstable disks. (a) The case of dissipativeless disk; (b) disk with small viscosity.

According to Hunter [9], the surface polytropic index for a self-gravitating disk can be related to the real 3D polytropic index γ_V as $\gamma_S = 3 - 2/\gamma_V$. Thus, in the region

$$5/3 < \gamma_S < 2, \qquad (4)$$

which is equivalent to $3/2 < \gamma_V < 2$, the nonlinear stabilization of the instability is possible at a certain level of wave amplitude

$$\varepsilon^2 = \frac{2v_0^2}{3(2-\gamma_S)(\gamma_S - 5/3)}. \qquad (5)$$

In a marginally unstable disk ($v_0^2 \ll 1$), stabilization is achieved at low wave amplitude. A soliton solution is possible in region (4) in the form of an envelop soliton (FIGURE 1). In the absence of viscosity, the soliton has a classical symmetric form like a normal distribution function (FIGURE 1a). But in the presence of low viscosity, the soliton transforms into a shock wave with an oscillating front [10] (FIGURE 1b). The shock wave which was predicted in a rotating stellar disk — a collisionless shock wave — has a similar form [11].

If $\gamma_S < 5/3$, that means $\gamma_V < 3/2$, then an explosive instability occurs with

$$\frac{\varepsilon}{\varepsilon(\tau=0)} = \frac{1}{\tau - 1/(A\varepsilon(0))}, \qquad (6)$$

where $A^2 = 3(2 - \gamma_S)(5/3 - \gamma_S)/4$.

SPECTRUM OF TURBULENCE OF CLOUDY POPULATION IN THE MILKY WAY

To date, numerous investigations have been devoted to measurements of the turbulent spectrum of the cloudy population in the Milky Way in both individual gaseous clouds and ensembles of clouds. In 1955, Kaplan found the correlation function $B_{rr} \sim r^{0.71}$, which is very close to the Kolmogorov spectrum. Systematic observations and the construction of correlation functions began in 1964. Larson's result [12] was close to that of Kaplan: $\Delta v \sim l^{0.38}$. More accurate investigations later resulted in a steeper spectrum. Mayers [13], Henriksen and Turner [14], and Vereschagin and Solov'ev [15] obtained the spectrum $\Delta v \sim l^{0.5}$. Sanders, Scoville, and Solomon [16] obtained $\Delta v \sim l^{0.62}$. These spectra are different from the Kolmogorov one and to explain them we must take into account anisotropy. Dolotin and Fridman [17] (see also [18]) have attempted to explain the observed spectrum as a result of the turbulence of Rossby waves in galactic gaseous disk.

Based on observed correlations the following was adopted: (1) correlation for the velocity fluctuations $\Delta v \sim l^{0.5}$; (2) correlation for the density fluctuations [13] $\rho \sim l^{-1}$; and (3) correlation for mass spectrum [19] $N(m)/N(m_0) \sim (m/m_0)^{-1.5}$. These correlations are consistent if we assume virial equilibrium on all scales.

The following general nonlinear equation can be derived in the low-frequency approximation, $\omega \ll \Omega_0$, for the solid body rotating self-gravitating cloud [17]:

$$\left(\frac{\partial}{\partial t} + \frac{1}{2\Omega_0}[\nabla_\perp \tilde{\chi}, \nabla_\perp]_z\right)\left(\Delta_\perp \tilde{\Psi} - \frac{\omega_0^2}{4\Omega_0^2}\Delta_\perp \tilde{\chi}\right) - \frac{(\omega_0^2)'}{2\Omega_0}\frac{1}{r}\frac{\partial \tilde{\chi}}{\partial \varphi} = 0. \qquad (7)$$

Here $\omega_0^2 \equiv 4\pi G\rho_0$, $(\omega_0^2)' \equiv d(\omega_0^2)/dr$, $\chi \equiv \mathcal{P} + \Psi$, Ψ is the gravitational potential, and $\mathcal{P}$ is the *pressure* function determined by the relation

$$\mathcal{P} \equiv \int dP_v/\rho, \qquad (8)$$

where P_v is the usual *volume* pressure and ρ is the volume density.

For small-scale perturbations corresponding to the case $\tilde{\Psi} \ll \tilde{\mathcal{P}}$, Eq. (7) can be reduced to

$$\left(\frac{\partial}{\partial t} + \frac{1}{2\Omega_0}[\nabla_\perp \tilde{\mathcal{P}}, \nabla_\perp]_z\right)\Delta_\perp \tilde{\mathcal{P}} - 2\Omega_0 - \frac{\rho_0'}{\rho_0}\frac{1}{r}\frac{\partial \tilde{\mathcal{P}}}{\partial \varphi} = 0. \qquad (9)$$

The latter equation is similar to the well-known Charney-Obukhov equation in hydrodynamics [20]–[22] and the Hasegava-Mima equation in plasma physics [23].

This analogy allows us to use results from well-researched fields to describe the dynamics of perturbations in gravitating molecular clouds. In particular, in accordance with Sazontov [24] and Mikhailovskii *et al.* [25] the nonstationary solution of Eq. (9) describes Rossby wave turbulence with energy spectra $w_k^{(1)} \sim k_y^{-3/2} k_x^{-2}$ and $w_k^{(2)} \sim k_y^{-3/2} k_x^{-3}$.

Passing to the limit $k_x \simeq k_y \simeq k_\perp$, and taking into account the characteristic property of Rossby waves $k_\perp \simeq k$, we have $w_k^{(1)} \sim k^{-3.5}$ and $w_k^{(2)} \sim k^{-4.5}$. According to the Hasegawa *et al.* numerical result [26], $w_k \sim k^{-4}$. The last relation [27] gives $v_k^2 \sim \int_k^\infty E_k dk = \int_k^\infty w_k k^2 dk \sim k^{-1} \sim \lambda$, so that:

(1) $v_\lambda \sim \lambda^{0.5}$.

However, $\mathbf{v}\nabla\mathbf{v} \sim \nabla\Psi$, so that $v_\lambda^2/\lambda \sim \Psi/\lambda \sim \lambda 4\pi G \rho_\lambda$. This yields a density spectrum

(2) $\rho_\lambda \sim \lambda^{-1}$.

Finally, as $P_\lambda \approx n_\lambda m_\lambda v_\lambda^2 \approx \rho_\lambda v_\lambda^2 = $ const and $m_\lambda \approx \rho_\lambda \lambda^3 \approx \lambda^2$ one obtains $n_\lambda \sim m_\lambda^{-1} v_\lambda^{-2}$. It follows that the mass spectrum satisfies

(3) $n_\lambda \sim m_\lambda^{-1.5}$.

We see that the turbulent spectra derived here correspond to the observed spectra, which is the evidence of weakly turbulent Rossby waves in the cloudy population in the Milky Way.

ON THE POSSIBILITY OF STUDYING THE DYNAMICS OF THIN DISKS IN A 2D APPROXIMATION

Before considering some essentially 3D nonlinear effects in thin astrophysical disks, we discuss the problem of 2D approximation and clarify why, in some cases, we must take into account finite, albeit small, disk thickness.

Traditional Conditions for a 2D Description of the Dynamics of Astrophysical Disks

The only cases in which a 3D description has been used in astrophysical disk dynamics are in the study of processes of disk bending and warping, and formation of bars and bulges.

Due to the fact that the semithickness H of the majority of astrophysical disks is much less than their radius R, $H \ll R$, the dynamical processes and structures in the disks have, as a rule, been studied in the framework of a 2D approximation. In so doing, two conditions are assumed to be fulfilled.

First: The structures and processes are symmetric about the disk plane.

Second: The typical scales of the processes and structures L are much greater than the semithickness of the disk, $L \gg H$.

However, it has never been proven rigorously that these constitute a complete set of sufficient conditions. Consequently, our first question is: What is the complete set of sufficient conditions required to justify describing the dynamics of astrophysical disks by 2D dynamical equations?

What Are the Initial Equations We Have to Write?

The 2D dynamical equations for astrophysical disks should be derived from general 3D equations. Analysis of this problem results in the following conclusions [28]. If the equation of state of the gaseous disk has:

1. the general form, $P_v = P_v(\rho, S)$, where P, ρ, and S are pressure, density, and entropy, respectively, then the derivation of 2D equations is problematic;
2. the barotropic form, $P_v = P_v(\rho)$, then we come to a closed 2D set of integro-differential equations, the solution of which is complicated;
3. the polytropic form, $P_v = A \cdot \rho^\gamma$, where A and γ are constants and γ is the polytropic index, then we come to a system of 2D partial differential equations, but with additional terms which have not been written before.

On the Sufficient Conditions of the Correctness of 2D Approximation

These conditions can be written in the following simple form [28], [29]:

$$\frac{H^2}{L^2} \ll 1, \quad \frac{H^2 R}{\zeta^2 L} \ll 1, \quad \omega^2 \ll \frac{c^2}{H^2}. \tag{10}$$

Here L and ζ are typical radial and azimuthal scales of perturbations; ω is the typical frequency of the process.

The first two strong inequalities are well-known conditions of large-scale approximation. *The third condition has not been used before.* In most cases this condition reduces to

$$\omega^2 \ll \Omega^2, \tag{11}$$

where Ω is the angular velocity of a disk. As a rule this condition has not been fulfilled in previous 2D studies of disk dynamics. In particular, as is easy to see, the majority of 2D theories of spiral structure do not satisfy this condition!

Typical Mistakes in Works Devoted to 2D Dynamics of Astrophysical Disks

Schematically, the reason for a mistake can be illustrated by a simple example. Let us consider the 3D equations of motion

$$\frac{dv_\perp}{dt} = -\nabla_\perp \chi, \tag{12}$$

$$\frac{dv_z}{dt} = -\frac{\partial}{\partial z}\chi, \qquad (13)$$

where the notations are given in the previous section. For low-frequency perturbations (11) the term dv_z/dt in Eq. (13) may be assumed to vanish, so that χ does not depend on z. Hence, the right side of Eq. (12) also does not depend on z. As a result, the left side does not depend on z and the equation of motion (12) can be reduced to the 2D form.

The latter argument is not true for the high-frequency perturbations, $\omega \sim \Omega$. In this case, a 2D approximation can be correct only in some special cases [29], including

(1) an isothermal disk in a strong outer gravitational field, and
(2) a self-gravitating disk with polytropic index $\gamma = 2$.

The Case of Low-Frequency Perturbations, $\omega \ll \Omega$

In this case, the 2D approximation is correct if the 2D equations are derived from 3D ones. Yet, as a rule the incorrect 2D dynamical equations are used where all functions have simply been integrated over z. Let us illustrate this with the simplest example of 2D equations of motion.

A Polytropic Disk in an External Gravitational Field

(a) Traditional form:

$$\frac{d\mathbf{v}}{dt} = -\frac{1}{\sigma}\nabla P_s - \nabla\Psi_c = -A_s\gamma_s\sigma^{\gamma_s-2}\nabla\sigma - \nabla\Psi_c, \qquad (14)$$

where $P_s = A_s\sigma^{\gamma_s}$ is the "flat" pressure, Ψ_c is gravitational potential in the plane $z = 0$, A_s and γ_s are constants, and γ_s is a "flat" polytropic index.

(b) Correct form [30]:

Substituting in (8) the expression for the 3D polytropic equation of state $P_v = A_v\rho^{\gamma_v}$, where A_v and γ_v are constants and γ_v is a "volume" polytropic index, we obtain $\mathcal{P} = A_v\rho^{\gamma_v-1}\gamma_v/(\gamma_v - 1)$. Thus, solving for ρ and integrating it over z from $-\infty$ to $+\infty$, we obtain the expression for the surface density

$$\sigma(r, \varphi, t) = \left(\frac{\gamma_v - 1}{A_v\gamma_v}\right)^{\frac{1}{\gamma_v-1}} \int_{-\infty}^{+\infty} [\chi(r, \varphi, t) - \Psi(r, \varphi, z, t)]^{\frac{1}{\gamma_v-1}} dz. \qquad (15)$$

Assuming that the disk has only infinitesimal thickness allows us to write a relationship between σ and χ for an arbitrary function Ψ, using its expansion in the neighborhood of the plane $z = 0$:

$$\Psi(r, \varphi, z, t) = \Psi_c(r, \varphi, t) + \Psi_c'(r, \varphi, t)z + \frac{1}{2}\Psi_c''(r, \varphi, t)z^2. \qquad (16)$$

Substituting (16) to (15) and integrating over z we obtain

$$\sigma(r, \varphi, t) = \sqrt{\pi} \left(\frac{\Psi_c'' }{2} \frac{\gamma_v - 1}{A_v \gamma_v} \right)^{\frac{1}{\gamma_v - 1}} \frac{\Gamma\left(\frac{\gamma_v}{\gamma_v - 1}\right)}{\Gamma\left(\frac{\gamma_v}{\gamma_v - 1} + \frac{1}{2}\right)} \left[\left(\frac{\Psi_c'}{\Psi_c''}\right) + \frac{2(\chi - \Psi_c)}{\Psi_c''} \right]^{\frac{\gamma_v + 1}{2(\gamma_v - 1)}}, \qquad (17)$$

whence

$$\chi = C\sigma^{2\lambda} + \Psi_c - \frac{(\Psi_c')^2}{2\Psi_c''}, \qquad (18)$$

where $\lambda \equiv (\gamma_v - 1)/(\gamma_v + 1)$, Γ is the gamma function, and

$$C \equiv \left[\frac{\Psi_c''}{2\pi} \frac{\Gamma^2\left(\frac{\gamma_v}{\gamma_v - 1} + \frac{1}{2}\right)}{\Gamma^2\left(\frac{\gamma_v}{\gamma_v - 1}\right)} \right]^{\lambda} \left(\frac{A_v \gamma_v}{\gamma_v - 1} \right)^{2/(\gamma_v + 1)}. \qquad (19)$$

As C depends on the external parameter (Ψ''), the actual "surface" equation of state is more general than the pure 2D equation of state, which has usually been adopted.

After substituting in (18), Eq. (12) takes the desired form of a 2D equation of motion for 2D functions

$$\frac{d\mathbf{v}}{dt} = -2\lambda C \sigma^{2\lambda - 1} \underline{\nabla \sigma} - \nabla \Psi_c - \sigma^{2\lambda} \underline{\nabla C} - \nabla \left[\frac{(\Psi')^2}{\Psi''} \right]_c, \qquad (20)$$

The terms in (20) different from those in (14) have been underlined.

A Self-Gravitating Disk

In this case, 2D dynamics can again be described by an equation similar to (14). The difference lies in the form of the expression for A_s.

(a) Traditional form [9]:

$$(A_s)_H = \frac{\pi^{(3/2 - 1/\gamma_v)} \Gamma(2 - 1/\gamma_v)}{2^{(2 - 1/\gamma_v)} \Gamma(5/2 - 1/\gamma_v)} A_v^{1/\gamma_v} G^{1 - 1/\gamma_v}. \qquad (21)$$

(b) Correct form [30]:

$$(A_s)_{\text{corr}} = \frac{2^{1/\gamma_v} \pi^{(1 - 1/\gamma_v)}}{3 - 2/\gamma_v} A_v^{1/\gamma_v} G^{1 - 1/\gamma_v}. \qquad (22)$$

Their ratio is not equal to unity:

$$\frac{(A_s)_H}{(A_s)_{corr}} = \frac{\pi^{1/2}}{2}\frac{\Gamma(\gamma_s/2+1/2)}{\Gamma(\gamma_s/2)}, \quad \gamma_s \equiv 3 - \frac{2}{\gamma_v}. \tag{23}$$

The reason for the demonstrated difference is very simple: on the right side of the correct 2D equation of motion stands the gradient of the volume pressure function, $\nabla \mathcal{P} = \nabla \int dP_v/\rho$, not the gradient of a "flat" pressure, $\sigma^{-1}\nabla P_s \equiv \sigma^{-1}\nabla \int P_v dz$, as in Eq. (14). The latter term cannot be interpreted physically as a force. Why?

Let us consider a tube of a conic section. If P_v is constant, the gas in the tube can exhibit no bulk motion, but the surface pressure P_s in different sections will still be different. In particular, if we compare two positions with different cross sections $z_1 > z_2$, we obtain

$$P_{S_1}(x_1) \equiv 2\int_0^{z_1} P_v(x_1)dz > P_{S_2}(x_2) \equiv 2\int_0^{z_2} P_v(x_2)dz. \tag{24}$$

According to Eq. (14) the gas must move from section S_1 to section S_2, which is clearly nonsensical!

SOLITARY VORTICES IN ASTROPHYSICAL DISKS

Using the vector equation of motion (20) for a disk in an external gravitational field for low-frequency perturbations (11), we derive [30] the following nonlinear dynamical equation for a solid-body rotating part of a disk, $\Omega = \text{const}$:

$$\frac{\partial}{\partial t}(\tilde{\chi} - a_R^2 \Delta \tilde{\chi}) + U_R \frac{\partial \tilde{\chi}}{\partial y} - \frac{c_s^2}{8\Omega^3} J(\tilde{\chi}, \nabla \tilde{\chi}) + \frac{(\ln C)_x'}{4\Omega}\frac{\partial \tilde{\chi}^2}{\partial y} = 0. \tag{25}$$

Here $\tilde{\chi}$ is the perturbation of χ (see (18)); $a_R \equiv c_s/2\Omega$ is the Rossby radius; $U_R \equiv -2a_R^2\Omega(\ln\sigma_0)_x'$ is the Rossby velocity; c_s is the sound speed determined by the "flat" functions

$$c_s^2 \equiv \sigma\left(\frac{\partial \chi}{\partial \sigma}\right)_0; \tag{26}$$

$$J(A, B) \equiv \frac{\partial A}{\partial x}\frac{\partial B}{\partial y} - \frac{\partial A}{\partial y}\frac{\partial B}{\partial x} \tag{27}$$

is a Jacobian; x, y are Cartesian coordinates with x along radius and y along azimuth; and the function C is given by (19).

Equation (25) contains both vector (third term) and scalar (last term) nonlinearities. It should be emphasized that the scalar nonlinearity is a consequence of the dependence of the "surface" equation of state on an external parameter (through the dependence on the external gravitational potential Ψ_c). If we start from the usual pure 2D hydrodynamical equations this term would be overlooked.

Derived for astrophysical disks, the nonlinear dynamical equation (25) is similar to the well-known Charney-Obukhov equation in hydrodynamics. In plasma physics, a similar equation was derived by Hasegawa and Mima [23]. Theoretical and laboratory modelling of these equations in hydrodynamics and plasma physics leads to the following results [30].

Equation (25) has two kinds of stationary solutions, which describe, respectively, two types of solitary vortices: single and double vortices. The sizes a of these structures are restricted: $1 < a/H < (R/H)^{1/3}$ where H is the disk thickness and R is the typical scale of the density inhomogeneity.

Single Vortices: Cyclones and Anticyclones

Due to the presence of the scalar nonlinearity, the existence of single Rossby vortices is possible in galactic disks. If the scalar nonlinearity $(\ln C)_x'$ has a fixed sign it is possible to form only one kind of solitary vortices: cyclones or anticyclones. A cyclone $((\ln C)_x' < 0)$ is characterized by minimum surface density. Anticyclone $((\ln C)_x' > 0)$ is characterized by a maximum surface density.

Double Vortices: Modons

Due to the presence of the vector nonlinearity, the existence of double Rossby vortices is possible in galactic disks. A double vortex–modon represents a cyclone-anticyclone pair and has one minimum and one maximum of perturbed surface density [17]. The astronomical application may be connected, for example, to the problem of double galactic nuclei formation and perhaps to other pairs of nearby objects, the scale of which is extremely different from the galactic nuclei.

On Strong Vortex Turbulence

The generation of several modons can result in their interaction and the formation of vortex turbulence [31]. In principle, vortex turbulence is distinct from wave turbulence. Wave turbulence is strong only if the amplitude of the perturbed values $\tilde{A}$ is comparable to, or larger than, the stationary one A_0. Vortex turbulence can be strong even if $\tilde{A} \ll A_0$, since the vortex–vortex interaction is much longer than that for wave–wave interaction.

On the Connection of Enhanced SFR with Large Areas of Solid-Body Rotation

Enhanced star formation has been found by Keel [32] for interacting galaxies with large areas of solid-body rotation. In particular, paired Seyfert galaxies show a striking number of solid body rotation curves — 80% of the 39 observed [33]. One of the conclusions of Keel's work [32] is: "The frequent presence of large areas of solid-body rotation, extending to a medial radius of 2.1 kpc, is not closely linked to the presence of optical bars."

It can be assumed that the reason for the dynamical activity of a gaseous disk in the absence of a bar, but with a large solid-body rotation area, is the development of strong vortex turbulence.

NONLINEAR RADIAL LAMINATED FLOW AS A MANIFESTATION OF 3D DYNAMICS OF ASTROPHYSICAL DISKS

In this section we will show that in the presence of a spiral density wave there is an observable manifestation of the three-dimensional nature of astrophysical disks in the form of a large-scale quasi-stationary radial flow laminar in z direction. The flow velocity has opposite directions in the central plane of the disk, $z = 0$, and on the disk periphery, near the planes $z = \pm H$. The streamlines are closed by vertical motions of smaller magnitude, $v_z \simeq v_r H/r$. As a whole, the flow has the form of four toroidal vortices separated by the vertical surface $r = r_c$, where r_c is a corotation radius, and the central plane of the disk (FIGURE 2). Therefore, the observation of this flow could provide a direct indication of the position of the corotation circle.

The characteristic velocity in the flow is comparable in size to the velocity perturbations in the density wave. For galactic disks it can be as large as several tens of

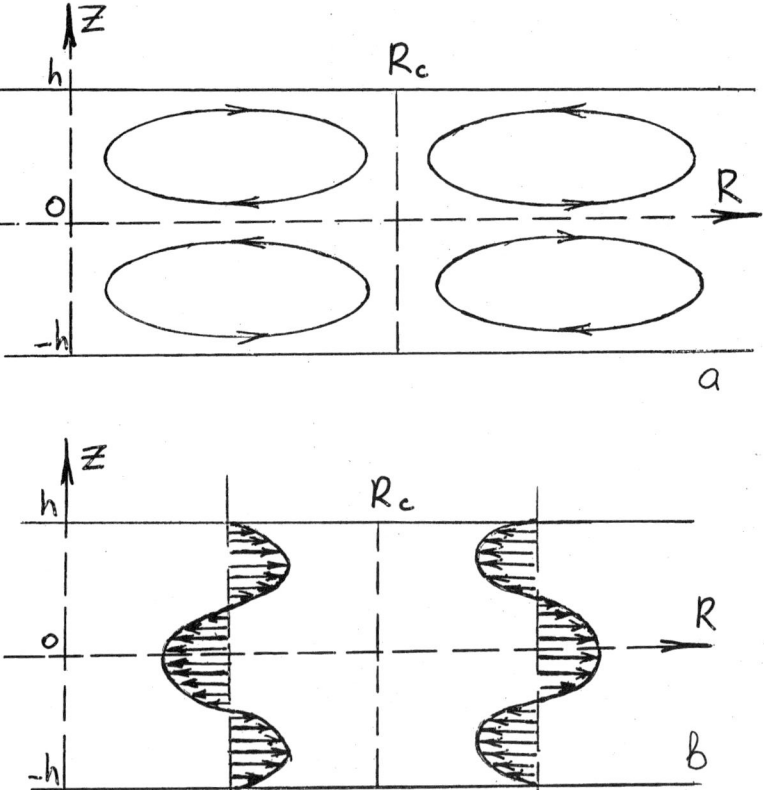

FIGURE 2. Schematic view of nonlinear radial flow induced by a quasi-stationary density wave. The radial cut of the disk is presented. The radial scale is squeezed. The flow is azimutally symmetrical and has a form of four tori. (**a**) Structure of streamlines in the flow; (**b**) The vertical profile of the flow radial velocity.

kilometers per second. The characteristic scale of the flow is comparable in size to the disk radius. Therefore, the existence of such convection can play a significant role in the overall dynamics of the disk.

By its nature the flow discussed above is a special kind of acoustic streaming. It is caused by the quasi-stationary component of nonlinear Reynolds stresses induced by the density wave. Classical acoustic streaming is caused by Reynolds stresses in strong acoustic waves (for a review see, e.g., [34]). It is a phenomenon that has been observed in hundreds of different laboratory experiments starting with Faraday's discovery [35] and described in numerous theoretical papers, starting with the pioneer work by Rayleigh [36].

Compared to the classical case, the main difference is a drift nature of any quasi-stationary flow in a rotating disk. This is a consequence of the dominant role of the Coriolis forces and means that the direction of the flow is perpendicular to the direction of the applied force. If the system is two-dimensional or uniform in the z direction, conservation of orbital momentum in the absence of dissipation will require the mutual annihilation of separate nonlinear azimuthal forces such as $\langle \mathrm{div}(\rho v_\varphi) \rangle$, $\langle \rho \partial \Phi / r \partial \varphi \rangle$, etc. But in a disk which is nonuniform in z direction, annihilation is only possible on the average. This means that locally the azimuthal force does not vanish and is zero only after an integration over the disk thickness. As a result, averaged over the disk thickness mass flux, the radial flow vanishes.

The Method

The method (for details see [37]) is based on the dividing each quantity entering into the dynamical equations into two parts:

- a "slow" quasi-stationary part;
- a "fast" pulsating part.

The former represents the dynamics of the system "averaged" over a period of time much longer than the period of pulsations. The latter represents the short-time dynamics of the system, but vanishes after averaging over time. This approach is well known in the theory of the turbulence, nonlinear optics, and acoustics, but has not been used previously to analyze the nonlinear self-action of astrophysical density waves.

This approach implies the following representations. For each value (physical parameter) f we can write:

$$f = \langle f \rangle + f_1, \qquad (28)$$

with

$$\langle f \rangle \equiv \frac{1}{\Delta} \int_0^\Delta f \, dt, \quad T \ll \Delta \ll t_{ev}, \qquad (29)$$

where T is a characteristic period of pulsations, t_{ev} is a characteristic time of the "slow" evolution, and

$$\langle f_1 \rangle = 0. \qquad (30)$$

For the product of two values, one can write:

$$fg = \langle f \rangle\langle g \rangle + \langle f_1 g_1 \rangle + \langle f \rangle g_1 + f_1 \langle g_1 \rangle. \tag{31}$$

For the case of a small but finite amplitude wave we can restrict our consideration to quadratic nonlinearities only. Simultaneously we can use the representation of the pulsations as a monochromatic wave,

$$f_1(r, \varphi, z, t) = \bar{f}_1(r, z, t)\exp[i(m\varphi - \omega t)] + \bar{f}_1^*(r, z, t)\exp[-i(m\varphi - \omega t)], \tag{32}$$

with the amplitude $\bar{f}_1(r, z, t)$ being a slow function of the time.

It follows that the quasi-stationary part of any nonlinear term has the form

$$\langle f_1 g_1 \rangle = \bar{f}_1 \bar{g}_1^* + \bar{f}_1^* \bar{g}_1. \tag{33}$$

This average is independent of the azimuthal angle, so that all effects described below are axisymmetric.

Nonlinear Quasi-Stationary Response of the Disk on the Density Wave

The total nonlinear quasi-stationary response of the disk to the density wave can be divided into the following three effects. Below we present the final results only; their derivation is the subject of a separate paper [38].

1. Nonlinear corrections to the potential function (enthalpy plus gravitational potential) of the disk: These are similar to the well-known "high frequency pressure" in plasma physics.

$$\frac{\partial \langle \chi \rangle}{\partial z} = -\frac{\partial}{\partial z}\left(\bar{v}_{z1}\bar{v}_{z1}^* + \frac{D}{\hat{\omega}^2}\bar{v}_{r1}\bar{v}_{r1}^* + \frac{m^2}{r^2\hat{\omega}^2}\bar{\chi}_1\bar{\chi}_1^*\right) \simeq -\frac{1}{D}\frac{\partial}{\partial z}\left(\frac{\partial \bar{\chi}_1}{\partial r}\frac{\partial \bar{\chi}_1^*}{\partial r}\right). \tag{34}$$

Here

$$\hat{\omega} \equiv \omega - m\Omega, \quad D \equiv \hat{\omega}^2 - \kappa^2, \tag{35}$$

and the functions $\langle \chi \rangle$ and χ_1 are defined as in Eq. (28).

Nonlinear corrections to $\langle \chi \rangle$ could influence the vertical equilibrium and the rotation curve of a gaseous disk. The former influence can be significant only for short-wave perturbations with $kH \simeq 1$. The contribution to the rotation law is of the order of the contribution of the gaseous pressure.

2. Nonlinear corrections to the rotation law:

$$\frac{\langle \rho \rangle V_\varphi^2}{r} = \frac{\partial \langle \chi \rangle}{\partial r} + \langle \rho_1 \frac{\partial \chi_1}{\partial r} \rangle + \frac{\partial}{\partial z}(\langle \rho \rangle \langle v_{r1} v_{z1} \rangle) + \frac{1}{r}\frac{\partial}{\partial r}(r\langle \rho \rangle \langle v_{r1}^2 \rangle) - \langle \rho \rangle \frac{\langle v_{\varphi 1}^2 \rangle}{r} \tag{36}$$

All nonlinear corrections to the rotation law are of the order of the gaseous pressure term. The calculation of the total effect requires accounting for corrections to $\langle \chi \rangle$ and thus can be performed as a 3D problem only.

3. The excitation of quasi-stationary "acoustic streaming": From the theory of classical acoustic streaming it is well known that the time-average flow of the medium is characterized by the average values of the mass flux components

$$\Pi_i \equiv \langle \rho v_i \rangle, \qquad (37)$$

rather than the average value of the velocity itself.

The radial (Π_r) and vertical (Π_z) components of the nonlinear quasi-stationary flow are the most interesting in describing the overall dynamics of a disk. It can be shown that the components of the quasi-stationary flow obey the continuity equation

$$\operatorname{div} \vec{\Pi} = \frac{1}{r}\frac{\partial (r\Pi_r)}{\partial r} + \frac{\partial \Pi_z}{\partial z} = 0, \qquad (38)$$

which means that the flow has closed streamlines.

Our analysis shows that an exact expression for the radial flow can only be obtained by taking into account the real three-dimensional structure of the disk and density wave:

$$\begin{aligned}\Pi_r &= -\frac{\partial}{\partial z}\left\{\frac{i\langle\rho\rangle}{\hat{\omega}D}\left[\frac{\partial\bar{\chi}_1}{\partial z}\frac{\partial\bar{\chi}_1^*}{\partial r} - \frac{\partial\bar{\chi}_1^*}{\partial z}\frac{\partial\bar{\chi}_1}{\partial r} - \frac{2m\Omega}{r\hat{\omega}}\left(\frac{\partial\bar{\chi}_1}{\partial z}\bar{\chi}_1^* - \frac{\partial\bar{\chi}_1^*}{\partial z}\bar{\chi}_1\right)\right]\right\} \\ &= -\frac{\partial}{\partial z}\left(\frac{\langle\rho\rangle}{\hat{\omega}^2}\langle v_{r1}\frac{\partial\bar{\chi}_1}{\partial z}\rangle\right)\end{aligned} \qquad (39)$$

The same is true for the vertical flow:

$$\Pi_z = -\int_0^z\left[\frac{1}{r}\frac{\partial(r\Pi_r)}{\partial r}\right]dz = \frac{1}{r}\frac{\partial}{\partial r}\left(\frac{r\langle\rho\rangle}{\hat{\omega}^2}\langle v_{r1}\frac{\partial\bar{\chi}_1}{\partial z}\rangle\right). \qquad (40)$$

It should be noted that in two-dimensional and cylindrical systems this effect vanishes. In the former case, $\partial/\partial z \equiv 0$. In the latter case, the perturbations are proportional to $\exp(ikz)$, so that terms like $f_1 g_1^*$ do not depend on z, and Π_r is equal to zero.

The Method of the Iteration

For an accurate analysis of the vertical structure of density waves, an iteration method was proposed [29]. This is based on reducing the system of linearized dynamical equations to an integro-differential equation for the potential function χ_1:

$$\chi_1(z) = \chi_1(z=0) - \hat{\omega}\int_0^z I(z)\frac{dz}{\rho}, \qquad (41)$$

where

$$I(z) \equiv -\hat{\omega}\int_0^z\left[\frac{m^2\rho\chi_1}{r^2\, D} - \rho_1 - \frac{1}{r}\frac{\partial}{\partial r}\left(\frac{r\rho}{D}\frac{\partial\chi_1}{\partial r}\right) + \frac{2m}{r\hat{\omega}}\frac{d}{dr}\left(\frac{\Omega\rho}{D}\right)\chi_1\right]dz. \qquad (42)$$

As a first approximation we use the solution obtained neglecting motion in the z-direction, which is equivalent to setting $I(z) = 0$, since $v_z \sim I$. Having determined the functions $\chi_1(r, z)$ and $\rho_1(r, z)$ of the first approximation in this way we can substitute them into the integral (42), and from (41) obtain these functions in the second approximation, and so on.

It can be shown that the iterations converge if the dispersion relation $I(\infty) = 0$ is satisfied. The rate of the convergence is determined by the parameter [39]

$$N = \hat{\omega}^2 H^2/c^2, \tag{43}$$

which characterizes the relation between $1/\hat{\omega}$, the time-scale of the wave, and H/c, the time required to establish vertical equilibrium. The greater the value N, the more complicated the vertical structure of the density wave.

If N is much less then unity, as for thin disks, combining the expressions above one can obtain

$$\Pi_r = \langle \rho_1(z=0) v_{r1}(z=0) \rangle$$
$$\times \frac{1}{\rho_1(z=0)} \left[\rho_1 + \frac{1}{r}\frac{\partial}{\partial r}\left(\frac{r\rho}{D}\frac{\partial \chi_1}{\partial r}\right) - \frac{m^2 \rho \chi_1}{r^2 D} - \frac{2m}{r\hat{\omega}}\frac{d}{dr}\left(\frac{\Omega \rho}{D}\right)\chi_1 \right], \tag{44}$$

where the first factor characterizes the magnitude of the flow and the second factor its dependence on z.

It is easily seen that the mass flux averaged over disk thickness is zero,

$$\int_0^H \Pi_r dz = 0, \tag{45}$$

and the characteristic velocity in the radial direction is about the peculiar velocity in the wave.

Structure of the Flow

The method described allows us to determine the vertical structure of the density wave and to calculate the structure of the nonlinear laminar flow. To illustrate this phenomenon let us consider a particular case of the tightly wound spiral-density wave generated in a disk with a Toomre parameter [40] $Q \gg 1$.

In this case the expression for the radial flow has the form

$$\Pi_r = -\langle \rho_1 v_{r1} \rangle_{z=0} \frac{\gamma+1}{2} \left(1 - \frac{z^2}{h^2}\right)^{\frac{2-\gamma}{\gamma-1}} \left(\frac{z^2}{h^2} - \frac{\gamma-1}{\gamma+1}\right), \tag{46}$$

and

$$\langle \rho_1 v_{r1} \rangle_{z=0} = A^2 \rho_0 \frac{\hat{\omega}}{k_r}. \tag{47}$$

Here, $A = \rho_1(z = 0)/\rho_0$ is the dimensionless amplitude of the density wave. Consequently, the flow has the form of four vortices separated by the vertical cylindrical surface $r = r_c$ and the central plane of the disk $z = 0$ (FIGURE 2).

For trailing spirals, the radial velocity in the central plane of the disk is negative inside the corotation circle and is positive outside. For leading spirals the situation is opposite. Therefore, the observational detection of such a flow could also provide an indication of the position of the corotation circle and the type of spiral.

Finally, we would like to note that the situation described above corresponds to a nondissipative disk. In the presence of dissipation, an additional nonlinear accretion (accretion drift) is caused by the density wave. This flow has a nonzero mass flux and can play a significant role in the redistribution of the surface density of a disk.

Specifically, Fridman, Khoruzhii, and Gor'kavyi [41] have shown that the excitation of such flow can give a reasonable explanation of the formation of gaps and ringlets in planetary rings. The radial mass flux in this accretion is negative inside the corotation and positive outside. Therefore, the accretion drift can also explain the secular flows found in most numerical simulations of gaseous galactic disks (see, e.g., [42]).

ACKNOWLEDGMENTS

The final version of this manuscript was written at the Newton Institute for Mathematical Sciences (Cambridge, UK), the hospitality of which we gratefully acknowledge.

REFERENCES

1. ZASOV, A.V. & S. SIMAKOV. 1988. Astrofizika **29**: 190.
2. BAEV, P.V. & A.M. FRIDMAN. 1989. Astron. Tsirk. No. **1535**: 1 – 2.
3. FRIDMAN, A.M. 1990. Sov. Phys. JETP **71**(4): 627–635.
4. FRIDMAN, A.M., V.L. POLYACHENKO & A.V. ZASOV. 1991. In Dynamics of Galaxies and Their Molecular Clouds Distributions, F. Combes & Casoli, Eds.: 109. Kluwer Acad. Publ., Dordrecht, Boston, London.
5. MIKHAILOVSKII, A.V., V.I. PETVIASHVILI & A.M. FRIDMAN. 1977. JETP Lett. **26**: 121.
6. MIKHAILOVSKII, A.V., V.I. PETVIASHVILI & A.M. FRIDMAN. 1977. JETP Lett. **26**: 341.
7. MIKHAILOVSKII, A.V., V.I. PETVIASHVILI & A.M. FRIDMAN. 1979. Soviet Astron. **23**: 133.
8. FRIDMAN, A.M. & V.L. POLYACHENKO. 1984. Physics of Gravitating Systems, Vols. 1 & 2. Springer-Verlag. New-York .
9. HUNTER, C. 1972. Ann. Rev. Fluid Mech. **4**: 219.
10. FRIDMAN, A.M. 1979. Pisma Astr. Zh. **5**: 325.
11. FRIDMAN, A.M., J. PALOUS & I.I. PASHA. 1981. Monthly Not. Rev. Astron. Soc. **194**: 705.
12. LARSON, R.B. 1981. Month. Not. RAS **194**: 809.
13. MAYERS, P.C. 1983. Ap. J. **270**: 105.
14. HENRIKSEN, L.N. & TURNER, I.R. 1984. Ap. J. **287**: 200.
15. VERESCHAGIN, S.V. & A.V. SOLOWEV. 1990. Astronom. Zh. **67**: 188.
16. SANDERS, D.B., N.Z. SCOVILLE & P.M. SOLOMON. 1985. Ap. J. **289**: 373.
17. DOLOTIN, V.V. & A.M. FRIDMAN. 1991. Sov. Phys. JETP **72**: 1
18. FRIDMAN, A.M. & O.V. KHORUZHII, O.V. 1996. In Chaos in Gravitational N-body Systems, J.C. Muzzio et al., Eds.: 207. Kluwer Acad. Pub. Netherlands.

19. MASEVICH, A.G. & A.V. TUTUKOV. 1988. Stellar Evolution: Theory and Observations, M.: Nauka.
20. CHARNEY, J.G. 1948. Geofysiske Publikasjoner Videnskab-akademi Oslo **17**: 3.
21. OBUKHOV, A.M. 1984. Izv. Akad. Nauk SSR, Ser.Geograph. & Geophys. **13**: 281.
22. OBUKHOV, A.M. 1988. In Turbulence and the Dynamics of the Atmosphere. Gidrometeoizdat. Leningrad. p. 409.
23. HASEGAVA, A. & K. MIMA. 1978. Phys. Fluids **21**: 87.
24. SAZONTOV, A.G. 1981. Preprint Inst. of Appl. Phys., N. 3, Gor'kii.
25. MIKHAILOVSKII, A.B., S.V. NOVAKOVSKII, V.P. LAKHTIN, S.V. MAKURIN, E.A. NOVAKOVSKAYA, O.G. ONISCHENKO. 1988. Preprint. Inst. of Space Res., N. 1356, Moscow.
26. HASEGAVA, A., C.G. MCLENNAN & I. KODAMA. 1979. Phys. Fluids **22**: 2122.
27. LANDAU, L.D. & E.M. LIFSHITZ. 1984. Fluid Mechanics. Pergamon Press.
28. FRIDMAN A.M., O.V. KHORUZHII & A.S. LIBIN. 1994. Appendix I to N.N.Gor'kavyi & A.M. Fridman, Fizika Planetnykh Kolets. Nauka Publ. Moscow. (See also A.M. Fridman & N.N. Gor'kavyi. 1999. Physics of planetary rings, Springer-Verlag. New York.)
29. FRIDMAN, A.M. & O.V. KHORUZHII. 1994. Appendix II to N.N. Gor'kavyi, A.M. Fridman, Fizika Planetnykh Kolets. Nauka Publ. Moscow. (See also A.M. Fridman & N.N.Gor'kavyi. 1999. Physics of Planetary Rings. Springer-Verlag. New York.)
30. FRIDMAN, A.M. & KHORUZHII, O.V. 1996. In Chaos in Gravitational N-body Systems, J.C.Muzzio et al., Eds.: 197. Kluwer Acad. Pub. Netherlands.
31. NEZLIN, M.N. & E.N. SNEZHKIN. 1993. Rossby Vortices, Spiral Structures, Solitons. Springer-Verlag. Berlin.
32. KEEL, W.C. 1993. Astron. J. **106**: 1771.
33. KEEL, W.C. 1995. Kinematic Instabilities, Interactions, and Fuelling of Seyfert Nuclei.
34. NYBORG, V. 1967. Physical Acoustics. **2B**: 265.
35. FARADAY, M. 1831. Philos. Trans. R. Soc. London **121**: 229.
36. LORD RAYLEIGH. 1884. Philos. Trans. R. Soc. London **171**: 1.
37. FRIDMAN, A.M. & O.V. KHORUZHII. 1994. Appendix V to N.N. Gor'kavyi & A.M. Fridman, Fizika Planetnykh Kolets. Nauka Publ. Moscow. (See also A.M. Fridman & N.N. Gor'kavyi. 1999. Physics of Planetary Rings. Springer-Verlag. New York.
38. FRIDMAN, A.M. & O.V. KHORUZHII. 1998. To be published.
39. FRIDMAN, A.M. & O.V. KHORUZHII. 1999. Introduction to Classical Gravity Physics, Gordon-Breach, accepted for publication.
40. TOOMRE, A. 1964. Ap. J. **204**: 1217.
41. FRIDMAN, A.M., O.V. KHORUZHII & N.N. GOR'KAVYI. 1996. Chaos **6**(3): 334.
42. LINDBLAD, P.A.B., P.O. LINDBLAD & E. ATHANASSOULA. 1996. Astron. Astrophys. **313**: 65.

Satellites as Probes of the Masses of Spiral Galaxies

LANCE K. ERICKSON,[a,d] S.T. GOTTESMAN,[b,d] AND JAMES H. HUNTER, JR.[b,d]

[a]*Department of Aeronautical Science, Embry-Riddle, Aeronautical University, Daytona Beach, Florida 32114*
[b]*Department of Astronomy, University of Florida, Gainesville, Florida 32611-2055*

ABSTRACT: We present atomic hydrogen (HI) observations and analyses of the kinematics of satellite-primary galaxy pairs. Two estimates for the masses of the primaries are available, one from their rotation curves and one from the orbital properties of the satellites. Defining χ as the ratio of these two mass estimates, it is a measure of the presence, or absence, of a significant halo. The χ distribution is presented and the selection effects are discussed. We show that our data, compared with the more numerous pairs identified by Zaritsky *et al.* [11], [12], have similar distributions for projected separations of less than 200 kpc, even though the selection criteria employed were quite different. Observational biases have a negligible effect; the biased and unbiased distributions are essentially identical. N-body calculations were executed to simulate the dynamical behavior of relatively low mass satellites orbiting primary disk galaxies with and without extended halos. In addition, we made a partially analytical analysis of the behavior of orbits in a logarithmic potential. We find that a "generic" model, characterized by a single disk-halo combination, cannot reproduce the observed $P(\chi)$ distribution. However, a simple two-component population of galaxies, composed of not more than 60% with halos and 40% without halos, is successful, if galaxies have dimensions of order 200 kpc. If galaxies are considerably larger with sizes extending to 400 kpc or more, no generic model can describe the full range of the observed $P(\chi)$, particularly if the distribution for $r_p < 200$ kpc is compared with that for $r_p > 200$ kpc. Regardless of the mix of orbital eccentricities, neither pure halo, nor canonical models (disk and halo masses are comparable within the disk radius) will work. A multicomponent approximation can be constructed; the canonical model must be mixed with a small fraction of systems essentially devoid of a massive dark halo. Only by including these complexities can the full range of $P(\chi)$ be modeled with any degree of success over all radial extents. We show that dynamical friction cannot be ignored in these explorations and that the average mass of a galaxy is in the range of $(1-5) \times 10^{12}$ $M_\odot$, with a mass-to-luminosity ratio of at most a few hundred. This is insufficient to close the Universe.

1. INTRODUCTION AND OBSERVATIONS

In an extension of earlier work [1]–[4], we present atomic hydrogen (HI) observations and analyses of the kinematics of 24 satellite-primary galaxy pairs with projected separations between 4.9 and 240 kpc. The satellites have masses of less than

---

[d]Addresses for e-mail, respectively: erickson@db.erau.edu, gott@astro.ufl.edu, hunter@astro.ufl.edu

3% of their primary spirals. Two estimates for the masses of the primaries are available, one from their rotation curves and one from the orbital properties of the satellites. Defining χ as the ratio of these two mass estimates, it is a measure of the presence, or absence, of a significant halo. Our selected systems were obtained primarily using the very large array (VLA) telescope of the National Radio Astronomy Observatory (NRAO). The resolution achieved at $\lambda 21$ cm was about 20″ (at a distance of 12.4 Mpc; this corresponds to 1.20 kpc). To ensure reasonable resolution of the disk, we observed spirals larger than 4′, and only those for which there was evidence of a rich HI content. To keep observing time to a minimum, we required that the primary and its satellites fill a field less than 30′ in diameter (defined by the primary telescopes of the array). Therefore, on average the projected field was 108 kpc in diameter, and the diameter of a typical spiral disk was 30 kpc.

The velocity range of possible satellites was restricted to ± 600 km s^{-1} with respect to the primary, a restriction imposed by instrumental considerations. Van Moorsel [5], [6] and earlier studies have used similar or even smaller values. At the average projected separation of our satellites, 63 kpc, $\Delta V = 600$ km s^{-1} gives an indicated mass of 5.3×10^{12} M$_\odot$ (ignoring all projection effects). Such systems were not observed.

To ensure that the satellites were minor constituents of the system they had to be at least 2 magnitudes fainter than, and at least 1–2′ smaller than the primaries. Indeed, in the groups employed, the satellites were typically more than 4 magnitudes fainter than the primaries, while the primaries had to be brighter than 13th magnitude.

The probability of physical association was established by an analysis of the background distribution of galaxies. (See Section 2 for a more detailed discussion.) For instance, groups for which there were more than 10 background objects within 1° of the primary were rejected. That is, to avoid contamination by nonphysical pairs, we searched for isolated systems of small angular size and large dispersed groups were rejected. Furthermore, systems were not accepted if there were nearby large galaxies, or if there was evidence of interactions (bridges and arms or other visible distortion).

Nine systems satisfied our computerized selection algorithm: NGC 1961, NGC 3893, NGC 3992, NGC 4111, NGC 4258, NGC 4303, NGC 4559, NGC 5371, and NGC 5689. Of these, NGC 3992 and NGC 1961 had been studied at a very early stage of this project [7], [8], before the algorithm based on our selection criteria had been developed. Five of these systems were observed at the old 300-ft telescope of the NRAO in order to ascertain their HI flux and redshift. Of these, two galaxies (NGC 3893 and NGC 4111) were detected and then observed at high resolution with the VLA telescope; for three other systems (NGC 4559, NGC 5371, and NGC 5689) insufficient HI emission was observed and they were not considered further. To supplement this list, seven other groups were added, either from the literature, or from data previously obtained by the Florida group for other projects. In total, 24 satellite primary pairs were employed in our analysis; their properties are listed in TABLE 1. An analysis of the fields around NGC 3893, NGC 4258, and NGC 4303 shows that roughly one-half of the observed satellites were detected in HI [9].

TABLE 1. Primary Groups[a]

Galaxy Group (1)	Primary Type (2)	V_{sys} (km s^{-1}) (3)	Distance to Primary (Mpc) (4)	Inclin. Angle (deg) (5)	R_{max} (kpc) (6)	$V(R_{max})$ (km s^{-1}) (7)	$m(V)$ ($\times 10^{10} M_\odot$) (8)	N_{sat} (9)	$\langle \chi \rangle$ (10)
NGC 0224	SbI-II	−301	0.7	78	28.0	230	34.4	4	0.5
NGC 1023	SBO$_1$ (5)	610	7.5	80	9.5	251	13.9	2	3.2
NGC 1961	SB(RS)IIP	3935	41.3	50	24.0	367	75.0	5	2.1
NGC 3359	SBc(S)I.8P	1009	11.0	51	20.8	140	9.5	1	0.3
NGC 3893	Sc(s)I.2	969	10.4	34	7.8	180	8.1	3	0.9
NGC 3992	SBB(RS)I	1046	14.2	53	19.0	240	25.4	3	0.1
NGC 4111	SO$_1$(9)	809	12.7	80	12.6	320	3.0	2	0.2
NGC 4258	Sb(s) II	465	5.2	72	19.0	208	19.5	1	0.0
NGC 4303	Sc(s)I.2	1561	12.9	19	12.2	221	14.0	1	4.8
NGC 4731	SBc(s)III:	1490	10.5	54	13.0	150	6.8	1	≪0.1
NGC 5084	SO$_1$(8)	1721	15.0	86	34.0	328	85.0	1	2.4

[a]*Notation:* Columns 1 to 2 are self-evident. Column 3 is measured heliocentric systemic velocity. Column 4 is distance to primary ($H_0 = 100$ km s^{-1} Mpc^{-1}). Column 5 is inclination angle of disk. Column 6 is radius of observed maximum rotation velocity. column 7. Column 8 is Keplerian, rotation mass of primary from columns 6 and 7. Column 9 is the number of satellites in a group, and Column 10 is mean value of (χ) for group.

2. METHOD OF ANALYSIS

Following the method of van Moorsel, we have treated each group as if it were a set of multiple satellite-primary pairs. Then, we may write the orbital mass, M, associated with the spiral in terms of the dimensionless function χ as,

$$M\chi = r_p V_r^2 / G \tag{1}$$

where r_p and V_r are the projected radial separation and radial velocity difference of a particular satellite. If each object is idealized to be a point mass, as shown in Paper I [1], the function χ depends on the true anomaly, v, the angle between the line of nodes and the major axis of the orbit, ω, the inclination of the plane of the orbit, i, and the eccentricity of the orbit e, viz.,

$$\chi \equiv \sin^2 i \frac{[\cos(v+\omega) + e\cos\omega]^2}{(1+e\cos v)} [1 - \sin^2(v+\omega)\sin^2 i]^{1/2}. \tag{2}$$

As a measure of the extent of any halo we have the observed value,

$$\chi_{obs} = r_p V_r^2 / G(m_1 + m_2), \tag{3}$$

TABLE 2. The Satellites[a]

Primary (1)	Satellites (2)	V_{sat} km sec^{-1} (3)	ΔV, km sec^{-1} $V_{sat}-V_{pri}$ (4)	d kpc (5)	$M(\Delta V)$ 10^{10} M$_\odot$ (6)	χ (M/m) (7)
NGC 224	NGC 147	−168	142	90.2	42.3	1.22
	NGC 187	−208	100	85.6	20.0	0.58
	NGC 205	−240	62	7.5	0.7	0.02
	NGC 221	−216	84	4.9	0.8	0.02
NGC 1023	north	905	295	34.2	85.9	6.12
	south	695	85	17.5	3.7	0.26
NGC 1961	A	4108	173	91	62.6	0.84
	B	3865	−70	106	3.9	0.05
	B1	3800	−135	122	52.4	0.70
	UGC 3342	3927	−8	157	0.2	0.00
	UGC 3349	4282	347	240	671.9	8.96
NGC 3359	dwarf	962	−47	48.5	2.5	0.26
NGC 3893	UGC 6797	963	−6	47.5	0.04	0.02
	UGC 6834	967	−2	30.3	0.23	0.04
	dwarf	1102	133	37.2	15.5	2.68
NGC 3992	UGC 6923	1062	16	60.6	0.4	0.02
	UGC 6940	1112	66	35.4	3.6	0.14
	UGC 6969	1115	69	45.3	5.0	0.20
NGC 4111	UGC 7089	789	−20	39.5	0.37	0.012
	UGC 7094	769	−40	36.	1.3	0.43
NGC 4258	UGC 7335	490	25	19.1	0.3	0.02
NGC 4303	UGC 7439	1270	−291	39.1	74.4	4.75
NGC 4731	RNGC 4731a	1505	15	32.1	0.2	0.03
NGC 5084	dwarf	2089	368	65.6	206.5	2.43

[a]*Notation:* Column 1, name of primary; Column 2, identification of satellites; Column 3, heliocentric velocity of satellites those, for NGC 224 are in the Andromeda-centric System); Column 4, velocity difference ($V_{sat}-V_{prim}$); Column 5 projected satellite-primary separation, if $H_0 = 100$ km s^{-1} Mpc^{-1}; Column 6, mass of primary based on satellite kinematics (the rotation mass (m) of the primary is taken from TABLE 1, col. 8); Column 7, value of χ_{obs} for each satellite -primary pair.

where m_1 and m_2 are the rotation masses of the primary and secondary, respectively. For our systems, $m_2/m_1 \ll 1$; hence, m_2 can be ignored and the subscripts dropped. χ_{obs} is related to χ by,

$$\chi_{obs} = (M/m)\chi. \qquad (4)$$

Thus, if the mass determined from the rotation curve measures the total mass of the system, the distribution of χ will be the same as that of χ_{obs}.

If values for the parameters of Eq. (2) are chosen randomly, the theoretical probability distribution, $N(\chi)$, is found to be sharply peaked for small values of χ. That is, there is a strong tendency to underestimate the orbital mass by almost an order of magnitude, because small values of r_p and V_r are highly favored. Furthermore, as long as the rotation curve reflects all the mass of the galaxy and the system is bound, $N(\chi) \to 0$ as $\chi \to (1 + e)$, which must be less than 2 for bound systems, assuming point masses.

In FIGURE 1a, we show the observed distribution of χ, $P(\chi)$, for the 24 pairs in our data base, of which 5 have values of $\chi > 2$. (These data and other information are listed in TABLE 2.)

Our galaxy groups were chosen from the Uppsala Catalog of Galaxies (UCG) [10]. We have simulated groups numerically, by creating a synthetic galaxy catalog and choosing members from it in the same fashion that galaxies were chosen from the UGC. Then, model $N(\chi)$ distributions were calculated, which provided information on biases. A comparison was made with distributions before and after selection criteria were applied (the unbiased and biased distributions). Our selection effects and biases are discussed in greater detail in papers III [3] and IV [4].

We are able to define a numerically consistent background distribution of separations between the satellites of our actual and synthetic galaxy for each interval in magnitude difference. As the background distributions (real and synthetic) at a particular difference in magnitude between primary and satellite(s) are modeled by Poisson statistics, an excess of neighbors implies a physical association. In modeling our selection effects, if the probability of association was greater than 0.5 the value of χ calculated for the synthetic pair was accumulated in both the biased and the unbiased distributions.

Therefore, in the absence of evidence identifying in an unambiguous fashion any pairs as optical rather than physical, we conclude that our data in FIGURE 1a represent the $P(\chi)$ distribution from a well-defined and representative group of primary galaxies, and any inferences drawn from these data should be uniformly applicable.

The biased and unbiased distributions of various quantities were compared in order to see how they are effected by the selection criteria. We applied the Kolmogorov-Smirnoff (K-S) test to gauge differences between the two distributions. We find that the biased and unbiased distributions are scarcely distinguishable, regardless of the choices for A(a), E(e), or for the choice of orbital phase. Therefore, χ is an ideal statistic to be employed in investigations of the mass distributions in small groups dominated by relatively massive primaries.

We have examined the EGH data for correlations between $N_s(\chi)$ and the projected separation (in both absolute terms and relative to the radius of the disk of the primary) of the primary-satellite pairs to discern any possible problems with the orbital masses because of the constraints imposed on the separations. Nothing was found at

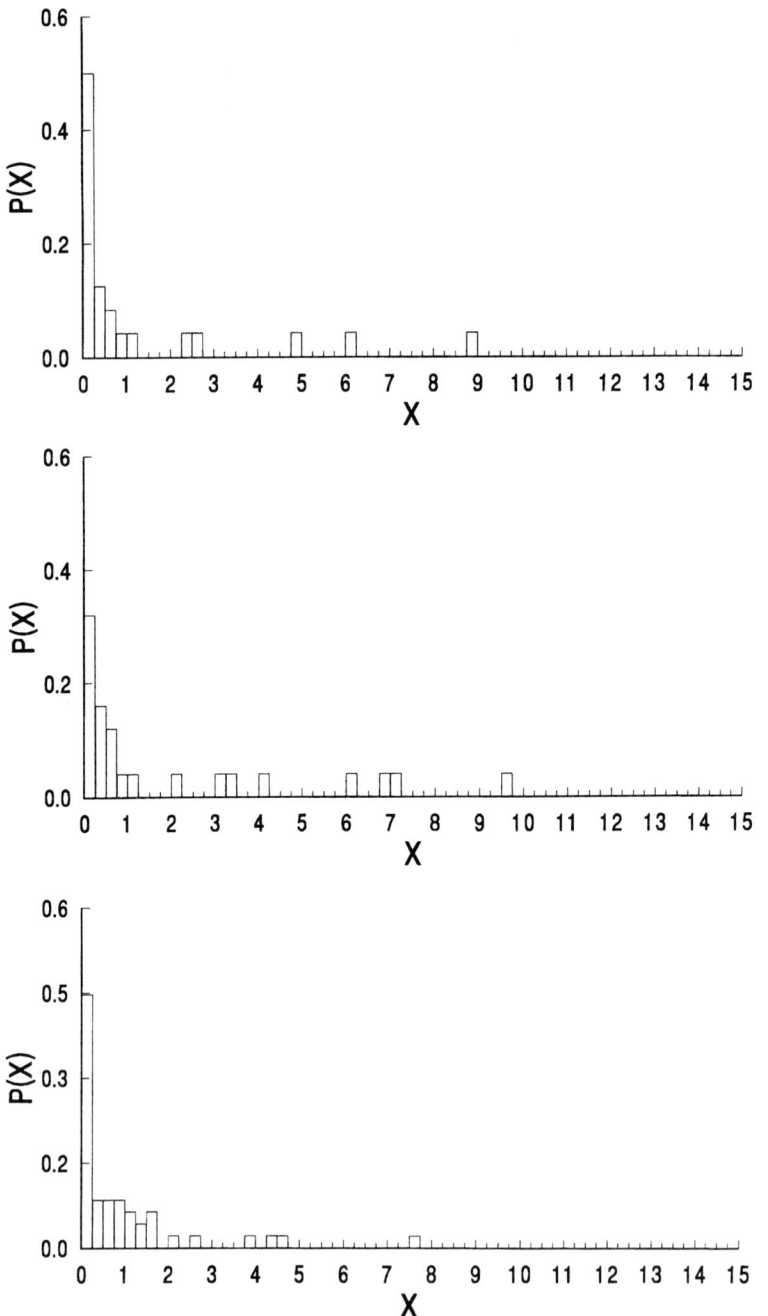

FIGURE 1. Plots of $P(\chi)$ for: **(a, top)** the data of EGH; **(b, middle)** the data of ZSFW1; and **(c, bottom)** the data of ZSFW2; $r_p < 200$ kpc in all three examples. The abscissa is plotted in units of 0.25 in χ, the ordinate is plotted in *percent* of the total population.

the 1% level of significance. However, the very largest values of χ are only observed from an admixture of orbits that have the very largest perigalactic distances (see Section 3). They are few in number and carry little weight in these correlations owing to the effect of projections.

In order to verify the conclusion that our data form a well-defined and representative sample, we have examined the data of Zaritsky et al. [11], [12] (ZSFW1, ZSFW2). The tabulated data of those papers allows us to use the method of Casertano and Shostak [13] to obtain the mass of each primary from the kinematics of the HI and the geometry of the disk; a mass estimate based upon a statistically viable rotation curve (the rotation mass). The kinematics of the satellites enable us to calculate the second estimate of the mass of the primary (the orbital mass) required to form $P(\chi)$.

For comparisons, we have considered the ZSFW1 and ZSFW2 data separately (although ZSFW1 is a subset of ZSFW2), and we have considered them within three radial regimes: all the data regardless of (projected) radial separation; the data for which the projected separations are greater than 200 kpc; and the data for which the projected separations are less than 200 kpc. In FIGURE 1b we show the data for $r_p < 200$ kpc taken from ZSFW1; FIGURE 1c shows the data for ZSFW2 for the same radial range. In other words, all three $P(\chi)$ distributions shown in FIGURE 1 are directly comparable. The data for EGH contain 24 points, that of ZSFW1 26 points, and that of ZSFW2 47 points. Owing to projection effects discussed earlier, all three are sharply peaked at $0 \leq \chi \leq 0.25$; in addition, all show a maximum value of $\chi < 10$. Thus, regardless of how the data were chosen, the distributions are in qualitative agreement.

There are some quantitative differences, particularly in the magnitude of the initial peak. However, as the values of χ are very small, small errors in the velocity difference between the primaries and satellites will have a significant effect on the population of this initial cell. Moving one point from the initial to the second histogram would reduce the peak for the data of EGH essentially to the same value as that of ZSFW2. In FIGURES 2a and 2b the data for ZSFW2 are shown for $r_p > 200$ kpc and for all projected radii ($r_p < 400$ kpc). The maximum value of χ is about 21. EGH do not have observations at these large radial separations.

From the data of ZSFW2, we conclude (1) that the larger radial separations ($r_p > 200$ kpc) include the larger values of χ. Furthermore, (2) at the smaller separations ($r_p < 200$ kpc), the $P(\chi)$ distribution has increased values for the inner region ($0 \leq \chi \leq 2$, 87% compared to 69% for $r_p > 200$ kpc), and for the initial peak (45% compared to 34%). Any model will need to describe these distributions and will need to reproduce the systematic trends (within the uncertainties) as a function of projected separation. We are strengthened in our belief that the differences in the distributions of the two radial regimes are real, because they are predicted by the models we will discuss later in this paper.

Finally, FIGURE 2c plots $P(\chi)$ for the combined data for NGC 1961 (5 satellites) and NGC 5084 (9 satellites) [14]. Both of these galaxies are known to be very massive before any consideration is given to satellites (see TABLE 1). NGC 1961 is common to the data of EGH and of ZSFW1 and ZSFW2. One satellite of NGC 5084 is included in EGH. The orbital extent of these data is less than 200 kpc and this figure should be compared with FIGURES 1a, 1b, and 1c. As we have seen, few points are

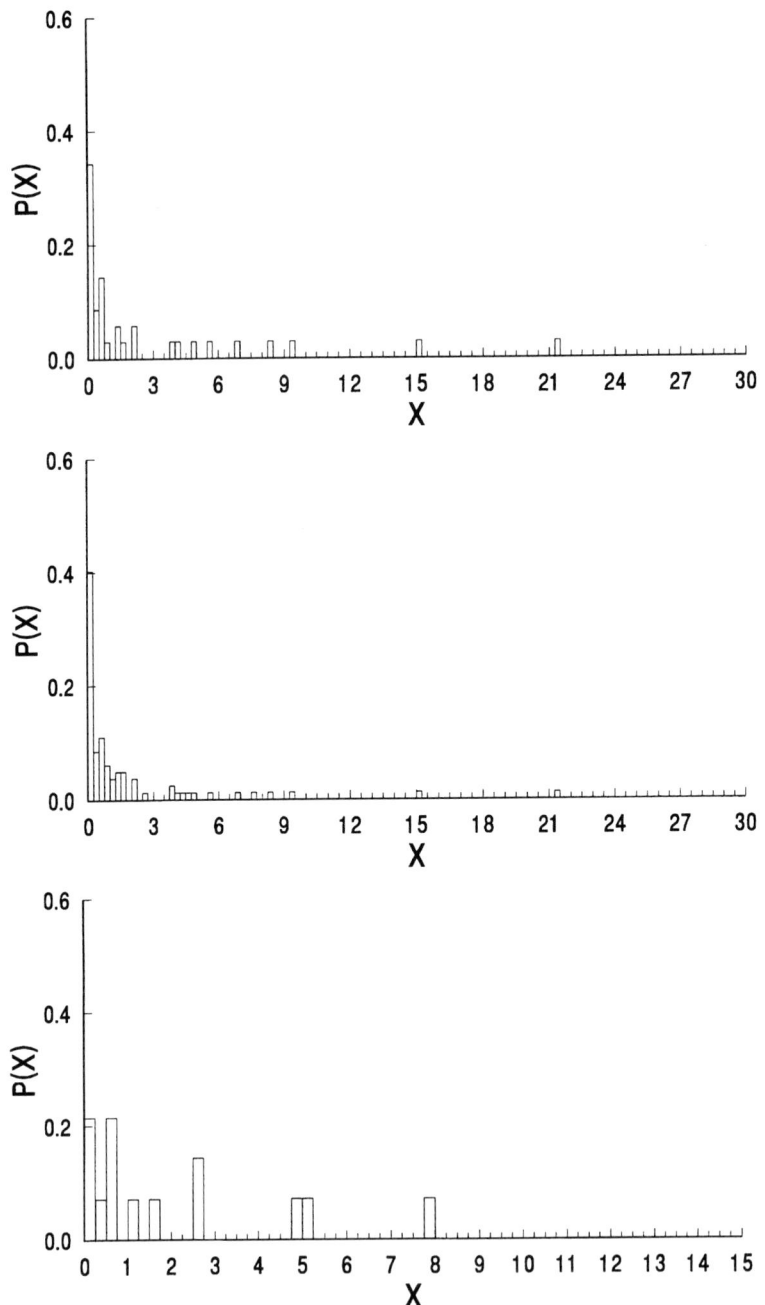

FIGURE 2. Plots of $P(\chi)$ for: **(a, top)** ZSFW2 for which $r_p > 200$ kpc; **(b, middle)** ZSFW2 for all projected radii ($r_p < 400$ kpc); and **(c, bottom)** for the massive galaxies NGC 1961 and NGC 5084 for which $r_p < 200$ kpc. The abscissa is plotted in units of 0.25 in χ, the ordinate is plotted in percent of the total population.

found at large values of χ; only five of the 14 points lie within $2 < \chi < 10$. An extended halo is certainly indicated for these massive disk systems. Unfortunately, the statistical uncertainties are large but the distribution appears to be much flatter than observed in FIGURES 1a, 1b, and 1c. As discussed in Section 5, models of galaxies completely dominated by their halos, for which the satellites are on circular orbits and for which dynamical friction is included, can reproduce a flatter $P(\chi)$ for $r_p < 200$ kpc. (We return to this point in Section 5.) Owing to their large disk mass (5.8×10^{11} $M_\odot$) the total mass within 200 kpc is comparable to massive halo systems discussed in Section 4.

The nature of NGC 3992 appears to differ substantially from these two galaxies. Employing the virial-like mass estimator of Bahcall and Tremaine [15], Gottesman and Hunter [7] argued that NGC 3992 showed no evidence for a significant halo. At that time there were three known satellites (see TABLE 2) within a projected separation of 61 kpc having a $\chi_m = 0.2$. Since that time, a fourth satellite has been found, UGC 06983, at a projected separation of 175 kpc having $\chi = 0.019$. This is a very large separation for such a small value. The rms velocity dispersion of these four satellites is 26 km s^{-1} (for NGC 3992, $V(R_{max}) = 240$ km s^{-1}). This is in striking contrast with an rms velocity dispersion of 324 km s^{-1} for NGC 1961 ($V(R_{max}) = 367$ km s^{-1}) combined with NGC 5084 $V(R_{max}) = 328$ km s^{-1}. Thus, our earlier conclusion is supported: "there is little latitude for a massive halo to be associated with this galaxy."

3. THEORY

In order to gain insight into the $N(\chi)$ distributions, we consider the general problem of orbit theory in massive halos. As a further idealization, we ignore dynamical friction and examine the idealized case of massless test particles orbiting in an isothermal halo without a primary disk. Defining the halo radius by r, letting r be any radius $\leq r_H$ and C the constant circular velocity in the halo, the halo potential $\phi = C^2(\ln x - 1)$, where $x = r/r_H \leq 1$. Since we are considering the orbit of a test particle subject to a central force, the total energy per unit mass, w, and specific angular momentum, j, of each test particle are constant. We define the following quantities:

$t_0 \equiv \dfrac{r_H}{C}$ = time required by test particle in a circular orbit at the halo radius traverse one radian;

$\tau \equiv \dfrac{t}{t_0}$ = dimensionless time;

$j \equiv r^2 \dot\theta$ where θ is the angular coordinate in the 2-body problem;

$r_{per} \equiv r_H x_p$ = perigalactic radius;

$r_{apo} \equiv r_H x_a$ = apogalactic radius;

In physical variables, conservation of energy and angular momentum read,

$$\frac{dr}{dt} = \pm\sqrt{2w - \left(\frac{j}{r}\right)^2 - 2\phi} \qquad (5a)$$

and

$$\frac{d\theta}{dt} = \frac{j}{r^2}. \tag{5b}$$

It is convenient to evaluate w at r_{per} and r_{apo}, where $\dot{r} = 0$. Hence,

$$\frac{2w}{C^2} = \frac{u_\theta^2}{x_p^2} + 2\ln x_p - 2 = \frac{u_\theta^2}{x_a^2} + 2\ln x_a - 2, \tag{6}$$

where angular momentum parameter $u_\theta \equiv j/r_H C$. In dimensionless variables, (5a) and (5b) become,

$$\frac{dx}{d\tau} = \pm \sqrt{u_\theta^2 \left(\frac{1}{x_p^2} - \frac{1}{x^2}\right) - 2\ln\left(\frac{x}{x_p}\right)} \tag{7a}$$

and

$$\frac{d\theta}{d\tau} = \frac{u_\theta}{x^2}. \tag{7b}$$

At $x = x_p$, Eq. (7a) changes sign from $-$ to $+$, and at $x = x_a$, the sign changes from $+$ to $-$. These equations were integrated using a Runge-Kutta algorithm. Although the orbits are closed only in the circular case (which does not require numerical integration), it is convenient to define eccentricity $e \equiv (x_a - x_p)/(x_a + x_p)$ exactly as in the Keplerian problem. Four sample orbits of different eccentricity are shown in FIGURE 3.

We have integrated ensembles of 10^4 orbits, subject to the following constraints:

(1) The number of orbits having apogalacticon distance x_a is proportional to x_a ($x_a \leq 1$); thus, the orbits follow the halo mass distribution.

(2) While no primary disks exist in this limiting model, we require $r_{per} \geq R$ (our average primary disk radius) = 15 kpc, or $x_p > 15/r_H$.

(3) The maximum orbital period $2\pi t_0 = 10^{10}$ yr, which limits both r_H and the total halo mass M_H. From Eq. (6),

$$u_\theta^2 = \frac{2x_a^2 \ln\left(\frac{x_a}{x_p}\right)}{\left[\left(\frac{x_a}{x_p}\right)^2 - 1\right]}. \tag{8}$$

Circular orbits are (trivial) special cases for which $x = x_a$ = constant, $u_\theta = r_{apo}/r_H = x_a$, and $\theta(\tau) = \frac{\tau}{x_a}$ + constant. We have projected $N(\chi)$ for ensembles of 10^4 orbits having varying ranges of e. The projection angles were selected at random, but the orbital phases were not randomly chosen. Instead, the phase angles for each orbit were evaluated at equal intervals of time before projection in accord with the law of

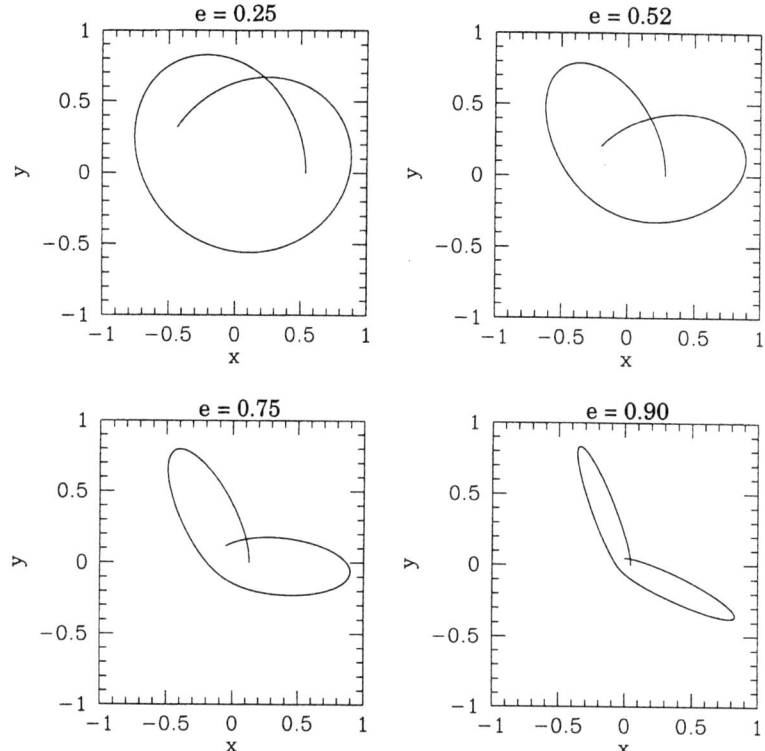

FIGURE 3. Plots of several orbits in the potential of a galaxy represented by a pure isothermal halo. The mass and radius of the system are $5.22 \times 10^{12}\, M_\odot$, and 390 kpc, respectively. The orbital eccentricity is defined as $\frac{(x_a - x_p)}{(x_a + x_p)}$, and the orbits are calculated over a Hubble time.

areas. The cases shown here were integrated for 10^{10} years with $C = 240$ km s^{-1}, $r_H = 390$ kpc, $2\pi t_0 = 10^{10}$ yr, and

$$\theta(\tau) = \int_0^\tau \left[\frac{u_\theta}{x^2 \tau}\right] d\tau + \theta_0.$$

where initial phase angle, θ_0, was selected randomly. The halo total mass $M_H = 5.22 \times 10^{12}\, M_\odot$. $N(\chi)$ distributions are shown in FIGURES 4a and 4b for circular orbits and for orbits with $e = 0.90$, respectively. The distributions are shown for all projected radii ≤ 390 kpc.

For an attractive inverse square law of force (Kepler's Problem) the maximum value of χ, χ_m, occurs at $x = x_p$. This fact follows from Eq. (2) with $i = \pi/2$ (orbital plane perpendicular to the plane of the sky) and $\omega = v = 0$; $\chi_m = 1 + e$, which is less

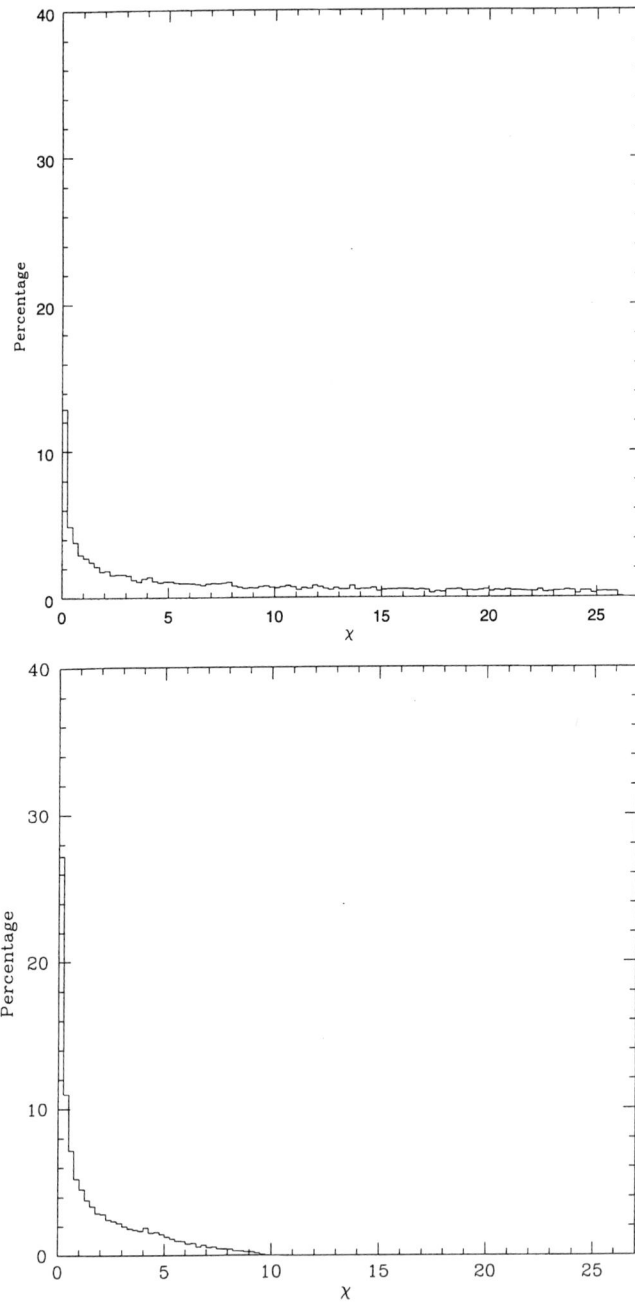

FIGURE 4. Plots of the expected $N(\chi)$ for the model described in FIGURE 3. **(a, top)** $N(\chi)$ is shown for an ensemble of orbits for which $e = 0$; and **(b, bottom)** for which $e = 0.9$. The plotted scale is as shown in FIGURES 1 and 2.

than two for bound orbits. However, in the isothermal halo model χ_m may not occur at perigalacticon. We restrict our discussion to orbits for which $i = \pi/2$ (a necessary condition for χ_m to be realized) and consider dimensionless radius vectors in the range $x_p \leq x \leq x_a$. We allow the radius vector to be oriented at any angle α with respect to the plane of the sky; if $\alpha = 0$, then x lies in the plane of the sky. For such orbits, the dimensionless radial velocity

$$u_r = \frac{dx}{d\tau}\sin\alpha + \left(x\frac{d\theta}{d\tau}\right)\cos\alpha,$$

$x\cos\alpha$ is the projection of x in the plane of the sky, and

$$\chi(x, \alpha) = \frac{r_H x}{R}\cos\alpha\left[\frac{dx}{d\tau}\sin\alpha + \left(x\frac{d\theta}{d\tau}\right)\sin\alpha\right]^2,$$

For each value of x there exists an α value, $\alpha = \alpha_{x_c}$, at which x has a maximum, x_c. Finally, defining $f(x) \equiv R \cdot \chi(x, \alpha)/r_H$, we plot $f(x)$ to establish the values of x_c and α_{x_c} at which χ has its absolute maximum, χ_m.

As an example, consider orbits in our isothermal halo model with $x = 0.95$. Upon examining the behavior of $f(x)$ over a range of eccentricities, we find that $\alpha_{x_c} = 0$ if $e < 0.7541$. That is, χ_m occurs at perigalacticon for $e < 0.7541$. Moreover, for the lower eccentricites,

$$\chi_m = \frac{2r_H x_a}{R}\left(\frac{x_a}{x_p}\right)\frac{\ln\left(\frac{x_a}{x_p}\right)}{\left[\left(\frac{x_a}{x_p}\right)^2 - 1\right]}. \tag{9}$$

TABLE 3. The Eccentricities of Satellite Orbits and the Maximum Values of χ

$\dfrac{r_{apo}}{r_{per}}$	e	x_c	α_{x_c}	χ_m	$\dfrac{r_{apo}}{R}(1 - \dfrac{2}{3}e^2)$
19	0.90	0.28000	46.310	9.7093	—
9	0.80	0.21556	34.550	12.470	—
7	0.75	$x_p = 0.13571$	0.000	14.019	15.438
$\dfrac{19}{6}$	0.52	$x_p = 0.30000$	0.000	19.974	20.247
$\dfrac{5}{3}$	0.25	$x_p = 0.57000$	0.000	23.658	23.671
1	0.00	$x_p = 0.95000$	0.000	24.700	24.700

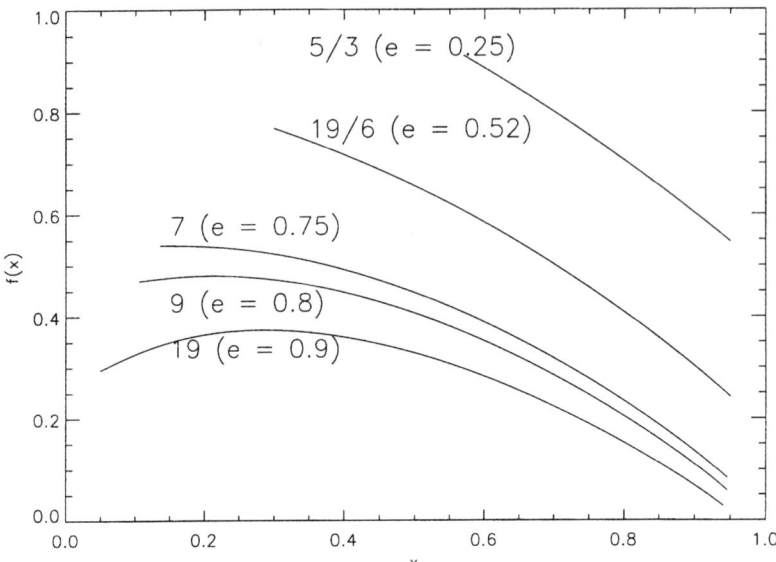

FIGURE 5. Various $f(x)$ curves to establish where each x_c, the maximum value of x, occurs when χ is at its absolute maximum. Results are shown in TABLE 3.

Also, as $e \to 0$, $\frac{x_a}{x_p} \to 1 + 2e$ and $\chi_m \to \frac{r_H x_a}{R}[1 - \frac{2}{3}e^2 + O(e^4) + ...]$. Plots of $f(x)$ for several eccentricities are shown in FIGURE 5 and numerical information is summarized in TABLE 3.

Thus, in a pure halo potential the large χ values originate in the vicinity of perigalacticon from orbits with large x_a ($r_{apo} \to r_H$) and with *small e*. However, in this model χ_m values as large as 20 can occur when e is as large as 0.5. (Recall that $\chi = 21$ is the largest value derived from the data of ZSFW2.)

4. SIMULATIONS

In order to include realistic disks, satellite-satellite interactions and dynamical friction, N-body calculations were carried out to simulate the dynamical behavior of relatively low mass satellites orbiting primary disk galaxies with, and without, extended halos. In the appropriate limit, these models do replicate the idealized models of the previous section. The models were "observed" in the same field of view and in the same fashion as were the actual groups. The numerical method used was the fourth-order scheme, developed by Aarseth [16], which was integrated for a Hubble time. Our model disks were assumed to have the following characteristics, derived from representative properties of our (EGH) selected systems: primary disk radius, $R = 15$ kpc, mass interior to disk radius $= 2 \times 10^{11}$ $M_\odot$, and the radius of the central bulge $= 4$ kpc. The mass and radius of the spherical halo surrounding an average primary disk were regarded as unknown quantities, to be determined, along with the disk mass, by the best fit of our models with the observations. The sum of the disk mass and radially distributed halo mass interior to the disk radius is subject to the observational constraint that the average circular velocity equals 240 km s^{-1} at disk radius R. Dynamical friction between the orbiting satellites and the spherical halo material, assumed nonrotating, also was included in our calculations. Our models were assumed to be randomly distributed within the interval 4–40 Mpc but constrained to have a mean distance of 12.4 Mpc, the average distance to our observed primaries if $H_0 = 100$ km s^{-1} Mpc^{-1}. These characteristics were scaled appropriately for models with different values of H_0. No significant differences were found for $50 < H_0 < 100$ km s^{-1} Mpc^{-1}. We have made no attempt to include active (evolving) halos in our present calculation; this is a problem we are investigating currently.

In this summary, we present results using a class of primary disk model, which represents less centrally condensed distributions of matter than are observed in galactic disks, and yet everywhere have masses and potentials that can be written in simple closed forms. In these models, we replaced the primary disks by very flattened, oblate spheroids of constant volume density (MacLaurin spheroids), having the same masses. Defining the semimajor axes in the x and y directions to be $a = b = R$ and the short, semiminor axis in the z direction by c, the total mass of the spheroid $M_D = \frac{4\pi}{3} \rho c R^2$. For all models discussed in this paper $\frac{c}{R} = 0.05$. Greater detail is given in references [3] and [4].

Dynamical friction can have important consequences for our calculations. In the end, after considerable numerical experimentation, we assumed that the model halos are nonrotating, i.e., that they are statistically at rest in our inertial (x, y, z) coordinate system, and used Chandrasekhar's [17] approximate expression for dynamical friction. Hence, we wrote the frictional deceleration as,

$$\vec{a}_f = -\Lambda \frac{G^2 M_s M_H(R)}{R r^2 V^3} \left[\text{erf}(X) - \frac{2X}{\sqrt{\pi}} e^{-X^2} \right] \vec{V}, \tag{10}$$

where M_s and $\vec{V}$ are, respectively, the mass and inertial velocity of a satellite moving through the halo matter and Λ is the coulomb logarithm. Defining the velocity dispersion of halo material at radius r from the origin by $\sigma(r)$, the dimensionless quantity,

$$X = V(r, t)/[\sqrt{2}\sigma(r)], \tag{11}$$

The halo velocity dispersion at each radius was assumed to be isotropic. In most of our models we calculated $\sigma(r)$ for our standard halo model with a point mass primary, M_p, by integrating the condition of hydrostatic equilibrium for the halo matter, viz.,

$$\frac{d}{dr}(\rho \sigma^2) = -\frac{G}{r^2}[M_p + M_H(r)]\rho, \tag{12a}$$

where $\sigma = \sigma(r)$, and $\rho = \rho(r)$ is defined by

$$\rho(r) = \rho_H \left(\frac{r_H}{r}\right)^2.$$

The solution of this equation is,

$$\sigma(r)^2 = \frac{GM_H(R)}{2R} + \frac{GM_p}{3r} = \frac{V_c^2}{2} + \frac{GM_p}{3r}, \qquad (12b)$$

where V_c is the circular velocity of the halo in the absence of the point mass. In order to prevent the velocity dispersions from becoming unrealistically large at small r, we assumed that the disk contribution to the halo velocity dispersion remained constant when $r < R$. In this regime, we experimented with several expressions for σ^2; e.g., $\sigma^2 = G[M_H(R) + M_p]/2R$ and $\sigma^2 = GM_H(R)/(2R) + GM_p/(3R)$ and concluded that any plausible choice of σ^2 has no discernible effect upon our simulated χ distributions. Moreover, in some model sequences we forced $\sigma^2 = G[M_H(R) + M_p]/2R$ for all r. Again, we found that this arbitrary procedure had little effect upon $N(\chi)$.

In most of our models, we let $\Lambda = \ln(900) \simeq 6.8$. This particular value of Λ was calculated using the following characteristics: $V = 240$ km s^{-1}, maximum impact parameter, $b_M = 67$ kpc, and satellite mass, $M_s = 1 \times 10^9$ M$_\odot$. Our relatively large b_M value might plausibly be reduced by a factor of 5, thereby reducing Λ to 5.2. Although we have not done so explicitly in discussing our models, for the purpose of computing the deceleration of dynamical friction, it is useful to define an effective satellite mass $\equiv M_{sat}\Lambda$.

Over a Hubble time, the relative importance of dynamical friction can range from being negligible to dominant. This effect is unimportant for low mass satellites, situated at relatively large radii, moving in low density regions of a halo. Such objects can have decay times $> 10^{11}$ yr. In contrast, the orbits of relatively massive satellites (e.g., $M = 5 \times 10^9$ M$_\odot$ or $\simeq 2.5\%$ of the primary mass), orbiting relatively close to the primary ($r \sim 30$ kpc, say) in a relatively dense halo, can decay in a few 10^9 yr. Some of our successful models with halos described in the next section had more than half of their satellites absorbed over a Hubble time. Circular orbits are effected more seriously at small radii where the halo density is greatest. Thus it is not surprising that the greatest effect of dynamical friction on eccentric orbits occurs at their perigalactic distance. In general the importance of dynamical friction depends on the mass of the satellite, the eccentricity of the orbit, the apogalactic distance and the orbital time scale (defined for this particular problem by the Hubble constant). One of our test problems was the successful simulation of the decay of the orbit of the Large Magellanic Cloud around our Galaxy, arriving at results that are essentially the same as those of Tremaine [18] and Murai and Fujimoto [19].

Dynamical models simulating our observed systems and the statistical distributions of their χ values were constructed to provide quantitative measure of the amount of halo material that is present. Because of the small number of objects in each system, the direct integration scheme of Aarseth [16], [20] was chosen for the N-body code. Regularization was not used, owing to the limited number of particles in these simulations. In addition, softening of the gravitational forces between the satellite and primary masses was unnecessary because we assumed (as noted previously) that any satellite coming within 4 kpc of the center of the primary (the nucleus) was devoured by the system and therefore removed from the computation. We recall that about one-half of our satellites are HI rich. Thus, they could not have passed through the disks of their primaries in recent times. If, during the last 10^9 years of the calculation, such an encounter occurred, the satellite was removed from

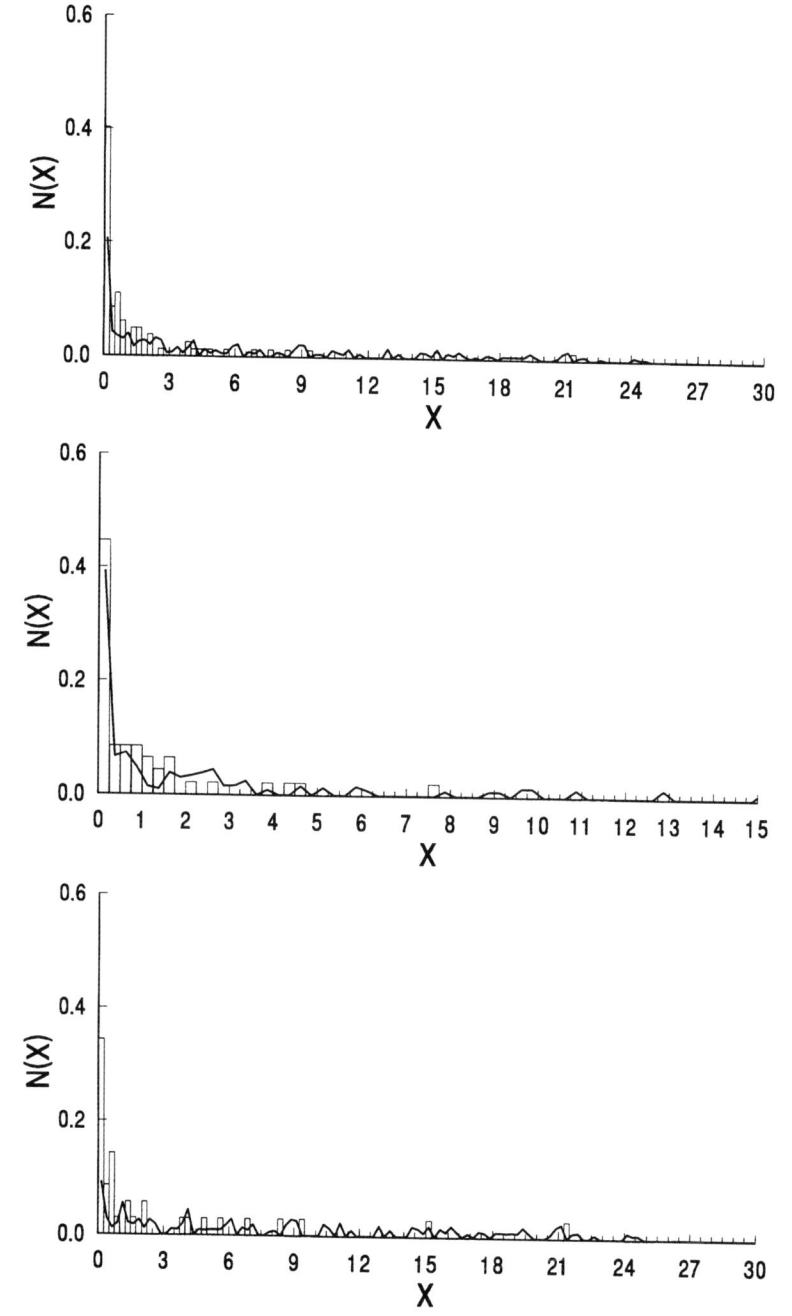

FIGURE 6. The $N(\chi)$ for bound satellites orbiting a galaxy represented by a massive, pure halo; $\beta = 1000$, $M = 5.3 \times 10^{12}\,M_\odot$, $r_H = 400$ kpc and $M_s = 1 \times 10^9\,M_\odot$. **(a, top)** all r_p; **(b, middle)** only $r_p < 200$ kpc; and **(c, bottom)** only $r_p > 200$ kpc. This is **Model 1**. The scales plotted in FIGURES 6–9 are as shown in FIGURES 1 and 2.

the experiment. Hence, the minimum impact parameter of our 24 observed, HI rich satellites is 15 kpc. The integration time for a complete simulation was the Hubble time (10^{10} yr, if $H_0 = 100$ km s^{-1} Mpc^{-1}).

In earlier studies, the accelerations were calculated using simple Newtonian forces between a point-mass primary and its satellites. However, in the present investigation (with the exception of one model shown in FIGURE 6) models of the primary galaxies were represented by the disk potentials described earlier, and the models included halos that dissipated orbital energy of the satellites via dynamical friction. The halo model densities were varied such that their masses within disk radius R ranged from ½ to 1000 times the disk mass. For convenience, we let $\beta \equiv M_H(R)/M_D$. A second model characteristic that was varied was r_H, the radius of the halo. Thus, the total halo mass $M_H(r_H) = \beta M_D r_H/R$, and the total mass of the system (disk + halo), $M = M_D(1 + \beta r_H/R)$. Recalling that $M(R) = M_D + M_H(R) = M_D(1 + \beta) = 2 \times 10^{11} M_\odot$, it follows that, in units of the solar mass, $M_D = 2 \times 10^{11}/(1 + \beta)$, and $M(r) = 2 \times 10^{11}(1 + \beta r/R)/(1 + \beta)$, where $R \leq r \leq r_H$. In addition, if $r > R$, $\sigma^2(r) = V_c^2[\beta + (2R)/(3r)]/[2(1 + \beta)]$, and $\sigma^2(r) = V_c^2[\beta + (2/3)/[2(1 + \beta)]$ when $r < R$. The circular velocity, $V_c = \sqrt{(GM(R))/R} = 240$ km s^{-1}. The halos cannot be larger than $\simeq 400$ kpc, the radius at which a circular orbit in a pure isothermal halo ($\beta \rightarrow \infty$) has an orbital period of $\sim H^{-1}$. Thus, for the models discussed in the next section, the maximum halo radius = 400 kpc.

5. MODELS

Each $N(\chi)$ distribution was built from fifty model integrations over a Hubble time, having different sets of initial conditions. For models with $H_0 = 100$ km s^{-1} Mpc^{-1}, each primary was assumed to be at a distance, randomly chosen, between 4 and 40 Mpc, with the mean distance being 12.4 Mpc, and the model was then projected at three randomly selected but orthogonal orientations. Model χ values were calculated from the projected separations and radial velocities of the satellite-primary pairs. At a mean distance of 12.4 Mpc for the observed systems, the 30' synthesized field of the EGH HI observations corresponds to a projected diameter of 108 kpc. The projected diameter of the field of view of a primary situated at a distance of l Mpc is 108 l/12.4 kpc. The most distant galaxy in our data, NGC 1961, was observed with several beam pointings. Effects of this restriction can be seen in our simulations of the EGH $N(\chi)$ distributions. If the model satellite orbits lie within this field, they are sampled completely, as is the case for the most distant objects in our sample. However, model satellite orbits that move beyond the "observational" window exhibit a correspondingly larger number of small χ values. Some of these orbits often move almost orthogonally to the line of sight and, consequently, their projected radial velocity differences relative to that of the primary are unusually small. Notwithstanding, any bias of this nature in our observations will be present in our models as well, because we "observe" them synthetically as if viewed with the VLA. In modeling the ZSFW data, no restriction was imposed upon the field of view. Our simulations of the ZSFW1 and ZSFW2 data are not subject to the "field of view constraint."

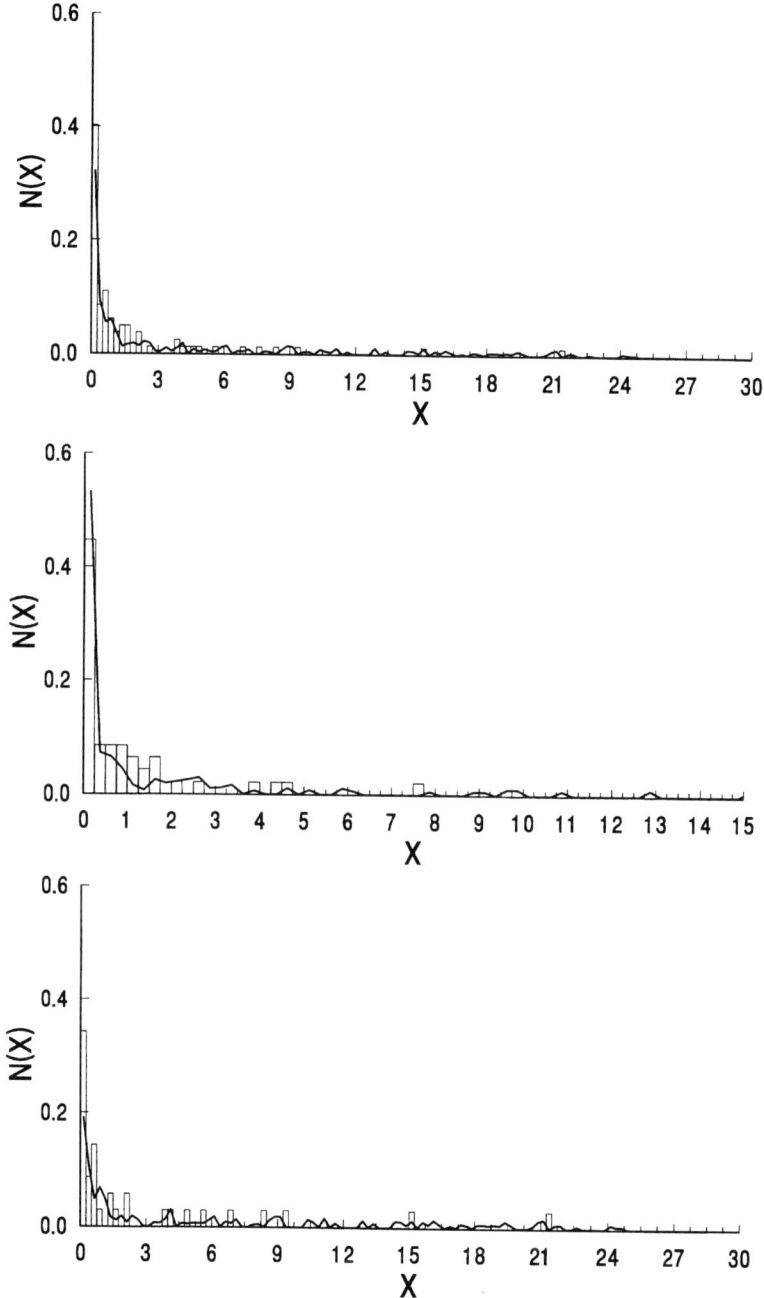

FIGURE 7. The $N(\chi)$ for bound satellites orbiting galaxies with two types of halos; 67% are represented by a $\beta = 1000$ model as shown in FIGURE 6, and 33% are galaxies with no halos: (a, top) all r_p; (b, middle) only $r_p < 200$ kpc; and (c, bottom) only $r_p > 200$ kpc. This is Model 2.

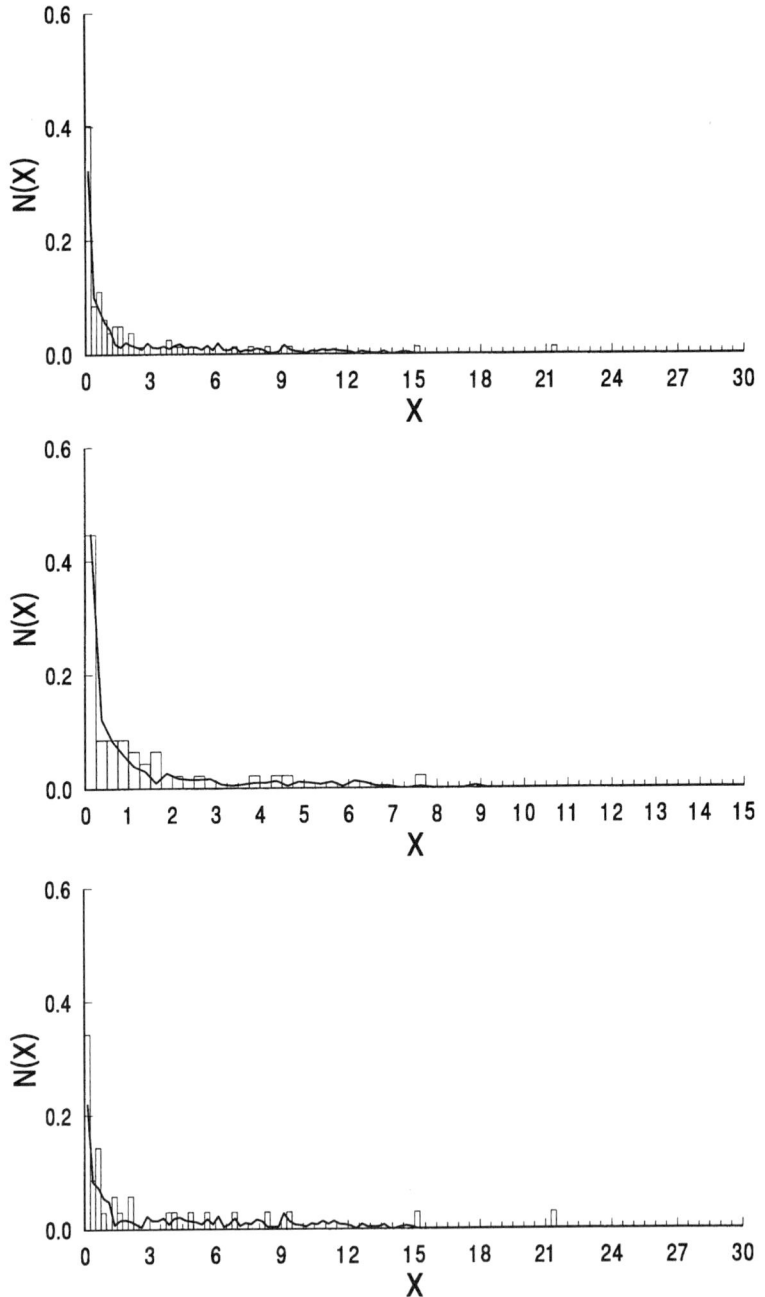

FIGURE 8. The N(χ) for bound satellites orbiting galaxies with two types of halos as in Model 2, except the 67% with-halo component is represented by a $\beta = 2$ model and dynamical friction is included: **(a, top)** all r_p; **(b, middle)** only $r_p < 200$ kpc; and **(c, bottom)** only $r_p > 200$ kpc. This is **Model 3**.

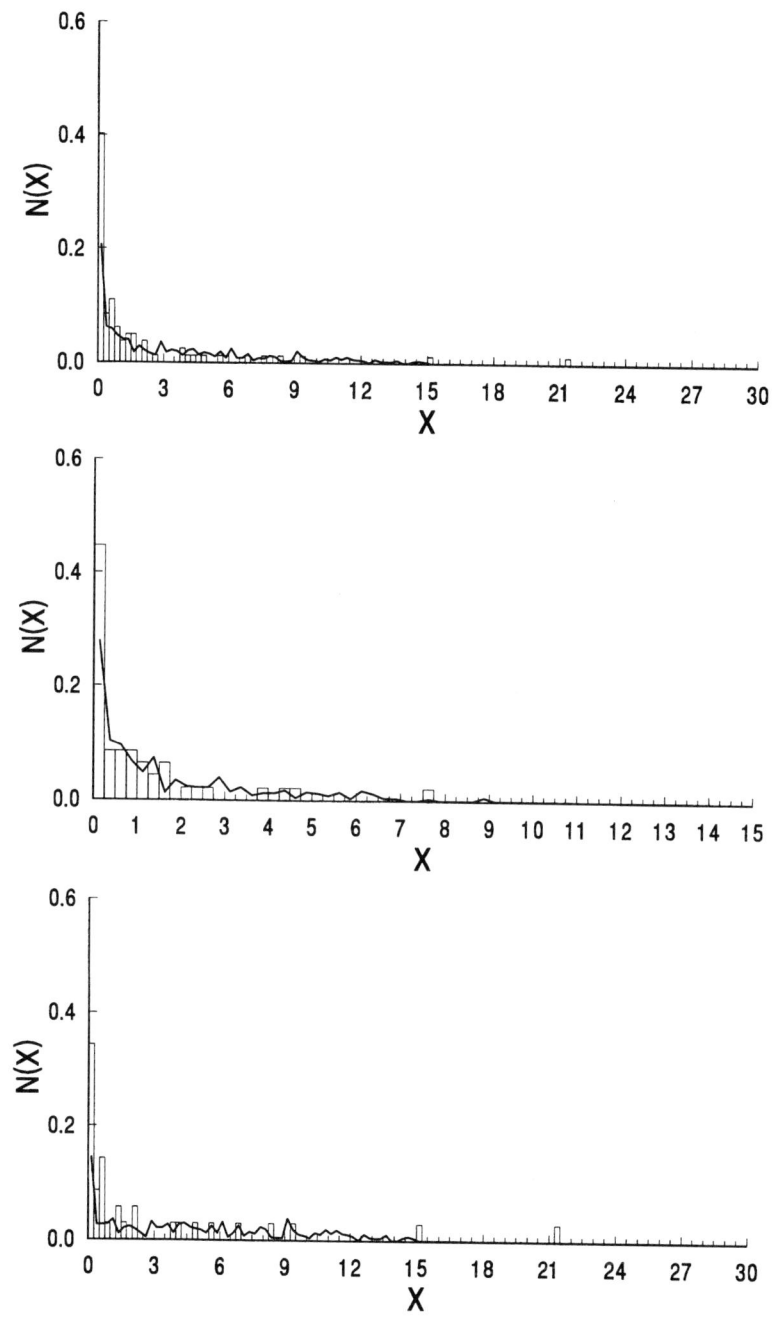

FIGURE 9. The $N(\chi)$ for bound satellites orbiting a $\beta = 2$ galaxy with an orbital mix consisting of 67% with $e = 0$, and 33% with $e = 0.9$. The model includes dynamical friction: **(a, top)** all r_p; **(b, middle)** only $r_p < 200$ kpc; and **(c, bottom)** only $r_p > 200$ kpc. This is **Model 4**.

We consider a model to be successful if it satisfies the following criteria:

(i) The predicted $N(\chi)$ must be consistent with the observed $P(\chi)$. In four of the more interesting models, this was evaluated by employing the chi-squared test. In addition, we also applied the K-S test to two of these models.

(ii) A significant fraction of the satellites ($\gtrsim 30\%$) must survive for a Hubble time. Thus, the halo characteristics and/or satellite masses cannot be such that most of the satellites spiral into, and are devoured by, the primaries through the action of dynamical friction. This requirement imposes constraints upon the halo masses and radii, as well as upon the satellite masses.

During the course of this work we have examined more than one thousand models, most of which were unsuccessful. The most obvious, potentially successful models, ones that occupied our thoughts for a long time, are generic. By this we mean that all of our galaxies may be represented by a single disk-halo combination. Thus, each generic system is characterized by "universal" values of β and r_H, with the only differences from case to case being due to differing selections of initial conditions for the satellites orbits. First we consider models of the ZSFW2 data, which are not constrained by the angular field of view.

Model 1 depicts the behavior of low mass satellites executing circular orbits in a massive isothermal halo with a negligible disk mass in the absence of dynamical friction. For this model, $\beta = 1000$, $r_H = 400$ kpc, $M(r_H) = 5.33 \times 10^{12}$ $M_\odot$, $M_s = 10^9$ $M_\odot$, and the number of satellites having initial radius r, $N(r) \sim r$. (The satellite numbers follow the halo mass distribution.) FIGURE 6a shows $N(\chi)$ distributions for all satellite orbits, whereas FIGURES 6b and 6c illustrate the $N(\chi)$ distributions for satellites having projected separations $r_p < 200$ kpc and $r_p > 200$ kpc, respectively. FIGURES 6a and 6c have satisfactory "tails" at large χ but display a paucity of small χ values; for $\chi < 0.25$ their respective χ distributions are roughly a factor of two and four too small. Only the predicted distribution for $r_p < 200$ kpc has a satisfactory number of small χ values. In an otherwise identical model including our expression for dynamical friction, the $N(\chi)$ distributions for all r_p and for $r_p < 200$ kpc are further depleted at small χ; for $\chi < 0.25$, $N(\chi)$ is less than half of the observed value when $r_p < 200$ kpc.

There are two very different types of limiting models that can retain the large χ tails and yet increase the percentage of small χ values: (1) a mixture of circular orbits (or orbits of relatively low eccentricity) with orbits of relatively high eccentricity, and (2) an admixture of haloless disks with heavy halo systems.

FIGURES 7a–7c show $N(\chi)$ for *Model 2* having 2/3 of its primaries with heavy halos and 1/3 of them with no halos. The heavy halo primaries and orbital characteristics are identical with those of Model 1. The approximate agreement of the observed and model distributions for all r_p and for $r_p < 200$ kpc is apparent. However, this composite model suffers from the obvious shortcoming of having heavy halo primaries that are essentially all halo and no disk. We have experimented with models similar to Models 1 and 2 that include significant disk masses. While their $N(\chi)$ distributions are qualitatively similar to those of Models 1 and 2, their maximum χ values, χ_m do not extend out to $\chi = 26$. For example, when $\beta = 2$, $\chi_m < 18$. The reason for this behavior is that the total mass within the disk radius is observationally constrained to be 2×10^{11} $M_\odot$. Hence, if $\beta = 2$, $M_D = \frac{2}{3} \times 10^{11}$ $M_\odot$, $M_H(R) = \frac{4}{3} \times$

10^{11} M$_\odot$ and $M(r_H) = \frac{2}{3} \times 10^{11}$ M$_\odot$ $[1 + 2\frac{r_H}{R}]$. Thus, the mass of the with halo component is $M = 3.62 \times 10^{11}$ M$_\odot$, and

$$\chi_m = \frac{\left(1 + 2\frac{r_H}{R}\right)}{3} \simeq 18$$

when $r_H = 400$ kpc and $R = 15$ kpc. On observational grounds [21], it is unreasonable to suppose $\beta > 2$; the canonical model, $\beta = 1$, has equal disk and halo masses within the disk radius. Further, as stated previously, r_H cannot exceed 400 kpc if the period of a circular orbit at that radius is to be ~H^{-1}.

Model 3 as shown in FIGURES 8a–8c is identical with Model 2 excepting that $\beta = 2$ and dynamical friction is included in the primaries with halos. This model fits the observations quite well for $\chi < 16$. (Of course it cannot match the single observed value at $\chi \simeq 21$.) We find somewhat similar results for models with mixed eccentricities, as shown in FIGURES 9a–9c, depicting *Model 4*. For this model $\beta = 2$ and the orbital mix consists of $\frac{2}{3}$ circular orbits, with the remaining $\frac{1}{3}$ having $e = 0.9$. Dynamical friction has been included in this simulation. The shortcoming of this model is that it fails to replicate the peaks at small χ. If we increase the fraction of highly eccentric orbits sufficiently to match the peaks at small χ, the amplitudes of the distributions for $\chi > 2$ become unacceptably low. However, if we have over estimated the importance of dynamical friction, it may be possible to find mixed eccentricity models that work. We emphasize that Models 2 through 4 are schematic, consisting of but two extreme components. When future observations have converged to particular distributions for all r, $r_p < 200$ kpc and $r_p > 200$ kpc, more realistic simulations will be warranted, including initial distributions of orbital elements, and/or halo masses and radii.

Although comparison of the data sets reveals significant variations (as we have noted), we have employed the chi-squared (not to be confused with our χ parameter) test to examine the compatibility of Models 3 and 4 with the ZSFW2 data. We have used the 5% critical value of the chi-squared distribution as our fiducial mark. Model 3 (mixed halo masses) is consistent with the data over all radial regimes (even allowing for the point at $\chi = 21$), and especially for $r_p < 200$ kpc. However, as indicated by visual inspection of FIGURES 9a–9c, Model 4 fails over all radial regimes. This model (mixed orbital eccentricities) cannot describe the full extent of the tail at large values of χ in the observed distributions, nor can it reproduce adequately the peak in the distribution at small values of χ.

In view of the essential agreement between the EGH, ZSFW1, and ZSFW2 $P(\chi)$ for $r_p < 200$ kpc, we thought it worthwhile to examine this regime more carefully. The models shown below compare simulated distributions with the EGH distribution, including the "field of view constraint." Similarities between the observed and calculated distributions have been gauged using the K-S test, as well as the chi-squared test for goodness of fit.

The $N(\chi)$ distribution of *Model 5*, shown in FIGURE 10a, is an optimized generic model with a small halo; its characteristics are: $\beta = 2$, $r_H = 60$ kpc, and total mass $M(r_H) = 6 \times 10^{11}$ M$_\odot$. This model is a borderline case in that it is the least massive halo ($M_H = 5.33 \times 10^{11}$ M$_\odot$) that barely passes our statistical tests most of the time. However, it is clear from a casual inspection of the figure that the fit is not very im-

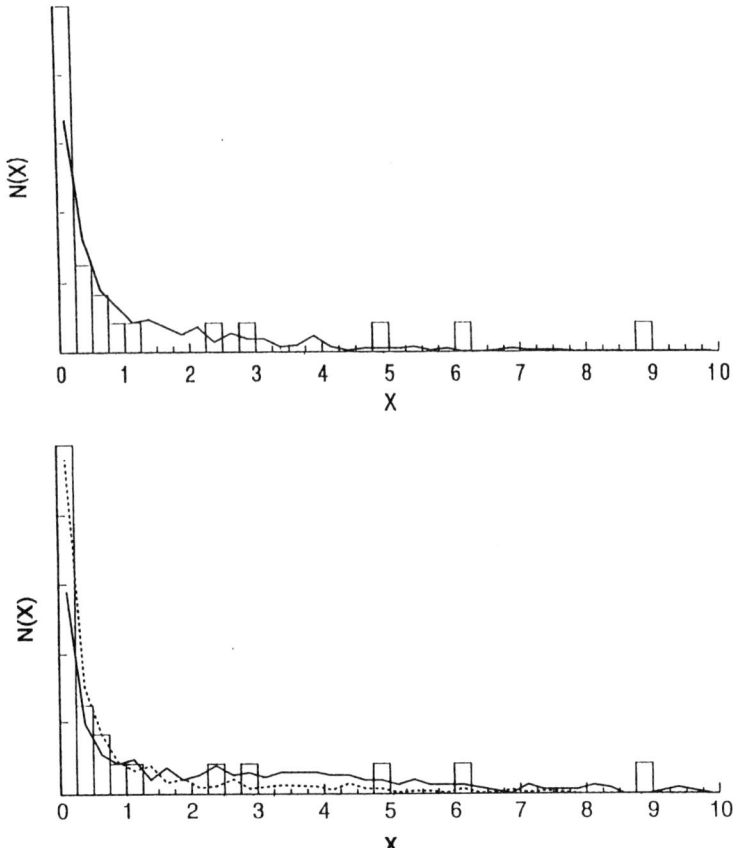

FIGURE 10. (a, top) Model 5. The $N(\chi)$ for a $\beta = 2$ galaxy with $r_H = 60$ kpc, and $M = 6 \times 10^{11} M_\odot$; the generic one component model. **(b, bottom) Model 6.** The $N(\chi)$ for a mixed model; 60% of the galaxies are represented by a $\beta = 2$ component, for which $r_H = 150$ kpc, and $M = 1.4 \times 10^{12} M_\odot$. The remaining 40% have no halo. The solid curve is for satellites with $M_s = 0.67 \times 10^9 M_\odot$. The dashed curve is for satellites with $M_s = 2 \times 10^9 M_\odot$. The models and data are normalized to the same area and the scale of the ordinate is not the same as in the earlier figures.

pressive; qualitatively speaking, the peak at small χ is too low and the "tail" at large χ does not extend out far enough.

Model 6 consists of 60% primaries with halos and 40% with no halos. The model characteristics are: $\beta = 2$, $r_H = 150$ kpc and $M(r_H) = 1.4 \times 10^{12}\ M_\odot$. The solid curve in FIGURE 10b shows that $N(\chi)$ when the satellite mass, $M_s = \frac{2}{3} \times 10^9\ M_\odot$, whereas the dashed $N(\chi)$ is for satellite $M_s = 2 \times 10^9\ M_\odot$. The differences between these distributions illustrate the crucial role that dynamical friction can play in the simulations. In this model, the simulation with the lower mass satellites passes the statistical tests, whereas the other does not. Both calculations were initialized with the satellite number density $\sim r$ and with a uniform mix of eccentricities, ranging up to $e = 0.9$. After integrating for a Hubble time, many of the relatively massive satellites have spiraled into small r values, thereby developing a peak at small χ at the expense of large χ values. Integration for an additional Hubble time leads to most of the satellites being devoured by their primaries. It should be noted that Models 5 and 6 were normalized to $N = 24$, the number of EGH satellites, rather than percentage of satellites as in Models 1 through 4. We emphasize that not only are Models 5 and 6 successful, but they do not require halos of 400-kpc extent (although such large halos cannot be ruled out by these data). The requirement for large halos depends very strongly on the reality of the two points in $P(\chi)$ at $\chi \simeq 15$ and at $\chi \simeq 21$. In all the models discussed above, satellite numbers are $\propto r$. We have examined similar models with satellite numbers $\propto r^2$, but we find these less satisfactory. In general, they produce relatively fewer points with small χ values.

5. DISCUSSION AND CONCLUSION

We have reviewed two observational studies of the kinematics of satellites to large spiral galaxies. The method we have employed allows the comparison of two mass estimates, one based on the rotation of the spiral disks and one based upon relative velocities of the satellites. As these measure the mass within different volumes, the ratio of the two estimates is sensitive to the gravitational effects of any halo composed of nonvisible matter.

The models consisted of an underlying disk embedded in a spherical, isothermal halo, the mass (relative to that of the disk) and size of which can be varied. The effects of this mass distribution have been explored by numerical experiments, and in the more simple case (orbits in a pure isothermal spherical halo) by a partially analytical approach. Considering a 400-kpc halo, constrained to have a circular velocity of 240 km s^{-1} at a radius of 15 kpc, the maximum value of χ varies from about 25 to about 15 as the eccentricities of the orbits increase from 0 to 0.75 (see TABLE 3).

Similarly, the strength of the inner peak ($\chi \leq 0.25$) in the observed $P(\chi)$ is sensitive to the radial extent of halos. However, in the pure halo model with circular orbits, dynamical friction suppresses the inner peak especially for the inner orbits. Such a model may be consistent with the observations for NGC 1961 and NGC 5084 shown in FIGURE 2c. Thus the strength of this inner peak and how it varies with radial extent, and the extent of the tail in $P(\chi)$ at large values of χ are important signatures. While large and massive halos, as expected, will adequately describe the extended tail, they do not describe the inner part of the $P(\chi)$ distribution. This disagreement is made worse with the inclusion of dynamical friction in the models.

Bahcall and Casertano [21] have emphasized that the data support models for which the mass of the halo is about equal to that of the disk within the radius of the disk. This reduction in the halo mass also reduces the extent of the tail to $\chi_m \leq 15$, whereas the data appear to extend to $\chi_m \approx 21$. At the same time the inner peak of a $\beta = 2$ model is too weak. Adding eccentric orbits to the mix of the model increases this inner peak but at the expense of the tail.

This is a classic example of the tail wagging the dog. If the χ values at 15 and 21 can be excluded from ZSFW2 then successful models can be constructed, such as those discussed in EGH III [3] and shown here by Models 5 and 6. However, we have no reason to exclude these points. Therefore, lacking further evidence a multicomponent model seems to be required. A haloless component only effects $N(\chi)$ for $r_p <$ 200 kpc, and a massive $\beta = 1000$ component adds most strongly to the tail of the distribution. We believe that the mass distribution of NGC 1961 and NGC 5084 are dominated by their halos, whereas NGC 3992 is an example of an (essentially) haloless system.

Models 3–6 explore the qualities of several forms of multicomponent distributions. Owing to the coarse nature of the data the models are crude, but they indicate the importance of the two largest χ values in the observed $P(\chi)$. The existence of these points requires a large mass and radius for at least some halos. Given current observations, we conclude, (1) regardless of model, that circular orbits alone will not describe $P(\chi)$ over all radial domains. (2) Adding an eccentric component improves the fit only to the inner part of $P(\chi)$. (3) Employing a more realistic $\beta = 2$ model is not successful unless the highest values of χ in $P(\chi)$ come from interlopers. (4) If those outlying values can be ignored, then galaxy halos may be of lower mass and lesser extent. (5) If these extreme values are included, a multicomponent model must be employed and, regardless of the distribution of orbital eccentricities, about 33% of the family of galaxies must be halo-poor compared to the standard model of Casertano and Bahcall.

We emphasize again that conclusion (5) is simplistic. The ratio of halo to disk masses most probably consists of a continuum of values. For instance, the addition of a small component of $\beta = 1000$ galaxies to Model 4 would easily reproduce the two (presumably very few) points observed beyond $\chi = 15$. Furthermore, such a component would have little effect at the smallest values of χ. However, the data as they exist are so sparse that Model 4 is successful without added complexities. A more complete exploration of parameter space is required. Individual galaxies must be observed in larger number that well represent the general population of spiral galaxies.

ACKNOWLEDGMENTS

Much of this project was completed in 1995–1996 while S.T. Gottesman was a visiting professor at the Instituto de Astrofisica de Canarias; he is grateful for the hospitality. This enterprise was partially supported by National Science Fondation Grants AST 81-16312 and AST 90-22827, and by the NASA/University JOVE Program. We are thankful to Mr. Chad Ellington, Mr. Dimitri Pourbaix, and Mr. Veera Boonyasait for their efforts on various aspects of this research. The comments by Dr. Dennis Zaritsky helped to clarify our presentation.

REFERENCES

1. ERICKSON, L.K., S.T. GOTTESMAN & J.H. HUNTER, JR. 1987. Nature **325**: 779. [Paper I]

2. ERICKSON, L.K., S.T. GOTTESMAN & J.H. HUNTER, JR. 1990. Ann. N.Y. Acad. Sci. **596**: 20. [Paper II]
3. ERICKSON, L.K., S.T. GOTTESMAN & J.H. HUNTER, JR. 1995. Ann. N.Y. Acad. Sci. **751**: 53. [Paper III]
4. ERICKSON, L.K., S.T. GOTTESMAN & J.H. HUNTER, JR. 1995. Ann. N.Y. Acad. Sci. **751**: 53. [Paper IV]
5. VAN MOORSEL, G. 1982. Ph.D. Dissertation, University of Groningen, Netherlands.
6. VAN MOORSEL, G. 1987. Astron. Astrophys. **176**: 13.
7. GOTTESMAN, S.T. & J.H. HUNTER, JR. 1982, Astrophys. J., 260, 65.
8. GOTTESMAN, S.T., J.H. HUNTER, JR. & G.S. Shostak. 1983. Mon. Not. R. Astro. Soc., **202**: 21.
9. ERICKSON, L.K. 1987. Ph. D. Dissertation. University of Florida.
10. NILSON, P. 1973. Ann. Uppsala Astron. Obs. **6**. [UGC]
11. ZARITSKY, D., R. SMITH, C. FRENK & S.D.M. WHITE. 1993. Astrophys. J. **405**: 464. [ZSFW1]
12. ZARITSKY, D., R. SMITH, C. FRENK & S.D.M. WHITE. 1997. Astrophys. J. **478**: 39. [ZSFW2]
13. CASERTANO, S.P.R. & G.S. SHOSTAK. 1980. Astron. Astrophys. **81**: 371.
14. CARIGNON, C., S. COTE, K.C. FREEMAN & P.J. QUINN. Astron. J. **113**: 1585.
15. BAHCALL, J.N. & S. TREMAINE. 1981. Astrophys. J. **244**: 805.
16. AARSETH, S.J. 1971. Astrophys. Space Sci. **14**: 118.
17. CHANDRASEKHAR, S. 1943. Astrophys. J. **97**: 251.
18. TREMAINE, S. 1976. Astrophys. J. **203**: 72.
19. MURAI, T. & M. FUJIMOTO. 1980. Pub. Astron. Soc. Japan **32**: 581.
20. AARSETH, S.J. 1987. *In* Galactic Dynamics, J Binney & S. Tremaine, Eds. Princeton University Press. Princeton, NJ.
21. BAHCALL, J. N. & S.P.R. CASERTANO. 1985. Astrophys. J. Lett. **293**: L7.

Chaos in the Centers of Galaxies

ISAAC SHLOSMAN,[a] CLAYTON HELLER,[b] AND INGO BERENTZEN[b]

[a]*Department of Physics and Astronomy, University of Kentucky, Lexington, Kentucky 40506-0055, USA*
[b]*Universitäts Sternwarte, Geismarlandstrasse 11, D-37083, Göttingen, Germany*

ABSTRACT: We compare diverging evolution of a two-component (gas+stars) galactic disk embedded in a "live" halo with that of an identical pure stellar disk. Our modeling supports the conjecture that the growth of central concentration in galaxies dissolves the main family of regular orbits in the stellar bar and assists in the formation of a galactic bulge.

INTRODUCTION

Evolution in disk galaxies is complicated by the dynamical action of stellar bars and other nonaxisymmetric features, like spiral arms, oval distortions, and general triaxialities on *all scales*. So far, we have acquired a basic understanding of stellar and gas dynamical processes excited by these global nonaxisymmetries. We are able, in principle, to explain the resulting large-scale (> 1 kpc) galactic morphologies in terms of prevailing families of stellar orbits and of gas dynamics in the background gravitational potential.

The circumnuclear regions of disk galaxies, on smaller spatial scales, remain largely unresolved by the ground-based observations. From the theoretical point of view, the dynamics of these regions, both stellar and that of the gas, is much more intricate and complex in comparison with with the large-scale dynamics. This complexity is a direct consequence of a number of different factors, most notably of (1) gas inflow toward the center and the resulting modification of mass distribution in the circumnuclear regions; (2) formation of the circumnuclear multiple resonance region due to the modified rotation velocity curve; (3) progressively increasing dynamic importance of gas (beyond the star formation) compared to its rather passive role elsewhere; and because of (4) thermal and dynamic effects of nuclear activity, stellar and/or nonstellar, induced by the above processes. In addition, the disk thickness cannot be neglected at these small radii. All these factors are expected to influence the circumnuclear morphology in terms of distribution of star forming sites, dust, formation of nuclear bars and rings, shapes and masses of the central bulges, isophote twists, lopsidedness, and more.

Recently, it became clear that major families of stellar orbits are profoundly influenced by dynamical and secular effects in the galactic disks, the latter on timescales shorter than the Hubble time. This evolution is driven in part by dynamical instabilities in the stellar component, such as the bar instability in the disk plane [1] and the bending instability out of the equatorial plane [2]–[4]. An additional driver of orbital evolution on the global scale is the disk gaseous component whose mass, although small compared to the overall galactic mass, can nevertheless exert dynamical effects on stars and substantially modify the galactic morphology. Because the cold interstellar medium is clumpy, with the mass spectrum which is weighted toward the higher masses [5], and with the high-mass cutoff around 10^7 $M_\odot$, it may efficiently and randomly scatter stars residing on periodic and semiperiodic orbits. This results in overall disk heating, and may even impede the bar instability, reducing the overall phase space available to these orbits [6].

In the circumnuclear regions of barred galaxies, the stellar dynamical evolution is expected to be even more gas-driven. Here the gas inflow from within the corotation radius, due to gravitational torques, can result in a substantial change in the galactic gravitational potential, and is likely to be accompanied by star formation and formation/fueling of an active galactic nucleus [7]. The orbit analysis in *non-self-consistent* potentials has shown that growing central mass concentrations, like compact bulges, nuclear star clusters, or supermassive black holes, all tend to destroy the main family of periodic orbits aligned with the stellar bar [8], [9]. This was claimed to weaken and dissolve the bar. Similar trends are expected when a massive nuclear molecular ring is present, preferentially destroying stable orbits between the inner Lindblad resonance(s) (ILRs) and the corotation [10].

However, the global effect of growing central mass concentration on the galactic dynamics was never tested in an evolutionary framework, i.e., when the background potential is adjusted to the changes in density distribution, which itself is closely tied to the evolution of major families of orbits. Below we perform an orbit analysis at selected times in an *evolving* and fully self-consistent galactic potential of a disk embedded in a live halo [11]. We provide a comparison between pure stellar and two-component stars+gas disks using identical initial conditions, so that the different evolutionary paths can be attributed to the presence of the dissipative gaseous component. The model axisymmetric galactic disk is chosen as to experience both bar and bending instabilities.

The numerical method employed in these calculations consists of an *N*-body algorithm to evolve the collisionless component, representing the stars and dark matter, combined with a smoothed particle hydrodynamics (SPH) to evolve the dissipative component, representing the gas. The algorithm, employs such features as a spatially varying smoothing length, a hierarchy of time bins to approximate individual particle timesteps, a viscosity "switch" to reduce the effects of viscous shear, and the special purpose GRAPE-3Af hardware to compute the gravitational forces and the neighbor interaction lists [12]. Further details and tests of this algorithm can be found in Heller and Shlosman [13] and Heller [14], and of the orbit analysis method (in Heller and Shlosman [10]). The potential is prepared by evaluating it on a rectangular 3D grid, from the particle model using the GRAPE hardware. The grid spacing is adjusted with position in order to give an appropriately smooth field without losing relevant features, such as the bar, gas ring, and inner disk. The potential is then symmetrized with eightfold symmetry.

MODELING THE MORPHOLOGICAL EVOLUTION

The initial model was taken as a relaxed configuration of a live disk-halo, derived from analytical model [15]. Details of the relaxation process can be found in Shlosman and Noguchi [6] and Heller and Shlosman [13]. A central object was added to absorb all particles within a radius of 40 pc with its mass growing at their expense. An isothermal equation of state with a temperature of 10^4 K was used for the gas. The units for mass, distance, and time are $M = 10^{11}\ M_\odot$, $R = 10$ kpc and $\tau = 4.7 \times 10^7$ yrs.

PURE STELLAR DISK

The model development, its dynamical instabilities and the overall secular evolution are shown in FIGURES 1 and 2. For the purely collisionless model A, the bar reaches maximum strength at $t \sim 20$, extending out to about 8 kpc with axis ratios of approximately $(x:y:z)$ $1.0:0.3:0.13$. The pattern speed of the bar decreases during the simulations, most rapidly during the bending instability, and then slows to a steady rate of about $\Delta\Omega_b = -0.002$ rad/τ^2, implying a slowdown timescale greater than a Hubble time. The nonlinear orbit analysis (below) confirms the absence of radial ILRs in this model, although linear diagrams erroneously claim a double ILR. There exists a single vertical ILR. At $t \sim 35$ a vertical buckling in the bar becomes visibly discernible as the bar has lost its symmetry with respect to the equatorial $z = 0$ plane. The amplitude of vertical asymmetry grows considerably, until the bar takes on a boxy appearance. The peanut shape of disk viewed along the bar's minor axis is clearly visible during this instability. After $t \sim 60$ the bar remains stable. The new axis ratios indicate both the weakening of the bar in the plane and the vertical thickening.

TWO-COMPONENT DISK

The two-component disk model (model B) experiences a similar bar instability which reaches maximum strength at about the same time as in A. At $t \sim 20$ the stellar distributions are nearly identical in the two models, while the gas in model B has formed a strong shock along the bar, offset in the leading direction and curving inward near the center. This gas morphology is indicative of the presence of an ILR [16], which is confirmed by our orbit analysis.

About 40% of gas, initially within 10 kpc, reside now within the central kpc, representing some 23% of the dynamical mass there. The central accreting object contains ~1.6% of the total mass within 10kpc. Following this large burst of accretion by the central object, it continues to grow linearly over time, reaching, at $t = 100$, $2.3 \times 10^9\ M_\odot$ or 2.2% of the mass inside 10 kpc. This corresponds to a gas density increase by a factor of ~6.5 within the central kpc. Unlike in model A, large spiral features in the gas persist throughout the run of B. The pattern speed of the bar decreases during this interval, after which it remains constant or possibly is even slowly increasing, in agreement with our previous unpublished work.

By $t \sim 30$ the gas circular velocities have developed a sharp discontinuity in the slope at about 1 kpc from the center. The gas rotation increases up to this radius and thereafter remains approximately flat. The central 300 pc are dominated by a growing oval gas disk, whose major axis leads the stellar bar by ~80°. The gas is accumulating close to the inner ILR (as the orbit analysis confirms). Outside this disk is a

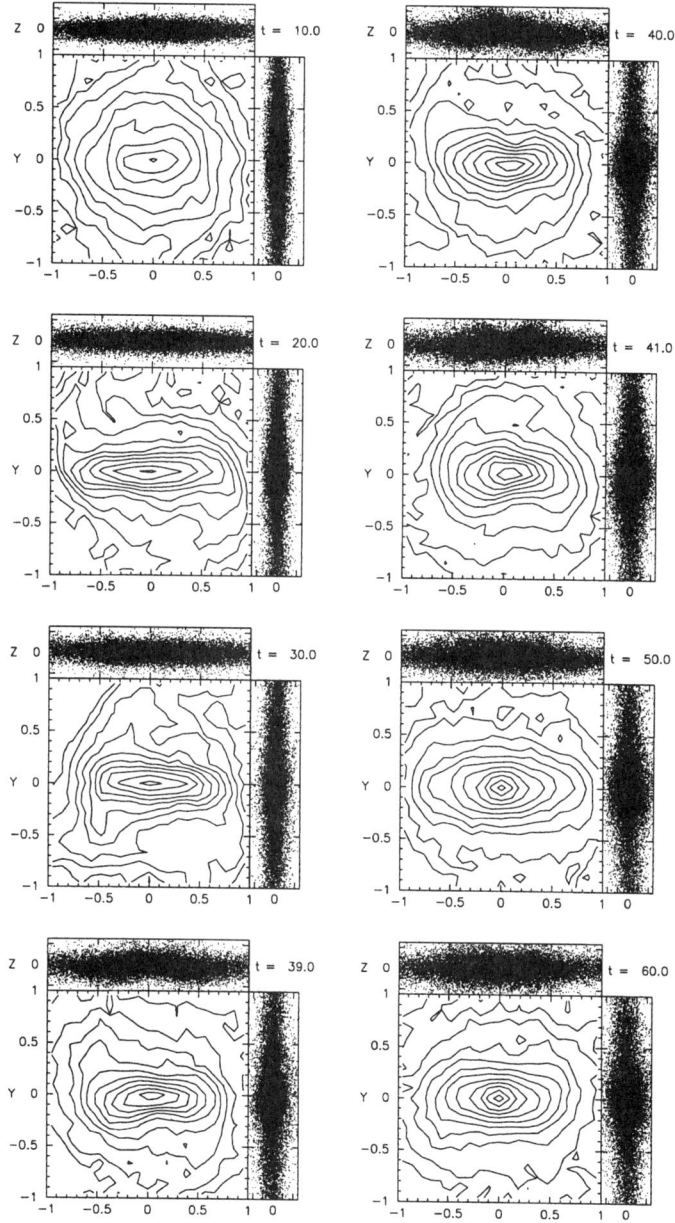

FIGURE 1. Evolution of stellar+gas disk in model B rotating counter-clockwise. Shown are face-on stellar contours, face-on gas distribution, and edge-on stellar distribution. Only ¼ of disk stellar particles are plotted. The bar is along the x-axis.

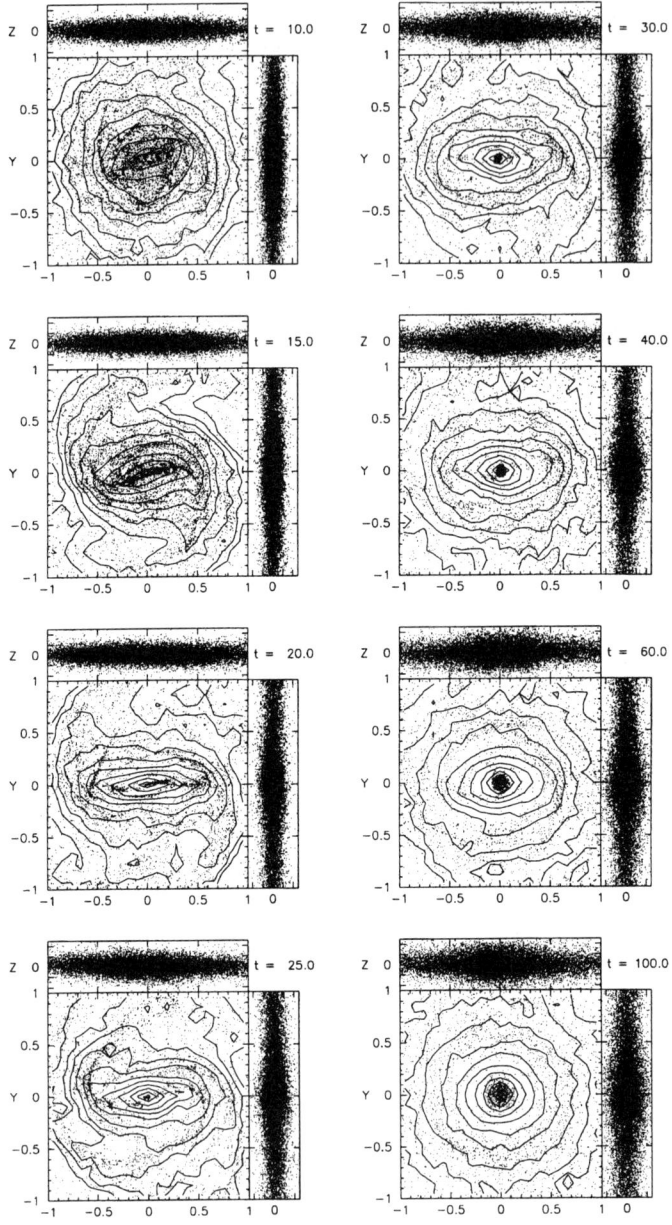

FIGURE 2. Evolution of stellar+gas disk in model B rotating counter-clockwise. Shown are face-on stellar contours, face-on gas distribution, and edge-on stellar distribution. Only ¼ of disk stellar particles are plotted. The bar is along the x-axis.

noticeable deficiency of gas in the bar, up to the radius of a forming oval gas ring which surrounds the stellar bar at about the position of the Ultra-Harmonic Resonance (UHR). Model A has the rectangular outer isophotes (FIG. 1) characteristic of a strong UHR, while in model B they are somewhat weaker (FIG. 2). The inner gas disk and the UHR gaseous ring remain throughout the run, connected by thin trailing gas spiral shocks, offset from the stellar bar in the leading direction. The inner gas disk continues to grow in size, being fed through the shocks and reaching a radius of approximately 1.2 kpc by $t = 100$. Similar gas morphology was obtained in the numerical modeling of M100 by Knapen *et al.* [17]. The weakening of the bar in the model B proceeds linearly with time. A vertical bending of the stellar disk also occurs, but earlier, near the time of maximum bar strength, and is not as dramatic as before. The gas acts to weaken the instability.

NONLINEAR ORBIT ANALYSIS

The linear (epicyclic) approximation gives erroneous results regarding the existense and positions of the IILR(s) even in the case of a moderate strength bar. Instead, we search and analyze the periodic orbits in the frozen potential at a given time. For simplicity, we restrict ourselves to only the lowest-order periodic orbits in the symmetrized potential within the corotation. We compare both the planar and 3D orbits, from each model at different times, at the bar's maximum strength (before the onset of the bending instability), and after the instability (when the evolution is quasi-static).

We test the stability of the simple planar 1-periodic orbits ($z = \dot{z} = 0$), that is planar orbits which are bi-symmetric with respect to the bar and close after one orbit around the center in the rotating frame. Subsequently, orbits which bifurcate in z and $\dot{z}$ from vertically unstable regions are located. The results are displayed in terms of a characteristic diagram, where the orbits are plotted with respect to their Jacobi integral, E_j, and either the y, z, or $\dot{z}$ intercept value with the $x = 0$ plane. The Jacobi integral is a conserved quantity along any given orbit in the rotating frame, and can be thought of as an effective energy. In the characteristic diagrams the orbits form curves or families. It is the study of these families, their properties and how these change during the evolution of the models, that concerns us here.

PURE STELLAR DISK

The characteristic diagrams for the (x, y) planar orbits of model A are shown at $t = 20$ (FIG. 3a) and $t = 65$ (FIG. 3d). The dashed curve denoted by ZVC (the zero velocity curve) delineates the accessible region in the plane based on energy considerations. Stable sections of the characteristics are represented with a solid line, while unstable are broken.

At $t = 20$, the corotation radius is at $E_j = -2.18$ or approximately 9.2 kpc, as defined by the Lagrange point in the orbit model potential. The family labeled x_1 are the orbits that predominately give the bar its structure. They are elongated along the bar and orbit in the same sense as the bar (direct). In the rotating reference frame of

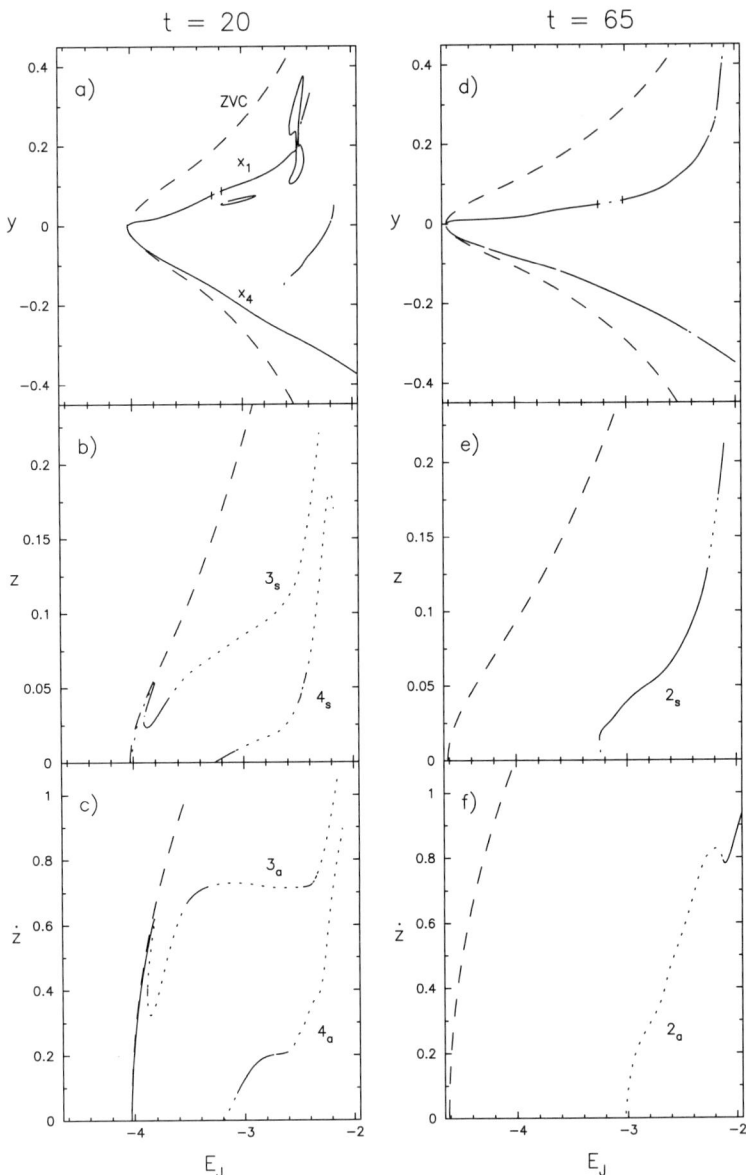

FIGURE 3. Characteristic diagrams for purely stellar disk model A. Stable sections: solid lines; unstable sections: broken lines. The dashed curve is the ZVC. The families consist of symmetric and anti-symmetric $2:n:1$ orbits, where the value of n is given in the diagram for each characteristic.

the bar they move in and out, twice for once around the center (2/1). As one follows the characteristic from the center out to higher energies, the orbits become more and more pointy, until at around $E_j = -3.62$ they develop small loops along the bar, and before the turn-up in the characteristic they shrink and disappear. At this point, the orbits extend to about 8 kpc along the bar and have an axis ratio of 1:4. Past the bend the orbits become increasingly rounded.

Bifurcating from the x_1 sequence near the bend is a bubble consisting of orbits which transition from 2/1 to 3/1 to 4/1 as one moves to higher energies. Most of the orbits have loops and the extreme 4/1 orbits have a pronounced diamond shape. The family labeled x_4 is present in FIGURE 3a (2/1 retrograde orbits). The characteristic is stable along most of the length shown in the figure, except for a few regions at lower energies where it is horizontally unstable. This family plays only a minor role at this early time, since a significant population of retrograde orbits would stabilize the disk against the bar formation.

Bifurcating from a vertically unstable section of the x_1 characteristic are two 3D orbit families. One family bifurcates in z from the low energy edge of the strip at $E_j = -3.26$, the other in $\dot{z}$ from the high energy side at $E_j = -3.20$ (FIGS. 3b, 3c). These orbits make two radial oscillations in the (x, y) plane and four vertical oscillations in z as the orbit closes one rotation about the center. The two families form a pair, with the bifurcations in z and $\dot{z}$ being, respectively, symmetric and anti-symmetric about the (y, z) plane. Using the notation of Sellwood and Wilkinson [18], these orbits are given as 2:4:1, with an s or a subscript added to indicate the sense of symmetry. These orbits have the same shape, in projection onto the (x, y) plane, as the x_1 orbit of the same energy, including the loops. The symmetric family is stable near the bifurcation, but becomes elsewhere. The anti-symmetric family is also unstable over most of its characteristic, except for a section between $E_j = -3.07$ to -2.64. The maximum z extent of the stable orbits of these two families is less than 300 pc.

A pair of 2:3:1 families apparently bifurcates from the origin. One should keep in mind that the gravity has been softened at the center. Both the symmetric and anti-symmetric family stay very close to the zero velocity curve until the orbits extend some 0.5 kpc above the disk plane, then move away, following an abrupt turn-down in their characteristics. The symmetric family is mostly unstable, except for a section following the turn-down. The anti-symmetric family is stable before the turn-down and mostly unstable following it. No x_2 orbits or symmetric 2:2:1 family corresponding to planar and vertical ILRs are present.

FIGURES 3d–3f show the same model at $t = 65$. Here the potential has deepened due to the mass redistribution in the system. The corotation has moved slightly out to $E_j = -1.92$ or about 11.5 kpc. Because the bar has weakened, the planar characteristics are simpler than before. The x_1 orbits are stable from the center out to $E_j = -3.24$, where a vertical instability strip begins. They extend to about 3 kpc along the bar, with an axis ratio of 1:6. At $E_j = -3.02$ the orbits once more become stable and remain so up to near where the characteristic turns upward toward the ZVC. The steep section of the characteristic has varying stability near the increasingly chaotic region of the corotation. No x_2 orbits have been detected, hence no radial ILR(s) exist.

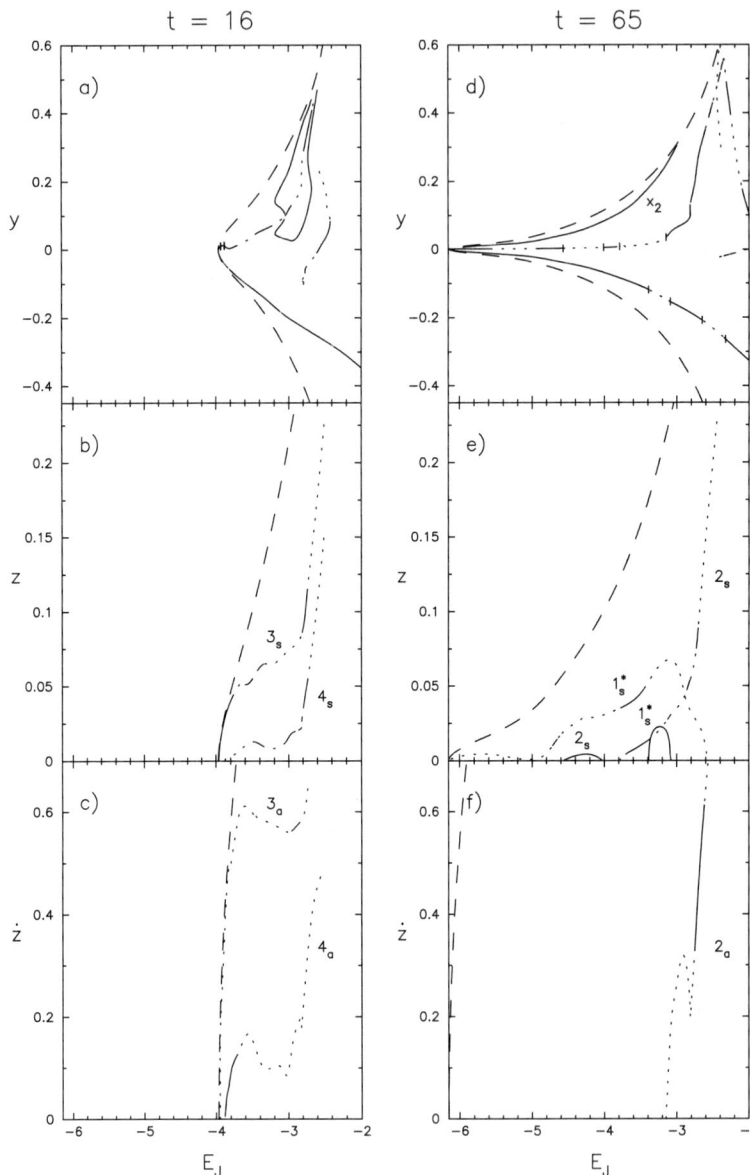

FIGURE 4. Characteristic diagrams for two-component (gas+stars) disk model B. Stable sections: solid lines; unstable sections: broken lines. The dashed curve is the ZVC. The families consist of symmetric and anti-symmetric $2:n:1$ orbits, where the value of n is given in the diagram for each characteristic. The asterisk indicates a retrograde family.

From the vertically unstable strip on x_1, bifurcate two families of 2:2:1 orbits: a symmetric family (BAN) in z at $E_j = -3.24$ and an anti-symmetric family (ABAN) in $\dot{z}$ at $E_j = -3.03$. The symmetric family is unstable near its bifurcation point, but becomes stable following a bend in the characteristic, and remains so up to $z \sim 2$kpc. In contrast the anti-symmetric family is unstable over most of the characteristic shown. The symmetric family bifurcating from x_1 at the instability strip are characteristic of a vertical ILR [3].

TWO-COMPONENT DISK

The disk is highly asymmetrical in the vertical direction when the bending instability is operating. The interpretation of the 3D-orbit families, computed from the symmetrized potential at such a time, would be highly ambiguous. For the full analysis we chose the time $t = 16$, at which the bar is at about 80% of its maximum strength and before the onset of the large gas inflow rate (FIG. 4a). The x_1 orbits between $E_j = -3.87$ and -3.63, or semi-major axes between 1–2v kpc are pinched perpendicular to the bar into a picklelike shape. Along the x_1 sequence there are several regions of instability. Three are horizontally unstable, from $E_j = -3.72$ to -3.43, $E_j = -3.25$ to -3.05, and covering the bend from $E_j = -2.97$ to -2.80. There are also two short vertically unstable sections near the center. Bifurcating from the plane near $E_j = -3.96$ is a pair of 2:3:1 orbit families. Similar to model A, these families at first stay very close to the zero velocity curve. The symmetric family is stable near the bifurcation point, but then alternates between stable and unstable sections, while the

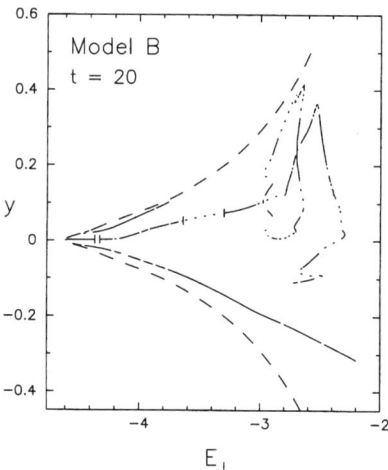

FIGURE 5. Characteristic diagrams for stellar+gas model B at $t = 20$. Stable sections: solid lines; unstable sections: broken lines. The dashed curve is the ZVC. At this time the bar is near maximum strength and planar ILRs exist.

anti-symmetric family is almost everywhere unstable. Also from near the center, at $E_j = -3.88$, bifurcate a pair of 2:4:1 families. The symmetric family is unstable over most of its characteristic, including near the bifurcation point, and has a complex unstable section following the sharp turn-up at $E_j = -2.82$. The anti-symmetric family is stable from it bifurcation point to $E_j = -3.72$.

For comparison with model A, we have also performed an orbit analysis of B at $t = 20$ (at the bar maximum strength; see FIG. 5). At this time the corotation is at $E_j = -2.38$ (~8.2 kpc). In contrast to model A, a family of x_2 orbits already exists due to the deepening of the central potential. The x_2 family are also direct 2/1 orbits, but elongated perpendicular to the bar. The characteristic covers a range in energy from $E_j = -4.45$ to -3.83, along which the orbits extend perpendicular to the bar from 0.18 kpc to 0.92 kpc, with axis ratios of 1:2 and 1:3.

By time $t = 65$ the potential well of model B has greatly deepened, a reflection of the steady accumulation of gas in the nuclear disk. The corotation has moved outward, though not as much as in A, to $E_j = -2.25$ or 9.2 kpc. The sections of instability have grown considerably and now the x_1 orbits are mostly unstable, except in the inner regions. The 4/1 bifurcation from the x_1 family (UHR) occurs at $E_j = -2.36$, at a point where the x_1 orbits have an extent along the bar of around $R_{UHR} = 6.4$ kpc and an eccentricity of ~0.5. This corresponds to the approximate location of the outer edge of the gas ring described earlier. Most of the 4/1 orbits are either unstable or have loops and the gas appears to have accumulated on x_1 orbits with E_j's just below the 4/1 gap. If we take R_{UHR} to be indicative of the maximum possible bar length, then the value of $R_{CR}/R_{UHR} = 1.4$ — within the limits of the empirical range [16].

The extent of the x_2 family has greatly increased from $t = 20$ and now covers a range of energy from essentially the minimum to $E_j = -3.0$, with an axis ratio of around 1:3. The family is stable along its entire length. The x_1 family has two sections of vertical instability before the sharp bend upward at around $E_j = -2.8$. The first has it's two ends at $E_j = -4.52$ and -4.03 connected by a loop of stable symmetrical 2:2:1 orbits (BAN), which bifurcate from the plane in z. These orbits are stable everywhere, span a range along the bar major axis of 0.06 — 1.1 kpc, have a maximum vertical extent of ~100 pc, and have loops in projection onto the (x, y) plane. From the second strip also bifurcate a pair of 2:2:1 families. The symmetric family, which bifurcates in z at $E_j = -3.78$, is unstable between its bifurcation point and $E_j = -3.07$, then alternates three times between stable and complex unstable sections until $E_j = -2.70$, after which it remains predominantly unstable, unlike in model A. The anti-symmetric family, which bifurcates in $\dot{z}$ at $E_j = -3.15$, is unstable between its bifurcation point and $E_j = -2.75$.

SURFACE OF SECTIONS ANALYSIS

The periodic orbits play an important role in galactic dynamics, in that they trap regions of phase space about them. Of course, most orbits in a realistic potential are not periodic. The trapped regions are referred to as regular regions and the orbits which reside in them are called regular orbits. The motions of these orbits are confined to a 2D surface called an invariant tori surrounding the parent periodic orbit. Orbits which are not trapped are called irregular. Unlike the regular orbits, they are not restricted to a subsurface and may wander throughout the nonregular regions of phase space, at least within energy considerations.

To determine the extent of trapping about the stable periodic orbits, we examine the surface of sections (SOS). These diagrams have been constructed by integrating orbits of a given E_j in the *unsymmetrized* potential, and marking a point in the $(y, \dot{y})$ plane each time it crosses the line $x = 0$ with $\dot{x} < 0$ (e.g., see the top left panels of FIGURE 6 in model A and 7 in B, for $E_j = -4.4$ and -5.4). The left side of each diagram, with $y < 0$, represents retrograde orbits, while the right side, with $y > 0$, represents the direct orbits. For the above frames chosen as examples, there are two regular regions in FIGURE 6, which from left-to-right, are associated with the x_4 and x_1 orbital families. In FIGURE 7, there are three regular regions, which from left-to-right, are associated with the x_4, x_1, and x_2 orbital families.

FIGURE 6 shows that the $y > 0$ regular regions of the phase space in the model A are dominated by the x_1 family which dissolves at around $E_j = -2.4$. In contrast, FIGURE 7 shows that most of the regular regions of phase space in B are dominated by the x_2 and x_4 families. At lower energies, where the x_1 family occupies a nonnegligible fraction of the available phase space, the fixed point for the family is at $\dot{y} > 0$, indicating that these orbits are inclined at an angle oblique to the bar in a trailing direction. The general trend of the offset angle is of increasing value as the center of the potential is approached. As one moves out in energy, the fraction of phase space which is regular decreases, with the x_1 family dissapearing by $E_j = -4.2$. At higher energies, as the corotation is approached, the stochastic regions continue to expand ($E_j = -3.8$ and -3.4), with the x_2 family completely dissolving by about $E_j = 3.0$. The regular retrograde orbits continue to fill up most of the available phase space for $y < 0$, so stochastic orbits are not as important here. No significant support for the bar, in terms of regular orbits, appears to exist at this time.

DIVERGING MORPHOLOGICAL AND DYNAMICAL EVOLUTION

The difference in the observed evolution of both numerical models is due to the gravitational effects of gas redistribution in the galactic plane. Although the models differ already at the time of dynamical instabilities, these differences become much more pronounced at the advanced stages of evolution. Dynamical parameters, such as the bar strength, rotation curves and density profiles, show growing differences delineating dissimilar underlying dynamics in both models. Broadly speaking, the trend is that the presence of the self-gravitating gas component in the disk makes both instabilities milder. The evolution is accelerated in the model B compared to A, even before the bar reaches its maximal strength (which happens simultaneously in both models). The bar pattern speed in B is substantially higher than in A after approximately one rotation and this difference increases with time. We note also that the pattern speeds of the stellar bars show different behavior after the initial quick decline — due to the continuous gas inflow to the center, the bar in model B slowly increases its speed, in contrast to that of the bar in A which continues its secular slowdown because of the interaction with the halo and the outer disk. Additional signature of the accelerated evolution of model B is the early onset of the bending instability.

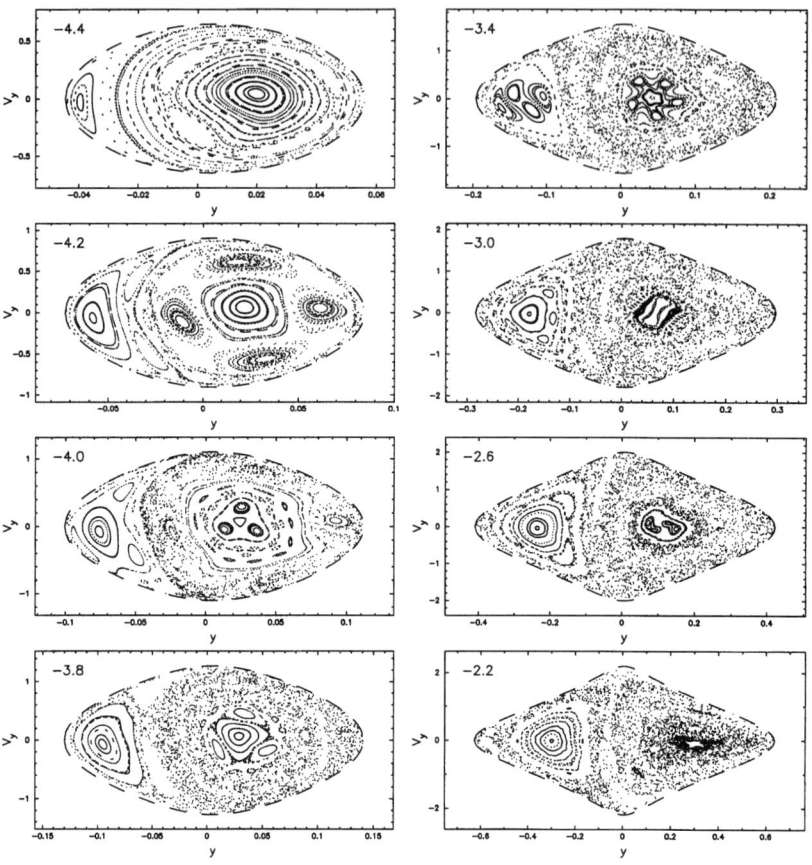

FIGURE 6. Surface of section diagrams for model A at $t = 65$. The direct x_1 and retrograde x_4 families dominate the phase space. At higher energies, as the corotation is approached, the fraction of phase space which is stochastic increases.

The central potential of model B substantially deepens due to the gas inflow, resulting in the formation of the x_2 family of orbits and the radial ILRs. FIGURE 4 shows that the inner resonance region is very broad and extends almost from the center (i.e., from IILR), to 3 kpc in radius (OILR), which encompasses the minor axis of the bar. The inner x_2 orbits in the vicinity of the IILR are heavily populated with gas. The stellar bar weakens as the anti-aligned x_2 orbits become more important. At the peak of its strength, the $m = 2$ mode amplitude is larger by ~50% in A, and the difference grows by a factor of two by $t = 65$, and by ten at $t = 100$. At the later time, the model B bar almost ceases to exist.

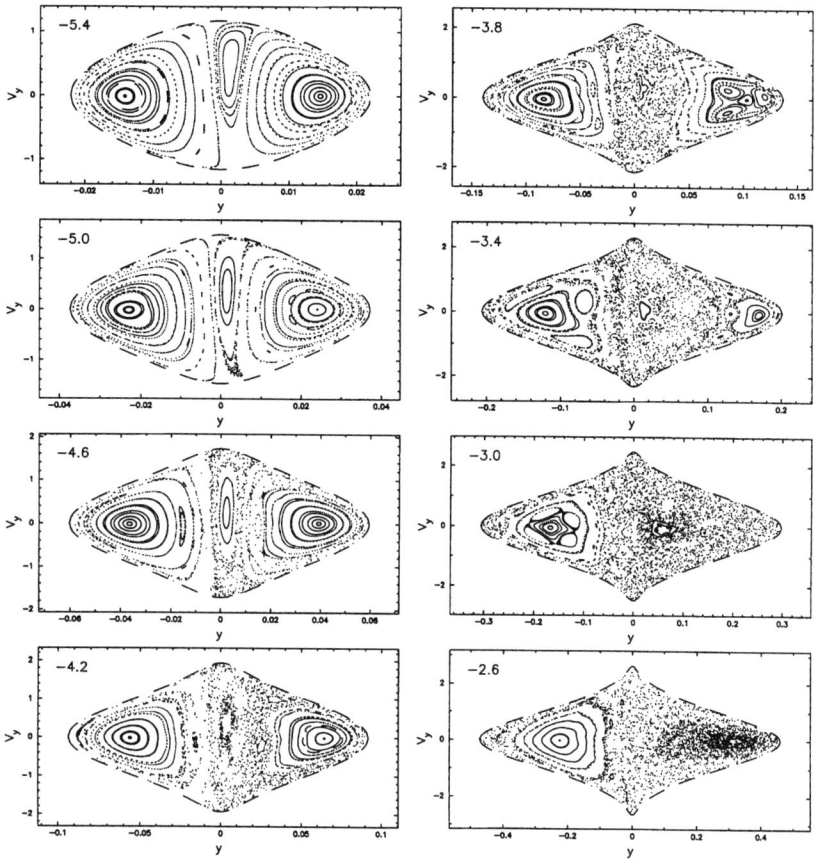

FIGURE 7. Surface of section diagrams for model B at $t = 65$. No significant support for the bar, in terms of regular orbits, exists at this time. The regular regions of phase space are dominated by the x_2 and x_4 families. At energies where the x_1 family traps a nonnegligible fraction of the available phase space (e.g., $E_j = -5.4$ and -5.0), the orbits are significantly inclined to the bar major-axis and provide little support. At higher energies the stochastic regions grow at the expense of the regular regions.

Because the gravitational potential is softened at the position of the IILR, the latter dynamical effects are limited, contrasting the modeling of inner disk in M100 by Knapen et al. [17]. In the latter work, the spacing between the ILRs is much smaller and the position of the IILR is further away from the center, leading to the formation of a nuclear ring just outside the IILR. Both numerical simulations underline the two main alternatives in the gas evolution in the circumnuclear regions of disks.

Additional important morphological difference between models A and B can be found in the properties of the central bulge which forms as a result of vertical buckling in the stellar bar. In particular, the bulge-to-disk ratio for model A is clearly outside the range found by observations, by a factor of ~3, while that for model B is much smaller and in agreement with observed values. Moreover, as the gas falls toward the center in B, the change in value of the bulge shape parameter is consistent with an evolution toward an earlier morphological type.

Morphologies of both models reflect fundamental differences in the orbital dynamics, mostly the change in the relative importance of various orbital families. First, as the SOS's show (FIG. 7), model B at $t = 65$ has a larger fraction of the phase space populated by chaotic orbits. This comes at the expense of the x_1 orbits, which are the most important orbits supporting the bar. The x_1 family itself becomes largely unstable (FIG. 4d), and although it remains present everywhere inside the corotation, it occupies only a small fraction of the phase space and is unable to trap substantial numbers of regular orbits. At these late times the x_1 family is dynamically unimportant.

SOS's of the unsymmetrized potential of model B have shown the presence of the innermost oblique x_1 orbits which dominate the phase space at the lowest energies during late evolution (FIG. 7). These orbits trail the major axis of the stellar bar and are mostly perpendicular to the major axis of the gaseous nuclear disk which leads the stellar bar by about 80°, not supporting the bar. The origin of this anomalous oblique family of x_1 orbits is explained by the fact that at these low energies (and small distances from the center), the nonaxisymmetric potential of the large-scale bar is barely felt, and the Laplace plane of the potential is instead controlled by the nuclear disk.

Most of the x_1 orbits in model B posses loops or appear "pointed" within the 5 kpc region of the bar, while x_1 orbits in A are round everywhere. Because intersecting orbits can not support a steady state gas flow, the gas should lose angular momentum and move inward. By changing the mass distribution in the galactic plane, the gas also affects the vertical structure of the disk by creating vertically unstable gaps in the 2D orbits as well as by destabilizing the 3D families bifurcating from the plane. The symmetric (BAN) and antisymmetric (ABAN) 2:2:1 families of orbits have the largest extensions away from the disk plane and appear to dominate the phase space (FIGS. 3 and 4). These families of 3D orbits bifurcate at the vertical resonance gap in the plane [3]. The vertical ILR is typically found in the neighborhood of the radial ILR [9]. In our case this is true for model B, which has the outer ILR at a distance of about 3 kpc from the center at $t = 65$. Model A has a vertical ILR in approximately the same place, but lacks the planar ILRs.

The orbit analysis reveals that there exists a substantial difference between the models in the stability of BAN orbits and in their extension above the plane and into the halo. Both BAN and ABAN appear with the bending instability and both are present in the models at $t = 65$ (FIGS. 3e, 3f, 4e, and 4f). The bifurcation points of these families moved inward during the evolution of model B, especially that of the BAN orbits. The BAN in model A, however, are stable up to ~2 kpc above the disk, while in the presence of the gas they are destabilized above $z \sim 500$ pc. The ABAN are unstable to large distances above the plane. The fact that in model B the stability region of these families shrinks and is limited to a narrow layer above the disk plane, is important if these families are related directly to the bending instability and the formation of the peanut shape bar profile. This means that the vertical bending in the bar will be (at least partially) damped in the presence of the gas and explains why the bending is substantially milder and reaches a smaller amplitude in the model B. The breaking of the z-symmetry of the bar, clearly observed in FIGURE 1 at $t \sim 39$–40, temporarily eliminates the planar and populates the BAN orbits, giving the bar its characteristic peanut shape. This effect is partially suppressed in B, both because the z-symmetry is maintained to a higher degree (FIG. 2), and because the BAN orbits are largely unstable (except in the immediate vicinity of the disk plane), and so the particles which leave the planar orbits and enter into the halo are populating chaotic orbits. The peanut shape is barely observed in this case.

CONCLUSIONS

We have tested numerically the diverging evolution of a two-component (gas+stars) galactic disk embedded in a live halo with that of an identical pure stellar disk. The orbit analysis confirms that the formation of the inner resonance zone of planar and vertical resonances modifies the circumnuclear orbital dynamics and leads to the weakening of the stellar bar. The major factor appears to be the appearance of the x_2 family of orbits oriented perpendicularly to the bar.

We find that the bar mode amplitude is substantially lower in the presence of the gas at all times, and the amplitude decays steadily on a time scale of $\sim 2 \times 10^9$ yrs. We also find that the bending instability which helps to populate the 3D $2:2:1$ orbits due to the temporary breaking of the z-symmetry in the disk, has a smaller amplitude in the presence of the gas. The characteristic peanut-shape of the inner bar is greatly weakened and "washed-out" due to destabilization of the 3D $2:2:1$ orbits in the more centrally concentrated model containing gas. The orbit analysis shows that the increased stability of the galactic disk should be attributed to the larger population of chaotic orbits following the growth of the central mass concentration to about 2% of the total mass within 10 kpc. Milder buckling of the stellar bar leeds to a much smaller bulge in better agreement with observations, and to possibly a diverging evolution between initially gas-rich and gas-poor barred galaxies.

Our modeling supports the conjecture that the growth of central concentration in galaxies dissolves the main family of regular orbits in the stellar bar and assists in the formation of a galactic bulge. Overall, it hints about evolution toward more axisymmetric disks and earlier morphological types. Taken at the face value, however, one can expect a lower frequency of stellar bars in S0's, compared to Sc's. Such correlation is not supported by optical and near-infrared data.

The characteristic time scale for bar dissolution appears to be short enough in comparison with the Hubble time to the extent that it is difficult to understand the large frequency, $\sim 2/3$, of barred galaxies in the Universe. It is implausible that this frequency can be explained by invoking only galaxy-interaction-induced transient bars, given that bar dissolution time scale is so short. Alternatively, recurrent stellar bars formed via intrinsically-driven instabilities seem to be equally implausible. The resolution of this paradox may lie in deeper understanding of stellar dynamics and, in particular, in the role of stochastic orbits in supporting the stellar bars.

ACKNOWLEDGMENTS

Support under NASA grants WKU-522762-98-6 and NAG5-3841, and HST AR-07982.01-96A is gratefully acknowledged. C.H. is supported by DFG grant Fr 325/39-1, 39-2.

REFERENCES

1. OSTRIKER, J.P. & P.J.E. PEEBLES. 1973. Astrophys. J. **186:** 467.
2. TOOMRE, A. 1966. Geophysical fluid dynamics. *In* Notes on the Summer Study Program at the Woods Hole Oceanographic Institute. Ref. no. 66-46, 111.
3. PFENNIGER, D. & D. FRIEDLI 1991. Astron. Astrophys. **252:** 75.
4. RAHA, N., J.A. SELLWOOD, R.A. JAMES & F.D. KAHN. 1991. Nature **352:** 411.
5. SANDERS D.B., N.Z. SCOVILLE & P.M. SOLOMON. 1985. Astrophys. J. **289:** 373.
6. SHLOSMAN, I. & M. NOGUCHI. 1993. Astrophys. J. **414:** 474
7. SHLOSMAN I., J. FRANK & M.C. BEGELMAN.1989. Nature **338:** 45.
8. HASAN, H. & C.A. NORMAN. 1990. Astrophys. J. **361:**69.
9. HASAN, H., D. PFENNIGER & C.A. NORMAN. 1993. Astrophys. J. **409:**91.
10. HELLER C.H. & I. SHLOSMAN. 1996. Astrophys. J. **471:** 143.
11. BERENTZEN, I., C.H. HELLER, I. SHLOSMAN & K.J. FRICKE. 1998. Mon. Not. R. Astr. Soc. **300:** 49.
12. STEINMETZ, M. 1996. Mon. Not. R. Astr. Soc. **278:**1005.
13. HELLER, C.H. & I. SHLOSMAN. 1994. Astrophys. J. **424:** 84.
14. HELLER, C.H. 1995. Astrophys. J. **455:** 252.
15. FALL, S.M. & G. EFSTATHIOU. 1980. Mon. Not. R. Astr. Soc. **193:** 189.
16. ATHANASSOULA E. 1992. Mon. Not. R. Astr. Soc. **259:** 345.
17. KNAPEN J.H., J.E. BECKMAN, C.H. HELLER, I. SHLOSMAN & R.S. DE JONG. 1995. Astrophys. J. **454:** 623.
18. SELLWOOD, J.A. & A. WILKINSON. 1993.Rep. Prog, Phys. **56:** 173.

Counterrotating Galaxies and Accretion Disks

R.V.E. LOVELACE[a]

Department of Astronomy, Cornell University, Ithaca, New York 14853-6801

ABSTRACT: Theoretical interest in astrophysical disks with counterrotating components of stars and/or gas has been stimulated by recently discovered counterrotating spiral and S0 galaxies. A variety of physical processes can occur in counterrotating disks. We have shown that a strong two-stream instability can occur for one armed ($m = 1$) tightly-wrapped spiral waves between co and counterrotating stellar components and/or between a corotating stellar component and a counterrotating gaseous component. The instability of counterrotating stellar components has been clearly seen in computer simulations. The unstable two-stream spiral waves can provide an effective viscosity for the gas causing its rapid accretion.

Accretion disks consisting of counterrotating gaseous components may exist with an intervening shear layer. Configurations of this type can arise from the accretion of newly supplied counterrotating gas onto an existing corotating gas disk. For example, the gas above the disk midplane can rotate with angular rate $+\Omega(r)$ while that well below has the same properties but rotates with rate $-\Omega(r)$. Using the Shakura-Sunyaev alpha turbulence model, we find self-similar solutions where a thin (relative to the full disk thickness) equatorial layer accretes very rapidly, essentially at free-fall speed. This type of accretion flow has now been observed in hydrodynamic simulations.

1. INTRODUCTION

A remarkable finding of recent high spectral resolution studies of normal galaxies is the occurrence of counterrotating gas and/or stars in galaxies of all morphological types, ellipticals, spirals, and irregulars (see reviews by Rubin [26] and Galletta [13]). In the ellipticals, the counterrotating component is usually in the nuclear core and may result from merging of galaxies with opposite angular momentum. Newly supplied gas with misaligned angular momentum in the nuclear region of a galaxy may be important in the formation of radio loud quasars if there is a rotating massive black hole at the galaxy's center [28]. In contrast, in a number of spiral and S0 galaxies, counterrotating disks of gas and/or stars have been found to coexist with the primary disk out to large distances (10–20 kpc). Examples include NGC 4550 [25], [24] , NGC 4826 [6], NGC 7217 [20], NGC 4546 [27], NGC 3626 [11], NGC 3593 [4], and NGC 4138 [15]. Another situation where there may be counterrotating disks is in low mass X-ray binary sources where the accreting, magnetized rotating neutron stars are observed to jump between states where they spin-up and those where they spin-down. Nelson *et al.* [22] and Chakrabarty *et al.* [8] have proposed that the change from spin-up to spin-down results from a reversal of the angular momentum of the wind supplied accreting matter.

[a]Address for telecommunication: rvl1@cornell.edu

Thakar and Ryden [31] discuss different possibilities for the formation of counterrotating galaxies, (1) that the counterrotating matter may come from the merger of an oppositely rotating gas rich dwarf galaxy with an existing spiral, and (2) that the accretion of gas may occur over the lifetime of a galaxy with the more recently accreted gas counterrotating. Important theoretical questions include: understanding the interactions between stars rotating in one direction ($+\Omega(r)$) and stars and/or gas rotating in the opposite direction ($-\Omega(r)$). We discuss in Section 2 the two-stream instability between coexisting counterrotating stars [19]. In Section 3 we discuss the theory of counterrotating gaseous accretion disks [18]. In Section 4 we present results of hydrodynamic simulations of counterrotating accretion disks [17].

2. TWO-STREAM INSTABILITY THEORY

We first give a brief summary of the linear WKB theory of tightly-wrapped spiral waves in a single component galaxy of stars rotating with angular rate $\Omega(r) > 0$ (see, e.g., [5] or [23]). We use an inertial, cylindrical (r, ϕ, z) coordinate system and assume a thin disk galaxy. In the midplane of the galaxy the perturbation of the gravitational potential is

$$\delta\Phi(r, \phi, 0, t) = C\exp\left[i\int^r dr' k_r(r') + im\phi - i\omega t\right] \quad (1)$$

where the radial wavenumber k_r satisfies $|k_r r| \gg 1$ for a tightly wrapped wave, $C(r)$ a slowly varying function of r, $m = \pm 1, \pm 2, \ldots$ is the number of spiral arms, and ω is the angular frequency of the wave. Only $m > 0$ need be considered because $\delta\Phi^*$ is a valid solution if $\delta\Phi$ is. Then, $k_r > 0$ (< 0) corresponds to a trailing (leading) spiral wave. The thin disk assumption requires $|k_r h| \ll 1$, where h is disk half-thickness. A well-known calculation (see [5] or [23]) of the linear perturbation of the dynamical equations leads to the dispersion relation

$$0 = \epsilon(\omega, k_r) \equiv 1 + \mathcal{P}_*(\omega, k_r), \quad (2)$$

where

$$\mathcal{P}_* = \frac{2|\bar{k}_r|\exp(-X_*)}{X_*} \sum_{l=1,2,\ldots} \frac{I_l(X_*)}{(s/l)^2 - 1}, \quad s \equiv (\omega - m\Omega)/\kappa,$$

$$\kappa^2 \equiv \frac{1}{r^3}\frac{d}{dr}(r^4\Omega^2), \quad \bar{k}_r \equiv k_r/k_{\rm crit}, \quad k_{\rm crit} \equiv \frac{\kappa^2}{2\pi G\Sigma_{\rm tot}},$$

$$X_* \equiv \left(\frac{k_r\sigma_r}{\kappa}\right)^2 = 0.28568(Q_*\bar{k}_r)^2, \quad Q_* \equiv \frac{\kappa\sigma_r}{3.3583 G\Sigma_{\rm tot}}$$

Here, $\epsilon(\omega, k_r)$ has the role of a dielectric function for the disk, and $\mathcal{P}_*$ the polarization function (the *–subscript denotes stars); s is a dimensionless frequency; $\kappa(r)$ is

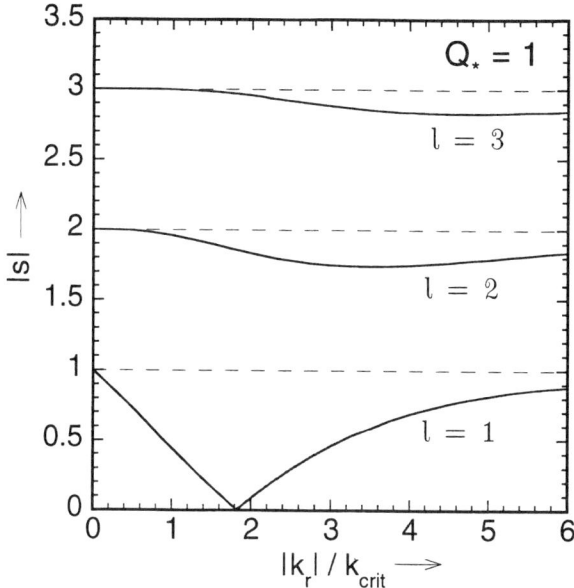

FIGURE 1. The figure shows the frequency $[s \equiv (\omega - m\Omega)/\kappa]$ – wavenumber (k_r) dependence of the tightly wrapped spiral wave modes in a disk of stars [19]. Here, Ω is the angular velocity of the stars, κ is their epicyclic frequency, k_{crit} is the critical wavenumber, and l labels different branches of the dispersion relation. The higher branches ($l \geq 2$) are analogues of the Bernstein modes in a magnetized plasma as discussed in the text.

the radial epicyclic frequency of a star; k_{crit} is a characteristic wavenumber, and $\overline{k}_r$ is the dimensionless radial wavenumber; $\Sigma_{tot}(r)$ is the total surface mass density of the disk; $Q_*(r)$ is Toomre's stability parameter; $\sigma_r(r)$ is the radial dispersion of the star velocities (for a stellar distribution function $f_* \propto \exp[-v_r^2/(2\sigma_r^2)]$); and I_l is the usual modified Bessel function of order l. The condition for axisymmetric ($m = 0$) stability is $Q_* > 1$ [32]. The influence of halo matter enters through $\Omega(r)$.

FIGURE 1 shows wavenumber (k_r) — frequency (s) plots for two values of Q_* obtained from Eq. (2). The corresponding plot in reference [5, Figure 6-14a] and in [23, Fig. 12.2] is *incomplete* because the higher-order branches, labeled $l = 2, 3, ...$, have been omitted. Further, the statement in [5, p. 367] that "a stellar disk has no pressure forces and therefore cannot support waves with $|s| > 1$" is incorrect. The branches $l = 2, 3, ...$, are the analogues of the well-known Bernstein modes which propagate across a uniform magnetic field in a collisionless plasma [3], [16]. Here, the coriolis force is the analogue of the Lorentz force in a magnetized plasma. The Bessel functions I_l arise from the sinusoidal particle motion in the sinusoidal wave field and the averaging over the Gaussian distribution function of v_r. The numerators for the different l terms can be written as $(3.74/Q_*)g_1(X_*)$ with $g_1 \equiv \exp(-X_*)I_1(X_*)/\sqrt{X_*}$. Note that $\sqrt{X_*}$ is the ratio of the amplitude of the radial epicyclic motion to the radial wavelength of the wave. The maximum of $g_1(X_*)$ is ≈ 0.222 (at $X_* \approx 0.584$), while that of g_2, ≈ 0.0665 (at $X_* \approx 2.3$), is smaller by a factor about 0.3, but this may be offset by a small denominator. Note that these higher-order modes give rise to a richer set of Linblad resonances corresponding to $s = \pm l$ or $\Omega_p = \Omega \pm l\kappa/m$, where $\Omega_p \equiv \omega/m$ is the pattern angular speed, $m = 1, 2, \ldots$, and $l = 2, 3, \ldots$.

The dispersion relation for a two-component galaxy of coexisting corotating ($+\Omega$) stars and a mass fraction $\xi_* \leq 1/2$ of counterrotating stars ($-\Omega$) is [19]

$$0 = \epsilon(\omega, k_r) \equiv 1 + (1 - \xi_*)\mathcal{P}_*(s, k_r) + \xi_*\mathcal{P}_*(s+w, k_r), \quad (3)$$

where $w \equiv 2m\Omega/\kappa$ and $\mathcal{P}_*$ is defined in equation (2). The corotating star modes are given by $0 = 1 + (1 - \xi_*)\mathcal{P}_*(s, k_r)$, while the counterrotating star modes are given by $0 = 1 + \xi_*\mathcal{P}_*(s+w, k_r)$. From Eq. (12), we find that the one-armed spiral waves ($m = 1$) can be strongly unstable [19]. The mode lines shown in FIGURE 2a do not cross but "attract" near the circled point to give instability as shown in FIGURE 2b. The exact symmetry $\phi \to -\phi$ for $\xi_* = 1/2$ makes it evident that the co- and counterrotating mode lines in FIGURE 2a must merge at $s = -\Omega/\kappa$ to give $\omega_r = 0$ (independent of Ω/κ) in the range of $\overline{k}_r$ where $\omega_i > 0$. Thus the $m = 1$ instability in this case is an *absolute* instability. In this respect it is similar to the $m = 0$ instability of Toomre [32]. Trailing and leading wave perturbations can be superposed to give a standing wave in place of Eq. (1), $\delta\Phi = C\cos(\int^r dr' k_r(r'))\sin(\phi)\exp(\omega_i t)$. FIGURE 3 shows the dependence of the maximum growth rate on Q_* for $\xi_* = 1/2$. Note that in a homoge-

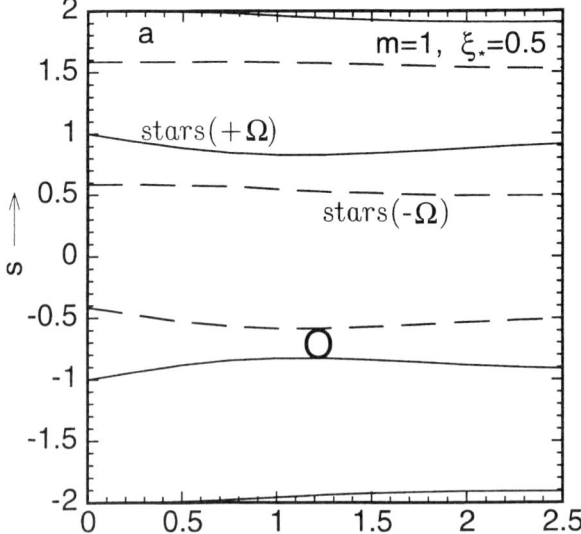

FIGURE 2. The figure shows the nature of the one-armed ($m = 1$) modes in a galaxy of corotating stars ($+\Omega$) and an equal mass ($\xi_* = 1/2$) of counterrotating stars ($-\Omega$) [19]. The horizontal axis is k_r/k_{crit}. (**a**) shows the mode lines, and the circle indicates the unstable near crossing of mode lines. (**b**) shows the real and imaginary parts of the wave frequency ω as a function of $\overline{k}_r$ obtained from Eq. (3). For both star components, $Q_* = 1.4$.

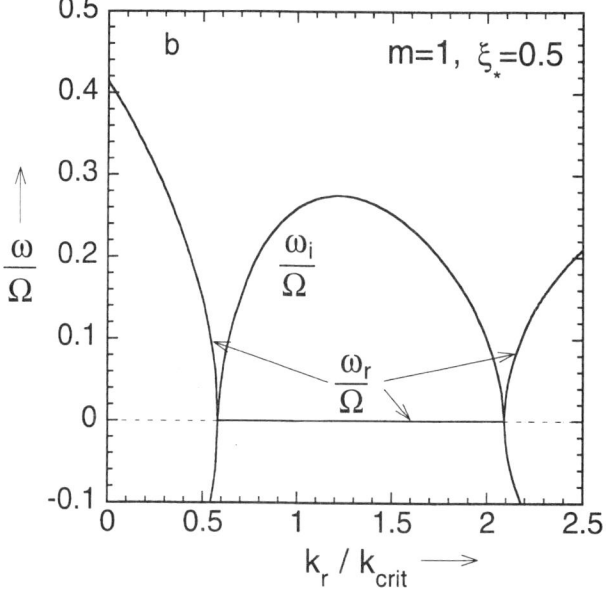

FIGURE 2b.

nous self-gravitating medium the two-stream instability is always dominated by the Jeans instability [1], [19].

With the help of Professor Contopoulos, I obtained an approximate analytic solution to Eq. (12) for the $m = 1$ instability for the symmetrical case $\xi_* = 1/2$ by neglecting the $l \geq 2$ terms. This gives

$$0 = \epsilon \approx 1 + \frac{A}{(\varpi - c)^2 - 1} + \frac{A}{(\varpi - c)^2 - 1}, \qquad (4)$$

where $\varpi \equiv \omega/\kappa$, $c \equiv \Omega/\kappa$ ($= 1/\sqrt{2}$ for a flat rotation curve), and $A \equiv |k_r| \exp(-X_*)I_1(X_*)/X_* = (1.871/Q_*)\exp(-X_*)I_1(X_*))/\sqrt{X_*}$. Expanding ϵ gives a quadratic in ϖ^2 which can be solved to give $\varpi^2 = 1 + c^2 - A \pm (4c^2 - 4c^2A + A^2)^{1/2}$ which shows that there is instability for $A > (1 - c^2)/2$ or $A > 1/4$ for a flat rotation curve. The maximum of $\exp(-X_*)I_1(X_*)/\sqrt{X_*}$ is ≈ 0.222 at $X_* \approx 0.584$ so that $A \approx 0.415/Q_*$. Thus for $Q_* = 1$, the growth rate is $\omega_i/\Omega \approx 0.38$ which is less than the value (0.49) of FIGURE 3 obtained from the full dispersion relation. The two-stream instability between counterrotating stellar components has been cleary identified and studied in simulations using ~10^5 particles by Comins et al. [10].

For a galaxy with both counterrotating stars and gas, the largest amplification is for the $m = 1$ leading spiral waves, and it occurs when the counterrotating mass fraction of gas plus stars is $\approx 1/2$ [19]. Possible nonlinear effects which may act to limit the wave growth and amplification include scattering of star orbits by the wave which increases Q_*, and heating of the gas. A residual level of $m = 1$ spiral waves

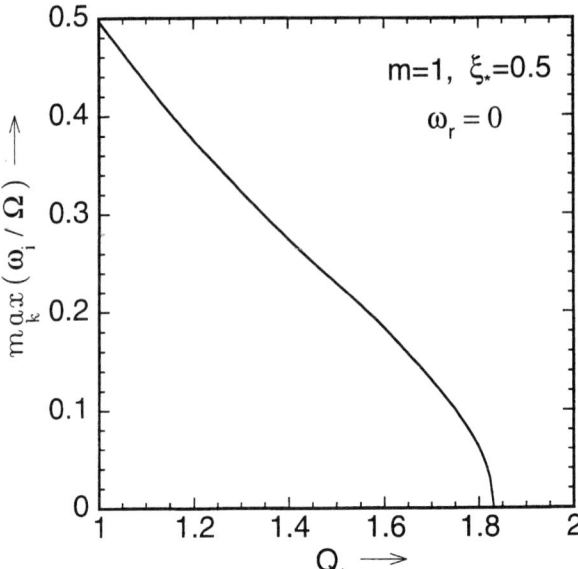

FIGURE 3. The figure shows the Q_* dependence of the maximum growth rate of the one-armed waves for a galaxy of equal mass co- and counterrotating components, $\xi_* = 1/2$ [19].

may remain excited which give an effective viscosity for the gas causing its enhanced accretion.

3. COUNTERROTATING ACCRETION DISKS

The widely considered models of accretion disks have gas rotating in one direction with a turbulent viscosity acting to transport angular momentum outward [29], [30]. The above mentioned observations point to more complicated disk structures on a galactic scale, in galactic nuclei, and perhaps in other accreting systems.

Here, we discuss accretion disks consisting of counterrotating gaseous components with gas at large z rotating with angular rate $+\Omega(r)$ and gas at large negative z rotating at rate $-\Omega(r)$ [18]. The interface between the components at $z \sim 0$ constitutes a supersonic shear layer and is sketched in FIGURE 4. A configuration similar to FIGURE 4 may arise from the accretion of newly supplied counterrotating gas onto an existing corotating disk. It might at first be thought that powerful Kelvin-Helmholtz instabilities would heat the gas to escape speed and rapidly destroy the assumed configuration. However, supersonic shear layers exist and exhibit gross stability in stellar and extra-galactic jets [14]. In the counterrotating disk, matter approaching the equatorial plane from above and below has angular momenta of opposite signs with the result that there is angular momentum annihilation at $z = 0$, the matter looses its centrifugal support and accretes at essentially free-fall speed. On the other hand, ac-

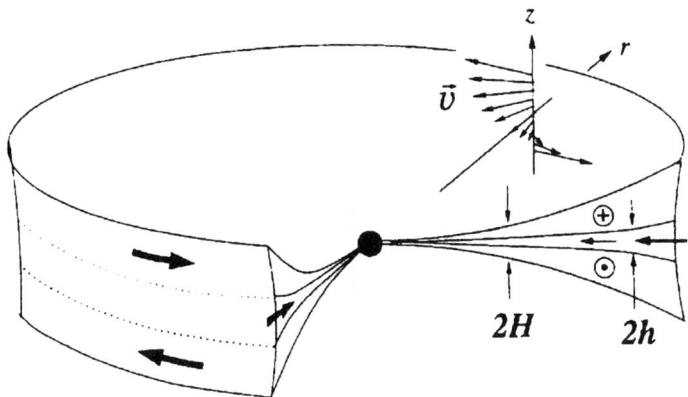

FIGURE 4. Structure of two apposed, counterrotating accretion disks and the midplane boundary layer. The inset shows the three dimensional view of the velocity field for the $n = 1/2$ case shown in FIGURE 5 [18]. The velocity variation is analogous to that in the Ekman layer of a rotating fluid (such as the ocean) where the Coriolis force balances the viscous force (see, e.g., [2]).

cretion disks rotating in one direction are modeled assuming a turbulent viscosity which is crucial for the outward transport of angular momentum [30]. The counter-rotating disks can also be expected to be turbulent owing in part to the Kelvin-Helmholtz instability, and turbulent viscosity can transport angular momentum outward in the large $|z|$ regions of the disk.

We consider a model of a stationary viscous disk consisting of co- and counter-rotating gas. For axisymmetric disk-like flows in cylindrical coordinates (r, ϕ, z) with $\mathbf{v} = [v_r(r, z), v_\phi(r, z), v_z(r, z)]$, the momentum and continuity equations are:

$$\rho \frac{\partial v_r}{\partial t} + \rho(\mathbf{v} \cdot \nabla) v_r = \frac{\rho v_\phi^2}{r} - \frac{\partial \tilde{p}}{\partial r} + \rho g_r + \frac{2}{r}\frac{\partial}{\partial r}\left(r\eta \frac{\partial v_r}{\partial r}\right) + \frac{\partial}{\partial z}\left[\eta\left(\frac{\partial v_r}{\partial z} + \frac{\partial v_z}{\partial r}\right)\right] - \frac{2\eta v_r}{r^2}, \quad (5)$$

$$\rho \frac{\partial v_\phi}{\partial t} + \rho(\mathbf{v} \cdot \nabla) v_\phi = \frac{-\rho v_r v_\phi}{r} + \frac{1}{r^2}\frac{\partial}{\partial r}\left(r^3 \eta \frac{\partial(v_\phi/r)}{\partial r}\right) + \frac{\partial}{\partial z}\left(\eta \frac{\partial v_\phi}{\partial z}\right), \quad (6)$$

$$\rho \frac{\partial v_z}{\partial t} + \rho(\mathbf{v} \cdot \nabla) v_z = \frac{\partial \tilde{p}}{\partial z} + \rho g_z + \frac{1}{r}\frac{\partial}{\partial r}\left[r\eta\left(\frac{\partial v_r}{\partial z} + \frac{\partial v_z}{\partial r}\right)\right] + 2\frac{\partial}{\partial z}\left(\eta \frac{\partial v_z}{\partial z}\right). \quad (7)$$

$$\frac{\partial \rho}{\partial t} + \frac{1}{r}\frac{\partial}{\partial r}(r\rho v_r) + \frac{\partial}{\partial z}(\rho v_z) = 0. \quad (8)$$

Here, $\rho(r, z)$ is the gas density, $\tilde{p}(r, z) \equiv p + \frac{2}{3}\eta(\nabla \cdot \mathbf{v})\delta_{ij}$ with p the pressure, $\mathbf{g} = -\nabla \Phi$ the gravitational acceleration with $\Phi(r, z)$ the potential, $\eta = \rho \nu$ the dynamic viscosity, and ν the kinematic viscosity. Because the microscopic viscosity is negligible for the conditions considered, we assume ν arises from small scale turbulence and can be approximated by the "alpha" prescription of Shakura [29] and Shakura and Sunyaev [30] as $\nu = \alpha c_s H$, where H is the full half-thickness of the disk which is assumed thin ($H \ll r$), c_s the sound speed, and α a dimensionless constant much less than unity. We also assume the second viscosity is zero so that the stress tensor $T_{ij} = \rho\, v_i v_j + p\delta_{ij} + T_{ij}^{\nu}$ with the viscous contribution $T_{ij}^{\nu} = -\rho\nu[\partial v_i/\partial x_j + \partial v_j/\partial x_i - \frac{2}{3}(\nabla \cdot \mathbf{v})\delta_{ij}]$.

We have found self-similar solutions to Eqs. (5)–(8) of the form[18]

$$v_r(r,z) = -u_r(\zeta)V_c(r), \quad v_\phi(r,z) = u_\phi(\zeta)V_c(r), \quad v_z(r,z) = -\frac{h(r)}{r}u_z(\zeta)V_c(r), \tag{9}$$

$$\rho(r,z) = \rho_0(r)Z\!\left[\frac{z}{H(r)}\right],$$

where $\zeta \equiv z/h(r)$ is the dimensionless vertical distance in the disk with $h(r)$ a length scale identified subsequently, and $V_c(r) = [r(\partial\Phi/\partial r)|_{z=0}]^{1/2}$ is the circular velocity of the gas. The potential is due in general to disk and halo matter and a central object. Here and subsequently we neglect the pressure force in the radial equation of motion. We consider cases where $V_c(r) = (GM/r_0)^{1/2}(r_0/r)^n$, with r_0 a constant length scale. The value $n = 1/2$ corresponds to a Keplerian disk around an object of mass M, and $n = 0$ to a flat rotation curve applicable to flat galaxies. We assume that the disk is not strongly self-gravitating and that it is vertically isothermal so that its overall half-thickness is $H \sim (c_s/V_c)r$ which is the same as for a disk rotating in one direction. Thus, for the "alpha" viscosity, $\eta = \alpha\, p/\Omega_c$, where $\Omega_c \equiv V_c/r$.

Substitution of Eq. (9) into Eqs. (5)–(8) gives

$$u_r'' = u_\phi^2 + nu_r^2 - 1 + k_h\zeta u_r u_r' - u_z u_r' + 2\delta^2\zeta u_r' + O(\epsilon^2), \tag{10}$$

$$u_\phi'' = (n-1)u_r u_\phi - u_z u_\phi' + k_h\zeta u_r u_\phi' + 2\delta^2\zeta u_\phi' + O(\epsilon^2). \tag{11}$$

Here, a prime denotes a derivative with respect to ζ, $k_h \equiv (r/h)(dh/dr)$, $\epsilon \equiv h/r \ll 1$, and $\delta \equiv h/H$. We have made the natural identification $h^2 = \nu r/V_c$ which implies $\delta = \sqrt{\alpha}$. We consider the ordering $\epsilon^2 \ll \delta^2 \ll 1$. For $\delta^2 \ll 1$, Eq. (7) simplifies to $\partial_z p = -\rho\partial_z\Phi$, which determines the isothermal density profile $Z(\zeta) = \exp(-\delta^2\zeta^2)$. From the assumed scaling $\rho_0(r) \propto r^{-\beta-1/2}$, we have

$$u_z' = [\beta + n - \tfrac{1}{2} - 2k_H\delta^2\zeta^2]u_r + k_h\zeta u_r' + 2\delta^2\zeta u_z, \tag{12}$$

where $k_H \equiv (r/H)(dH/dr)$. For consistency of the self-similar solutions, δ and k_h must be independent of r which implies $k_H =$ constant.

Equations (10)–(12) constitute a closed system for the dimensionless functions $u_r(\zeta)$, $u_\phi(\zeta)$, and $u_z(\zeta)$. Applying (10)–(12) to the flow suggested in FIGURE 4, we infer that $u_r(\zeta)$ is an even function of ζ, while $u_\phi(\zeta)$ and $u_z(\zeta)$ are odd functions. For $|\zeta| \gg 1$, we impose $u_r \to 0$ and $u_\phi \to \pm 1$ so as to have $u_r'' \to 0$ and $u_\phi'' \to 0$. In a more complete treatment with the $O(\epsilon^2)$ terms retained in Eqs. (10) and (11), the solution at large $|\zeta|$ approaches that of Shakura and Sunyaev [30] with $u_r = O(\epsilon^2)$. FIG-

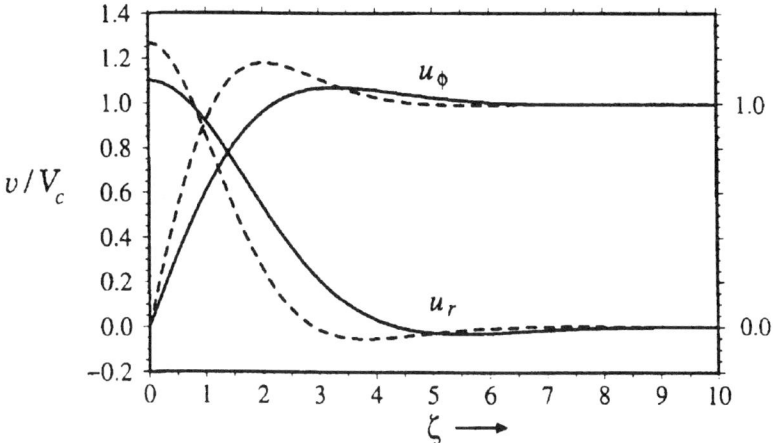

FIGURE 5. Solutions to Eqs. (10)–(12) for $\delta = 0.1$ as a function of $\zeta = z/h(r)$. The solid lines correspond to a Keplerian disk ($n = 1/2$, $\beta = 1$) with $u_r(0) = 1.10011267$ and $u_\phi'(0) = 0.66265042$. The dashed curves correspond to $n = 0$ (flat rotation curve) and $\beta = 3/2$ and have $u_r(0) = 1.26728062$ and $u_\phi'(0) = 1.12573559$. These numerical values are obtained by a shooting method [18].

URE 5 shows solutions to Eqs. (10)–(12) for $\delta = h/H = 0.1$. Both Keplerian ($n = 1/2$, $\beta = 1$) and galactic ($n = 0$, $\beta = 3/2$) solutions are shown. Note that for $3 \lesssim |\zeta| \lesssim 7$, the solutions have a region with $u_r < 0$ indicating spiraling outflows on both sides of the midplane. For large ζ, the density fall off gives $\rho v_z \to 0$. The solutions remain qualitatively the same for small δ and small, nonzero ϵ.

For an alpha disk rotating in one direction, the mass accretion rate is $\dot{M}_{SS} = 2\pi r \Sigma|v_r|_{SS}$, with the accretion speed $|v_r|_{SS} \sim \alpha c_s H/r \ll c_s$, where $\Sigma = \int dz \rho$ is the disk's surface mass density [30]. In contrast, for a counterrotating disk $\dot{M}_{CR} = 2\pi r \Sigma|v_r|_{CR}$, with accretion speed (averaged over z) $|v_r|_{CR} \sim (h/H)V_c$. For the same α, we have $|v_r|_{CR} \gg |v_r|_{SS}$: the accretion speed of the counterrotating disk is *much larger* than that of the disk rotating in one direction.

4. HYDRODYNAMIC SIMULATIONS OF COUNTERROTATING DISKS

Axisymmetric accretion of counterrotating disks have been simulated using the inviscid hydrodynamic (Euler) equations in spherical, inertial coordinates (R, ϕ, θ), with $\theta = 0$ the equatorial plane [17]. Of course, the finite grids used give a numerical viscosity which acts like alpha viscosity in essential respects. So far we have investigated only Keplerian disks around central mass M. An explicit Lax-Friedrichs finite-difference scheme with flux correction in Chakravarthy-Osher [9] form was used. This scheme is first order of approximation in time and third order of approximation in space and is oscillation-free near discontinuities. For simulations, we used a non-homogeneous (R, θ) grid, where radial grid increment δR increases outwards exponentially: $\delta R = \delta R_0 a^{R/R_0}$, where $a = 1.05$. The θ grid $\delta\theta$ increases in a

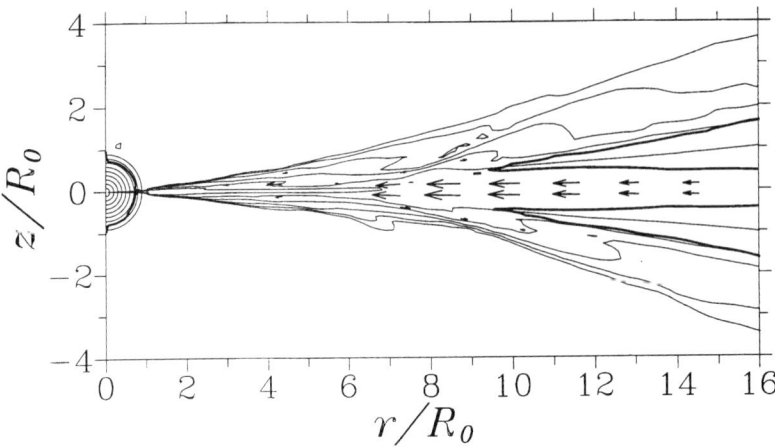

FIGURE 6. Iso-density contours (solid lines) and mass flux vectors ($\rho\mathbf{v}$) in counterrotating accretion disk at time $t = 14t_0$, where t_0 is the rotation period at the inner radius of the disk (from [17]). The poloidal flow near the midplane of the disk is supersonic, excluding the small regions separated by the bold lines.

similar way going away from the equatorial plane. In simulations we used grids of $N_R \times N_\theta$ of 100×100 or 200×200 in some cases.

The case we describe here is almost isothermal and is close to the analytical model of Lovelace and Chou [18]. We took equilibrium initial conditions for the density and velocity of the disk, but put the angular velocity in the lower-half of the disk ($z < 0$) opposite to the angular velocity of the upper-half ($z > 0$). FIGURE 6 shows that matter starts to rapidly flow radially inward in a narrow region near the midplane of the disk. The poloidal flow near the midplane is supersonic, and the radial inflow velocity increases as r decreases in approximate agreement with the analytical model.

The change from spin-up to spin-down of X-ray pulsars in binary systems may results from a reversal of the angular momentum of the wind supplied accreting matter [22], [8]. The matter inflowing to the pulsar is assumed to be supplied at a rate $\dot{M}$ at a radius r_{out} in the equatorial plane of the binary system. Initially, the pulsar has a conventional, corotating disk. Then, due to some disturbance, the sense of rotation of the supplied matter reverses so that it is counterrotating. The "new" counterrotating matter encouters the "old" corotating disk. Interaction between the "new" and "old" matter leads to fast boundary layer accretion with average accretion speed $u_{\text{CR}} \sim 3.25(h/H)V_K$, where the 3.25 factor is estimated from the simulations. The time-scale for this fast accretion to reach the pulsar is

$$t_{\text{CR}} = \int_0^{r_{\text{out}}} \frac{dr}{u_{\text{CR}}} \sim 2\text{d}\left(\frac{M_\odot}{M}\right)^{\frac{1}{2}}\left(\frac{r_{\text{out}}}{10^{12}\text{cm}}\right)^{\frac{3}{2}}, \qquad (13)$$

where we have assumed $h/H = \text{const} = 0.1$, where h is the boundary layer half-thickness and H is the disk half-thickness. The sign of the angular momentum carried by the boundary layer flows at the much smaller Alfvén radius $r_A \ll r_{out}$, which determines the spin-up or spin-down of the pulsar, is uncertain from the present work. If this sign has the sense of the counterrotating material supplied at r_{out}, then a rapid switch between spin-up and spin-down of the pulsar would occur with time scale of order $t_{CR} \sim$ days.

5. CONCLUSIONS

Recent observations of counterrotating galaxies disturb the conventional picture of spiral galaxies consisting of stars and gas rotating in one direction. New physical processes are important in these systems and are just coming under detailed investigation. In particular, the two-stream instability can be important for giving an effective viscosity and inward radial transport of gas in counterrotating galaxies [19]. New accretion flow geometries are possible in counterrotating gaseous disks where the co and counterrotating components are separated by a thin boundary layer [18]. Other new accretion flow geometries have been found in hydrodynamic simulations of counterrotating disks [17].

ACKNOWLEDGMENTS

The author is indebted to Professor Contopoulos for many, many discussions and continuous encouragement over the past twenty years. The author thanks his collaborators on the work described here, Drs. T. Chou, V.M. Chechetkin, N.F. Comins, M.P. Haynes, K.P. Jore, O.A. Kuznetsov, M.M. Romanova, P. Shorey, and T. Zeltwanger. This work was supported in part by NSF grant AST-9320068.

REFERENCES

1. ARAKI, S. 1987. Astronomy J. **94**: 99.
2. BATCHELOR, G.K. 1967. An Introduction to Fluid Dynamics: 197. Cambridge University Press. Cambridge.
3. BERNSTEIN, I.B. 1958. Phys. Rev. **109**: 10.
4. BERTOLA, F., P. CINZANO, E.M. CORSINI, A. PIZZELLA, M. PERSIC & P. SALUCCI. 1996. Astrophys. J. **458**: L67.
5. BINNEY, J. & S. TREMNAINE. 1987. Galactic Dynamics. Chap. 6. Princeton University Press. Princeton, NJ.
6. BRAUN, R., R.A.M. WALTERBOS & R.C. KENNICUTT, JR. 1992. Nature **360**: 442.
7. Buta, R., Van Dricl, W., Braine, J., Combes, F., Wakamatsu, K., Sofue, Y., & Tomita, A. 1995, Astrophys. J. 450, 593.
8. CHAKRABARTY, D., L. BILDSTEN., M.H. FINGER, J.M. GRUNSFELD, D.T. KOH, R.W. NELSON, T.A. PRINCE, B.A. VAUGHAN & R.B. WILSON. 1997. Astrophys. J. **481**: L101.
9. CHAKRAVARTY, S. & S. OSHER. 1985. AIAA Paper No. 85-0363.
10. COMINS, N.F., R.V.E. LOVELACE, P. SHOREY & T. ZELTWANGER. 1997. Astrophys. J. **484**: L33.
11. CIRI, R., BETTONI & G. GALLETTA. 1995. Nature **375**: 661.
12. COPPI, B., M.N. ROSENBLUTH & R.N. SUDAN. 1969. Ann. Phys. **55**: 207.

13. GALLETTA, G. 1996. *In* Barred Galaxies (IAU Coll. No. 157), R. Buta, D. Crocker & B. Elmegreen, Eds. Astronomical Society of the Pacific. San Francisco, CA.
14. HARDEE, P. E., M.A. COOPER & D.A. CLARKE. 1994. Astrophys. J. **424**: 126.
15. JORE, K.P., A.H. BROEILS & M.P. HAYNES. 1996. Astronomy J. **112**: 438.
16. KRALL, N.A. & A.W. TRIVELPIECE. 1973. Principles of Plasma Physics. Chap. 8. McGraw Hill. New York.
17. KUZNETSOV, O.A., R.V.E. LOVELACE, M.M. ROMANOVA & V.M. CHECHETKIN. 1998. Astrophys. J. To appear.
18. LOVELACE, R.V.E. & T. CHOU. 1996. Astrophys. J. Lett. **468**: L25.
19. LOVELACE, R.V.E., K.P. JORE & M.P. HAYNES. 1997.
20. MERRIFIELD, M.R. & K. KUIJKEN. 1994. Astrophys. J. **432**: 575.
21. MERRITT, D. & J.A. SELLWOOD. 1994. Astrophys. J. **425**: 551.
22. NELSON, R.W., L. BILDSTEN., D. CHAKRABARTY, M.H. FINGER, D.T. KOH, T.A. PRINCE, B.C. RUBIN, D.M. SCOTT, B.A. VAUGHAN & R.B. WILSON. 1997. Astrophys. J. **488**: L117.
23. PALMER, P.L. 1994. *In* Stability of Collisionless Stellar Systems. Chap. 12. Kluwer. Dordrecht.
24. RIX, H.W., M. FRANX, D. FISHER & C. ILLINGWORTH. 1992. Astrophys. J. **400**: L5.
25. RUBIN, V.C., J.A. GRAHAM & J.D.P. KENNEY. 1992. Astrophys. J. **394**: L9.
26. RUBIN, V.C. 1994. Astronomy J. **108**: 456.
27. SAGE, L.J. & G. GALLETA. 1994. Astronomy J. **108**: 1633.
28. SCHEUER, P.A.G. 1992. *In* Extragalactic Radio Sources: From Beams to Jets. J. Roland, H. Sol & C. Pelletier, Eds.: 368. Cambridge University Press. Cambridge.
29. SHAKURA, N.I. 1973. SvA. **16**: 756.
30. SHAKURA, N.I. & R.A. SUNYAEV. 1973. Astron. Astrophys. **24**: 337.
31. THAKAR, A.R. & B.S. RYDEN. 1996. Astrophys. J. **461**: 55.
32. TOORNRE, A. 1964. Astrophys. J. **139**: 1217.

Global Spiral Patterns in Galaxies: Complexity and Simplicity

C.C. LIN

Department of Mathematics, Room 2-330, Massachusetts Institute of Technology, Cambridge, Massachusetts 02139 and Florida State University, Tallahassee, Florida 32306

ABSTRACT: A coherent exposition of the density wave theory of galactic spirals is presented in a recent monograph [3]. It is centered on the working hypothesis of quasi-stationary spiral structure, a possibility first proposed by Bertil Lindblad [23]. This hypothesis has since been found to be widely applicable in a number of physical contexts, including the explanation of the Hubble classification system and other categorical classes. Direct empirical support of this hypothesis has been provided especially by the regularity of the infrared images frequently observed in a number of galaxies (e.g., NGC 309), and by the observed amplitude modulation along the spiral arms (e.g., M51, M81, and NGC 1300). The present paper is a brief review of this theory with further clarification of the fundamentals and of certain specific issues raised in the literature.

On the theoretical side, the likelihood for the validity of this hypothesis has been supported by modal studies. Emphasis is placed on the widely observed coexistence of a single regular structure in the Pop II objects and the more complex irregular structures in the Pop I objects, a contrast first discovered by Zwicky many years ago in the main disk of *M*51. It is pointed out that, in both barred and nonbarred spirals, this basic phenomenon may be understood by noting the fact that the microscale of the collisionless system of Pop II stars — i.e., the diameter of the epicycle — is typically on the same order of magnitude as the observed spacing between the spiral arms. The spiral pattern is thus a very compact structure which is unlikely to respond readily to internal and external disturbances of moderate magnitudes. This apparent robustness in structure is suggested by its observed regularity despite the impact from the coexisting Pop I objects with their strong irregular turbulent motions.

This physical picture supports Oort's conjecture [28] of a limited role for tidal interaction in most of the spiral structure of galaxies. It also enables us to place in proper perspective a controversy between two schools of thought, with different emphasis placed on intrinsic mechanisms and on tidal interaction. It is primarily a matter of applicability, or frequency of occurrence of the different scenarios proposed. There is, as yet, little observational evidence presented in the literature that would support the need of an interpretation of the global spiral structure in the galactic disk in terms of a fast evolving structure recently generated from a featureless initial state through tidal interaction. Dynamically, the likelihood for realizing such a scenario is also estimated to be quite low [3, p. 130].

The above discussions are placed in the context of a general point of view much advocated in current scientific literature; namely, the need to analyze alternative mathematical models for diverse physical contexts in the study of "complex" systems and to judge *separately* the merits of each model on the basis of its empirical confirmation in an appropriate physical context.

1. INTRODUCTION

I am privileged to have this opportunity to honor our good friend of many years, George Contopoulos, upon his retirement. I wish him well with this relief of some of his official duties. I expect to continue to hear from him and to see him from time to time just as before. I present to him and to the Gainesville Workshop my third and final report on global wave pattern in galaxies. My first full report of our early work in this area was presented, with George in the audience, some 33 years ago, in 1965, at the beginning of my work on this subject, in collaboration with Frank Shu. The title of my paper was "On the mathematical theory of a galaxy of stars." I recently read the paper again, and found it could still serve as an adequate introductory exposition to our theory (even though there are minor improvements to be made in view of subsequent developments). In that paper, the physical basis of the theory was quite fully described, and the *complexity* of the galaxy as a physical system stands out very clearly. Indeed, it is also quite clear that the research efforts, beginning with Hubble's morphological classifications of galaxies, were devoted to the search for *simpler* concepts and *simpler* models (cf. [19]). The main objective of this paper is to clarify certain important issues under continual discussion over the decades, including the controversy between two schools of thought on the generation, evolution, and long-term maintenance of global structures in spiral galaxies. We shall see that even though there is considerable complexity of the issues, there is a certain degree of simplicity in that the set of issues often raised could be addressed with one single perception: the coexistence of *different* spiral structures dominated respectively by the behavior of Pop II and Pop I objects. This is a phenomenon first discovered by Fritz Zwicky in the main disk of M51. Recent infrared data in that galaxy [30] and in other galaxies (e.g., NGC 309; see [6], [7]) give further evidence in support of this perception. The controversy between different schools of thought will be examined in the context of Oort's conjecture of 1969 [28]. It will be shown that his conjecture is fully supported by observational data and by dynamical studies. (See Sections 4 and 5 below.)

All these issues will be presented in a general perception described as *complexity* in recent literature on mathematical modeling. It has been suggested that complex systems generally require the study of *alternative* theoretical models and their confirmation by *empirical data* obtained in diverse physical contexts that are appropriate to each school of thought. They generally exhibit different aspects of the total complex phenomena. A good example of complexity is the study of weather forecasting. Different models are needed for numerical forecasting on different scales, from local to global. The appropriate matching of alternative mathematical models with diverse physical contexts holds the key to the resolution of controversies. (See Section 3 for further discussions of the present case.)

1.1. Some Basic Issues

An exposition of a density wave theory for the interpretation of the observed global spiral structures in galaxies was recently published [3]. It is semiempirical and based on the working hypothesis[1] that the *global* spiral structure observed in a gal-

---

[1]One of the definitions to the word *theory* in Webster's International Dictionary is a *working hypothesis* (definition 4b, p. 2371 of the 1966 edition).

axy is generally *quasi-stationary*; that is, in most cases, the global spiral structure is currently maintained in a slowly evolving state. This working hypothesis has been successfully applied to a wide variety of observational data. As mentioned above, there are a few subtle points that need to be addressed more fully. These issues and their answers will be presented in this paper. (For more detailed coherent discussions of certain specific issues, see Bertin [2] and Lin and Bertin [16]). A description of the issues will be presented first (Subsections 1.2–1.4), to be followed by a brief sketch of the answers (Section 2). The full details will be given in subsequent sections. As mentioned above, it turns out that the answers are all to be based on one single theme: that is, Pop II objects and Pop I objects behave differently, with *regularity associated with the Pop II objects* and irregularity associated with Pop I objects (cf. Section 5). The issues will now be described with the help of three quotations from the published literature.

1.2. Complexity in the Morphology of Spiral Structure

The morphology of observed spiral structure is far more complex than that represented by Hubble's diagram [10] for describing his classification (FIGS. 1 and 2). Nevertheless, the diagram does catch the main trends and serves as a successful framework for classifying galaxies, and its formulation was an enormous achievement, as reflected in the following description of Jeans [11].

> The great nebulae exhibit an enormous difference of structural detail, but Hubble, who has devoted much skill and care to their classification, finds that most of the observed forms can be reduced to *law and order*. They fall primarily into the two great classes of nebulae of regular shape which exhibit no spiral arms, and true "spirals," consisting of a rather vaguely-defined central region from which two spiral arms emerge. Hubble finds that the true spiral forms, as regards their main features at least, can be specified in terms of three distinct quantities, which he describes as (1) the relative size of the unresolved nuclear region, (2) the extent to which the arms are unwound (the openness or angle of the spiral), (3) the degree of condensation in the arms. When these quantities are estimated or evaluated observationally for different nebulae, they are found to vary very approximately in unison, with the result that practically all observed nebular configurations can be arranged to form linear series.

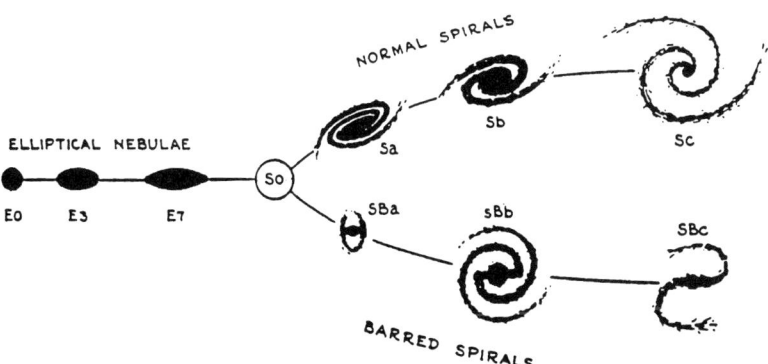

FIGURE 1. Hubble classification of galaxies.

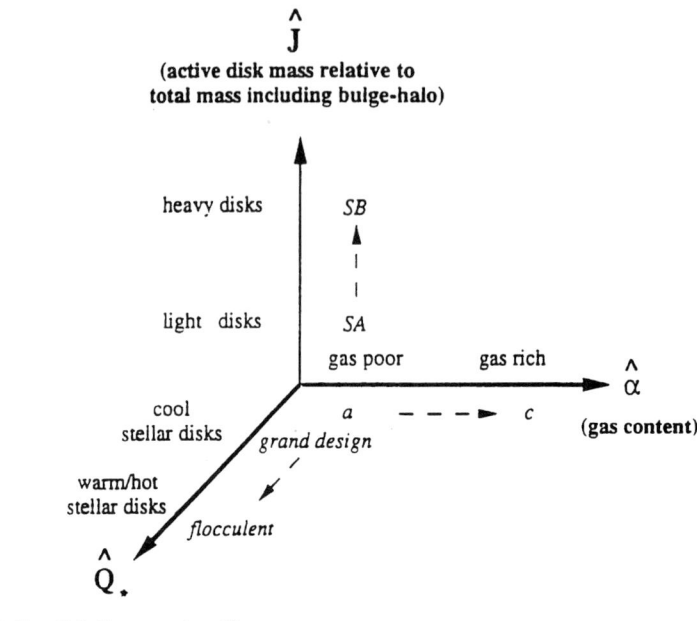

FIGURE 2. Theoretical framework for the classification of the morphologies of spiral galaxies on the basis of their intrinsic modal characteristics. (See [3, Sect. 4.7] for detailed explanation.)

1.3. Irregularity in Outer Arms

The sketch of Hubble [10], in the form of two-armed spirals is obviously an oversimplification. For there are multiarmed spirals, and there is generally irregularity in the outer spiral structure. The concern for this complexity is natural to the theorists, as reflected in the following quotation from the review of the Bertin-Lin monograph by Lynden-Bell [27], who raised this issue in the context of the working hypothesis of *quasi-stationary* spiral structure.

> Perhaps the very irregularities of the observed galaxies are to blame for the slightly blurred impression left on me by the earlier chapters. This is typified in the following passage on the justification for the quasi-stationary spiral structure hypothesis (p. 82): "In any case, the dynamical basis of the hypothesis lies in the existence of global spiral modes, and this in turn depends sensitively on the nature of the axisymmetric basic state which may or may not support global spiral modes. When global spiral structures are observed they may be expected to evolve at a much slower rate in a rotating framework appropriate to the wave pattern, provided the wave pattern is largely composed of a single dominant mode or a very small number of such modes rotating at approximately the same angular velocity ... Naturally the wave pattern itself may evolve substantially in the course of time and in general it does."

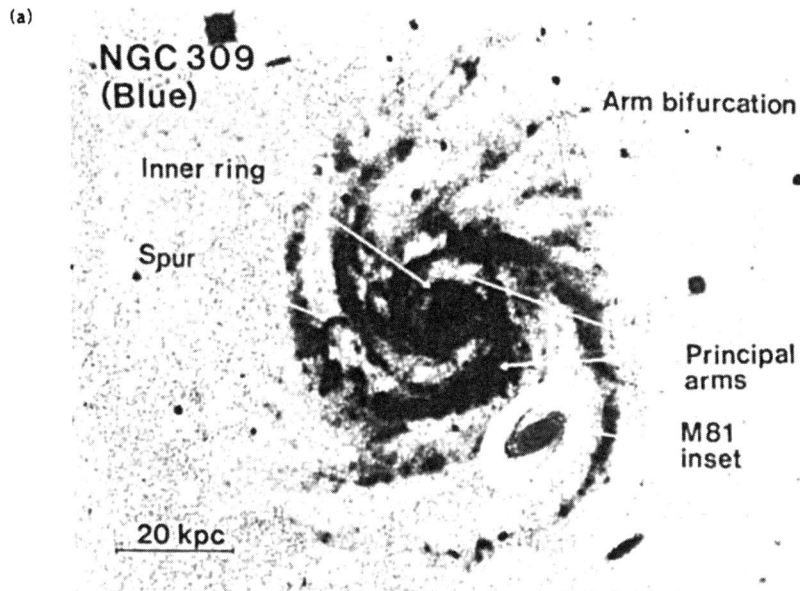

FIGURE 3. Optical and infrared appearance of the multiarmed Sc galaxy NGC 309. The infrared image **(b)** reveals a smooth bisymmetric structure with a prominent bar, while the optical picture **(a)** is characterized by multiarmed morphology (see [3, Sect. 5.2.4]).

Note that the long-term evolution in the last sentence of this quotation is essentially the quasi-periodic evolution suggested by P.O. Lindblad [25]. The response to Lynden-Bell's comment is to be based on the theme outlined in Section 1.1 and discussed in detail in Section 5. Quasi-stationary behavior does not hold in the irregular structure of Pop I objects.

1.4. Interacting Galaxies: Oort's Conjecture

Another set of issues concern the general regularity and irregularity of spiral structures, not just in the outer part. These issues were raised by Oort [28]. The following is a quotation from Lin and Bertin [16].

> A deeper issue on global structure of galaxies was also discussed by Oort almost exactly 25 years ago. At the 38th symposium of the International Astronomical Union (IAU), he raised the following issue of *regularity*: "A more serious problem seems that of the long-term permanence of the spiral wave — can they continue to run around during 50 revolutions without fatal damage to their regularity?" In response to this issue, he stated that "in most cases, we cannot invoke interaction. Moreover, if there had been interaction in the past it might well have contributed to *disturb* the regular wave pattern, but less likely to rebuild it." He continued by pointing out that a statement on the existence of the global structure should be "supplemented by two essential additions. First, that in about half the spirals, the structure is either unclear, or there are more than two arms. Second, that even in the half that can be classed among the two-armed spirals there are invariably important additional features *between* the two principal arms."

These two supplementary points were also stressed by Lin [15] in terms of the *coexistence* of all kinds of spiral features. Three-armed features and interarm bridges are now usually attributed to contributions from the gaseous component; the dominant contribution to the two-armed global spiral gravitational field is believed to be associated with the "older" evolved stars. This point receives direct observational support from recent infrared observations (see; e.g., FIG. 3 adopted from Block *et al.* [7]), where the two-armed grand design stands out clearly despite the irregularities present in optical and radio observations. There are thus two *different* scenarios proposed — or two different schools of thought advocated — for the same observed global spiral structure in the case of interacting galaxies. The spiral structure is considered to be either (1) recently *generated* by an interaction from a featureless initial state, or (2) *disturbed* (perhaps only to a limited extent) by the interaction from a preexisting state that is primarily maintained by internal dynamical mechanisms. Oort conjectured that the latter scenario holds in most cases. The resolution of such a debate is obviously to be based on comparison with empirical data, and the data are found to support Oort's conjecture. Indeed, advocates of the alternative suggestion (see, e.g., [37], [9]) still have to face the difficulty of offering a fully successful case of comparison with observational data without "failures." (See [9] and Section 4 below.)

2. ANSWERS IN BRIEF OUTLINE

The complex issues raised above concerning the regularity of global structure and the mechanisms for their generation and maintenance has been extensively studied from both the observational and the dynamical points of view. (See Chapter 4 of monograph by Bertin and Lin for a coherent presentation; see also [16] and Section 4 below.) Both of the scenarios (1) and (2) just mentioned above are considered.

These studies fully support Oort's conjecture. Indeed, detailed dynamical arguments are given (in Section 4.8 of the book) to support the conclusion that there is still "a wide gap between the [simple] empirical suggestion of tidal interaction and a quantitative demonstration that the process is a viable general explanation for regular spiral structures."

As we shall see in the following discussions, the key to the resolution of all the issues raised above may be traced to the perception formed from the discovery of Zwicky [44], who found that, in the main disk NGC 5194 of the M51 interacting system, there are two coexisting spiral structures: the "red" picture is far more regular than the "blue" picture. This discovery is confirmed by the recent multiwave-band observation of Rix and Rieke [30]. Thus, we have empirically discovered simplicity in the infrared, instead of the complexity shown in the optical.

One may therefore make the general suggestion that perhaps in every galaxy with a global spiral structure, there is a *regular two-armed* structure in the principal internal part that matches the general sketch made by Hubble. It would form the *backbone* for the support of the observed global spiral structure in all cases. Such a conjecture is definitely encouraged by all recent infrared data. The galaxy NGC 309 is, indeed, observed to have a regular two-armed barred image in infrared, even though the outer structure suggests a multiarmed Sc classification (FIG. 3). There is also no obvious nearby companion. The infrared structure thus appears to be clearly *self-sustained*. (See [3, p. 146].)

Dynamically, the following picture emerges. A *single* wave pattern of Pop II stars forms a *robust* backbone for the global spiral structure (cf. Section 5.2 for the discussion of the dynamical basis of this statement). The Pop I objects, which are dynamically much more responsive and prone to develop turbulent motions on small scales and mesoscales, would respond to the large scale forcing of the spiral gravitational potential of the Pop II stars, and thereby reinforce this large scale structure, resulting in a generally self-sustained pattern.

With such a robust backbone, Oort's conjecture on galactic encounters appears to be fully supported from the dynamical point of view, at least when the interaction is not too strong to destroy or totally distort the inner stellar disk. (See further detailed discussions in Sections 4 and 5 below.) The Pop I objects in the outer disk would in general yield readily to tidal forcing. Before presenting the details described above, let us first consider some general points concerning the development of semiempirical theories for understanding and describing complex physical systems. Naturally, there are difficulties that often lead to the development of different schools of thought. The perception that should be adopted for the resolution of the resultant controversies has been examined and discussed in recent years under the term *grappling with complexity* (see article by Lee Segel [34]). The main themes of these discussions will now be briefly presented as a background for the discussion of the conjectures and controversies sketched above.

3. COMPLEXITY

A consideration of all the above diverse issues in perspective indicates that the spiral structures observed in galaxies is a set of *complex* phenomena. Theorists are challenged to understand and to describe this wide range of phenomena on the basis of relatively *simple* models and scenarios based on the principles of stellar dynamics (Jeans equations) and fluid dynamics (Navier-Stokes equations). As noted above, it is natural that such difficult challenges would lead to different schools of thought with emphasis placed on different theoretical scenarios or different physical contexts. Sometimes it may turnout to be difficult to determine, directly from the observational data for given *individual* objects, which of the scenarios are significant (see [3, p. 7]).

Such controversies can only be resolved by careful analyses of observational data obtained in the relevant physical contexts. *For each theoretical scenario proposed, there must be empirical confirmation in appropriate physical contexts.* Note that two *different* scenarios may *both* be valid because they apply to *different physical contexts*. Each scenario must stand or fall on its own merits.

This requirement for empirical check can be quite a frustrating experience, especially when accurate and reliable empirical data are difficult to obtain. Indeed, this frustration is shared by scientists in all areas of research, especially in newly developed areas such as theoretical biology, where empirical data are often not yet well organized. This experience is described by Segel [34] in the paper cited above. The following is a quotation from this article:

> One might think that all of the individual models could somehow be agglomerated into one giant model, but generally this will spoil the simplicity required for understanding and will render calculations infeasible. Indeed, we assert that *a complex system can be characterized by the fact that it must be attacked via many models.* No single model will ever suffice. Moreover, *the more models required, the more complex the system must be.*

The need for different models is well illustrated in atmospheric sciences. Different models and different methodologies are used for treating phenomena in local scales, mesoscales, and the global scale: predicting the motion of an intensive storm moving at 100 km/hr is very different from that for a hurricane. Different models are required to answer this often posed question: how is global warming related to El Niño?

One has to keep in mind that the simpler models are often imperfect approximations to reality. Some investigators might even label them as "wrong" by taking strict standards. Nevertheless, they are usually useful approximations, well within the limits of observational error. Even in extreme cases, when the models are too oversimplified to be good approximations, their analyses can help with general understanding. (See discussion of "artificial complex structures" beginning with page 23 of [34].)

Finally, it should be noted that we are dealing with semiempirical theories that are supported by theoretical analysis of both the mathematical models and the empirical data obtained from observations. The former can only demonstrate the likelihood of the occurrence of the scenarios proposed. It is the analysis of the empirical data that would furnish the proof of their physical realization.

3.1. Simplicity in the Galactic Context

Although "many" models are mentioned in the above quotation on the study of complex systems, the actual number needed is often quite small. This happens when simplicity is exhibited in the empirical data; i.e., when the data already exhibit *law and order.* Indeed, in such cases, the development of mathematical models is more likely to be successful. In the present context, this simplicity is provided by Hubble's discovery of his classification system. This impression of simplicity is further strengthened by the recently observed regularity of infrared images — which could be anticipated from Zwicky's discovery in the M51 system and our general understanding of the physical nature of the Pop II and Pop I objects.

We may thus make the following conjecture. In an essentially isolated galaxy, the Pop II objects form a stable standing wave pattern, even though it is constantly being disturbed by the turbulent dynamical processes in the Pop I objects. This stable structure forms the backbone of the global spiral structure as a whole. There are spiral structures on smaller scales in the Pop I objects, but they have only limited influence on the behavior of the Pop II system. We are thus led to the highly successful *modal approach* to the global structure of galaxies. In this approach, the global structure is regarded as supported by the intrinsic dynamical mechanisms (cf. [3, Sect. 4.7] and FIG. 2 of this paper). Naturally, the question to be answered is this: why do the two kinds of objects behave so differently? The answer, as it turned out, is to be traced to the collisionless nature of the stellar system of Pop II stars (cf. [3, p. 14] and Section 5.2 below.)

3.2. Two Schools of Thought

Different mathematical models are generally needed to account for the phenomena observed in different physical contexts. This point is obvious when one compares the cases of essentially isolated galaxies and interacting galaxies. In different schools of thought, the attention is focussed on different physical contexts. In the present paper, we consider global spiral patterns in the galactic disk. We study their maintenance, generation, and modification by tidal forcing from external galaxies whenever (which is infrequent) close neighbors are present. Others (see Toomre and Toomre [45]) focus their attention on the impact of (generally time-dependent) tidal forcing, where the outskirts of the galaxy take on greater prominence than the principal part of the galactic disk. (See Sections 4 and 6.3 for further discussions).

In this latter approach, it is proposed that the global spiral structure is produced by the effect of tidal forcing from a recent encounter with a companion (or satellite). We are thus dealing with two *different types* of physical contexts, each requiring its own model. One should note, however, that the scenario of interacting galaxies, which is suggested by the observational data in the case of the pair of galaxies NGC 5194/5, has a number of alternative subcases: (i) Are we dealing with the case of a spiral structure generated from an initially featureless galactic disk? (ii) Did the most recent interaction take place when there was already a preexisting spiral structure? (iii) If so, was the spiral structure greatly modified or only somewhat disturbed by this tidal interaction? All are possible scenarios. An investigation should aim at deciding the applicability of these scenarios in the observed galaxies (i.e, the frequency of realization of these individual scenarios) supplemented by theoretical considerations of the likelihood of their physical realization.

In the following sections, we shall emphasize considerations of the all important step of comparison with observational data. We first consider interacting systems

(Section 4), to be followed by essentially isolated galaxies (Section 5). It will become clear why many galaxies do not have a global spiral structure and why such structures, when they do exist, would generally be dominated by a *single* two-armed structure, as sketched by Hubble (FIG. 1).

3.3. Semiemperical Theories

A theory based on a working hypothesis is semiempirical (cf. Section 1.1, footnote 1). Such a theory must eventually be supported by accumulation of both theoretical and empirical evidence. Indeed, this is generally true of the modeling of all physical systems, simple and complex. Self-consistency is naturally essential in the mathematical theory, and the theorist should also demonstrate the likelihood of occurrence of the particular scenario favored. But the "proof" of the theory can only be provided by empirical confirmation. It is therefore important to emphasize the proper matching of the mathematical models with appropriate physical contexts.

In the present context, the QSSS hypothesis should be applied to essentially isolated galaxies, while Oort's conjecture should be applied to interacting galaxies with moderate interaction. Furthermore, it is important to note that we are focusing our primary attention on the spiral structure of the Pop II objects and the associated global spiral structure of the Pop I objects which have been brought under the direct influence of this spiral gravitational field. There are also spiral structures on smaller scales in the Pop I objects which evolve at faster rates.

Optical and radio observations of the distribution and kinematics of Pop I objects would present both the global spiral structure and the spiral structure on smaller scales. The backbone of the global spiral structure is conclusively revealed in the Pop II objects by infrared data, essentially free from the complication of the structures on smaller scales.

The 21-cm data of an interacting pair of galaxies, to be discussed in Section 4.1, show various spiral structures, partly in the intergalactic space and partly inside the galactic disk; the latter is an illustration of the above discussions.

4. INTERACTING GALAXIES

We now examine the merits of the different schools of thought, including Oort's conjecture on interacting galaxies, in light of observational data available so far. We first focus on the two well-studied systems of interacting galaxies, NGC 3031/3077 (or the M81 system) and NGC 5194/5195 (or the M51system). This will be followed by a discussion of the likelihood of occurrence and the frequency of observation of the various possible scenarios (cf. [16, Sect. 4]).

4.1. The M81 Interacting System

To get a good overall picture of interacting galaxies, including a substantial amount of quantitative data, let us examine the 21-cm data of neutral hydrogen in the M81 interacting system.

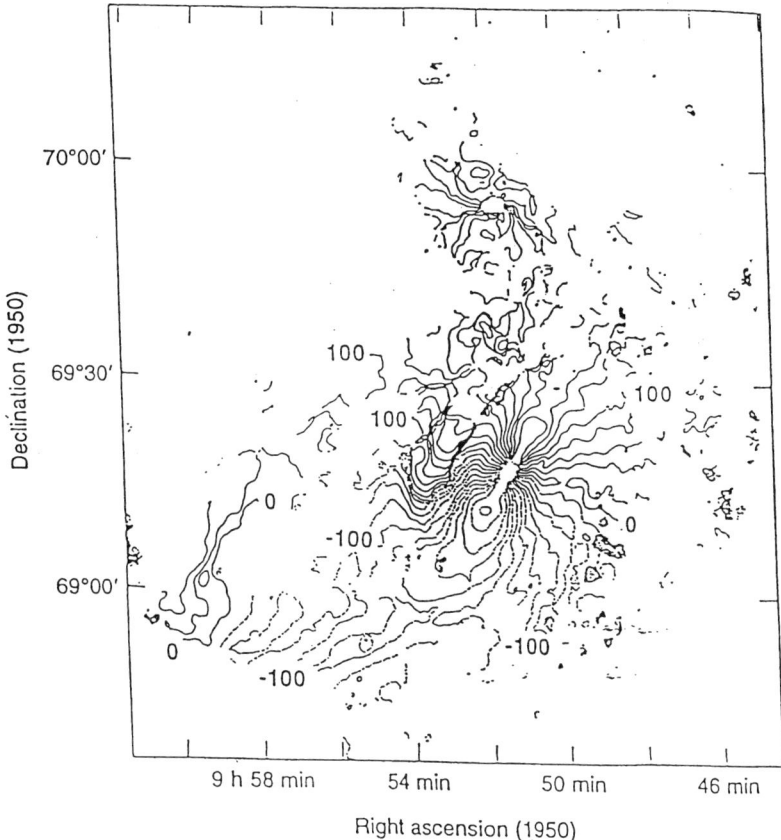

FIGURE 4. Kinematics of neutral hydrogen in the interacting system of NGC 3031/ 3077 (see [42], where the distribution of HI is also given).

In FIGURE 4, we show the data for the kinematics of neutral hydrogen in the intergalactic medium as well as that in the interstellar medium [42]. The latter reminds us of the familiar diagram analyzed in great detail by Visser [38] (see FIG. 5). While the latter shows the dominant influence of the stellar component in the galaxy M81, the former is substantially influenced by the joint gravitational field of both of the interacting galaxies, with the history of the passage of NGC 3077 very much in evidence.

The picture that emerges is that of the close passage of a satellite galaxy (NGC 3077) in the distant past, that *disturbed* the structure of a main galaxy (NGC 3031) only to a limited extent, as conjectured by Oort. The detailed studies of the current state by Visser indicate that the kinematics observed in M81 is consistent with that of a standing wave pattern described in the hypothesis of quasi-stationary spiral

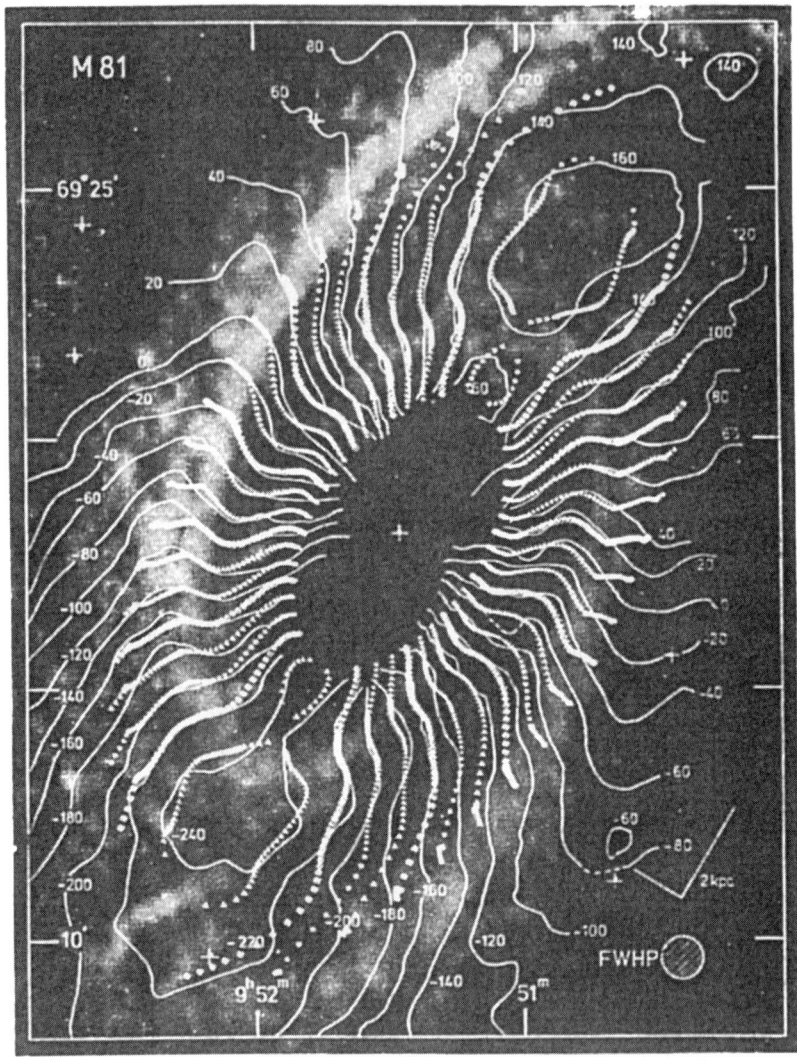

FIGURE 5. Kinematics of neutral hydrogen in the main part of the galaxy M81, according to Visser (cf. FIG. 9 for the distribution showing amplitude modulation).

structure (cf. Section 5). Indeed, a vivid scenario of such a passage is provided by the "photograph" of the locations of the various objects at both the optical (see, e.g., [35, pp. 92–300, color plates 13, 24, 25] for the galaxies M51, M101, and M81) and radio frequencies (see [35] cited above, where the distribution of HI is also given). Thus, in this rather typical case, the whole set of data lends strong support to the scenario proposed by Oort as the most likely to occur and to be observed. Note also that the generation scenario fails to provide amplitude modulation observed and predicted in the modal theory (see FIG. 8).

4.2. The M51 interacting system

In the interacting pair of galaxies NGC 5194/5195, there are indications of strong distortion from tidal interaction in the *outer* arms (FIG. 6). But there is no clear indication of the state of the galactic disk of NGC 5194 *prior* to the interaction. Was it featureless or was there a preexisting spiral structure? There is also no clear indi-

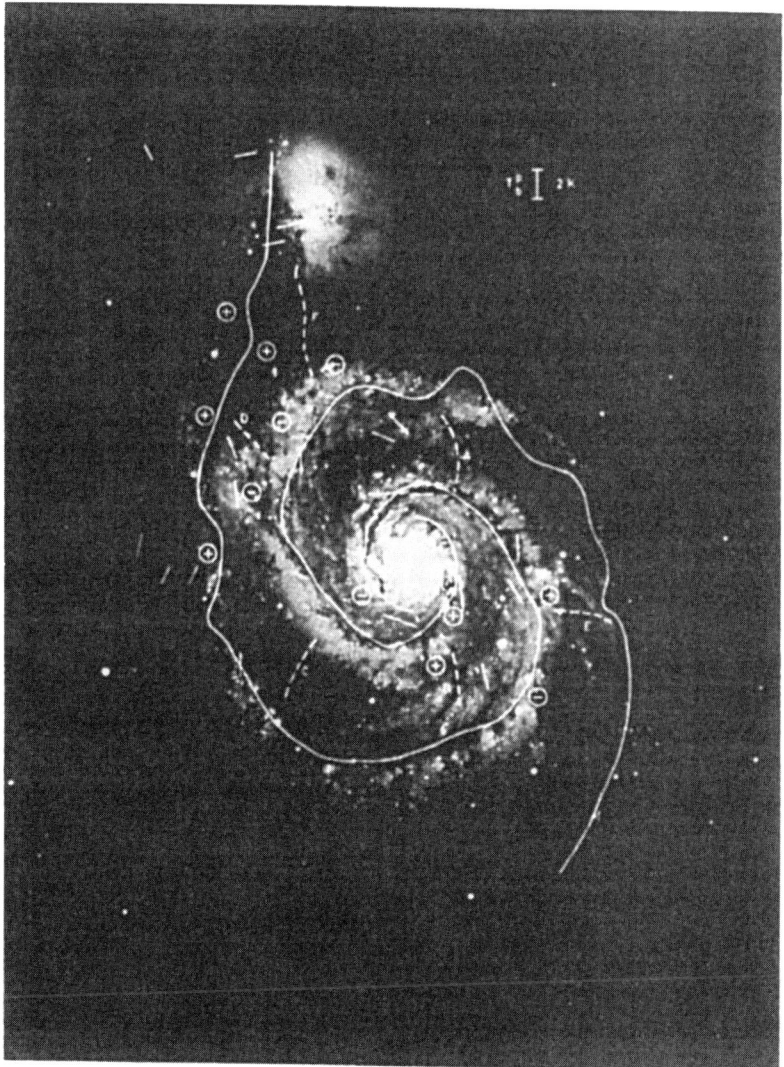

FIGURE 6. Peak of synchrotion emission in M51 and other characteristics of Pop I objects showing irregular structures. (See relevant discussion in [3, Chap. 5].)

cation whether or not the system is in a state of continual interaction with repeated close encounters (cf. [31]). These authors successfully studied the peculiar system Arp 86 in great detail both from the observational point of view and through N-particle simulation with repeated close encounters. They suggested that Arp 86 is a typical interacting pair similar to the M51 system.

N-particle simulation of this latter pair has been attempted by several authors. Yet all the attempts to simulate it with a *featureless* initial state has so far not been entirely satisfactory. In a concluding section of a paper by Hernquist [9], with the heading "Successes and Failures," he described how discrepancies would remain in the inner spiral structures when the outer arms were well simulated. He concluded his paper with a rather pessimistic note: "To paraphrase Lord Ross: Although many of our conjectures now seem at least plausible, a thorough understanding of galaxies like M51 remains elusive."

We now understand the system much better through the observational data in infrared, obtained and analyzed by Rix and Rieke [30] (see [16, p. 133] for a more de-

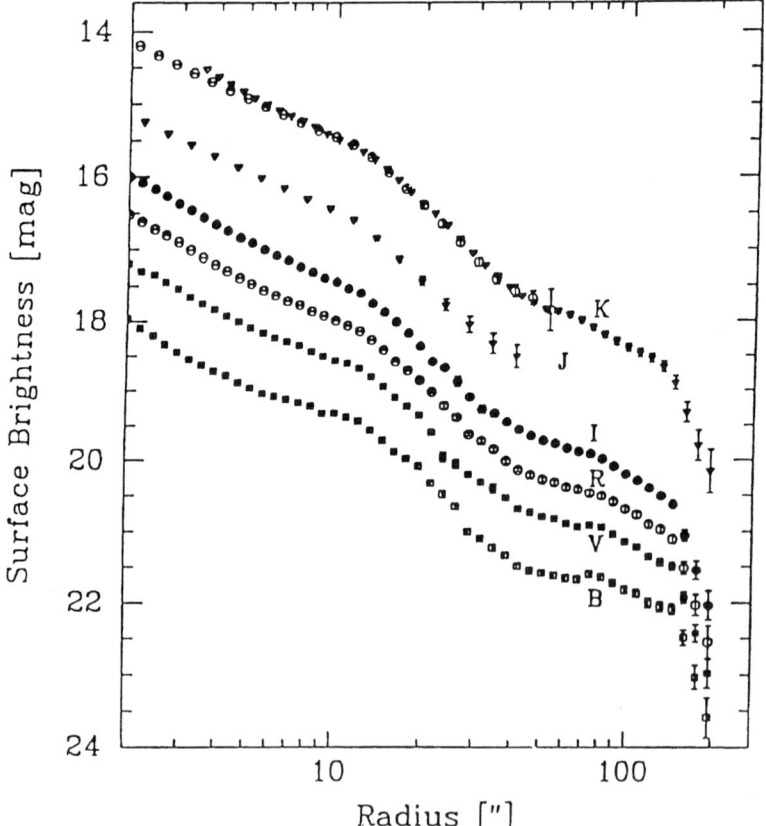

FIGURE 7. Amplitude modulation in the gravitational potential of the spiral structure of M51 is reflected in the distribution of objects observed at several wavelengths (after [30]).

tailed discussion). These indicate a complex picture: "the spiral arm amplitude clearly shows smooth, strong radial variations ... (FIG. 7). These variations may arise from interference of a preexisting spiral pattern with tidally induced spiral arms." This is clearly supportive of the conjecture of Oort quoted above. Indeed, the very regularity of the two-armed infrared image indicates that the "disturbance" visualized by Oort might well be quite limited. (See also [6, p. 10]). This limited disturbance is also indicated by an estimation of the order of magnitude of the tidal forcing (cf. [16, p. 142]). The smooth strong radial variations thus would primarily be the amplitude modulation intrinsic to the preexisting pattern of standing waves (cf. Section 5).

4.3. Frequency of Occurrence of Alternative Scenarios

Let us now consider the likelihood of the occurrence of the various scenarios of interaction from the dynamical point of view.

In the above discussions, we emphasized the possibility of a limited impact on the stellar disk, even when the outer Pop I disk is highly distorted. To be sure, there are also cases of very strong interaction in which there is almost total destruction of the galaxies. We should therefore focus our attention on the frequency of occurrence of the various scenarios. If we consider the fact that the peculiar galaxies form only a small fraction of all galaxies, we see that strong interactions are rare events. Our interest in spiral structure is likely to be centered on cases where any *recent* interaction does not exceed the intensity experienced in the case of M51 or Arp 86. Thus, there is a significant likelihood for a quasi-stationary structure to exist in the Pop II objects, and in the overall gravitational field on a large scale, even in interacting galaxies.

The scenario of an initially featureless state has been considered by Bertin and Lin [3, p. 130] from the dynamical point of view. It is pointed out that such an initial condition implies a stable stellar disk (not just with respect to axisymmetric disturbances) and hence a disk difficult to excite into a prominent spiral structure. It is also pointed out that, as shown by numerical simulations of tidal forcing, an encounter must be *"strong"* and *"rapid"* in order to have a sizable impact. Such encounters are infrequent. Consequently, there appears to be a wide unresolved gap between the simple empirical suggestion of tidal generation and a quantitative demonstration that the process provides a viable general explanation for spiral structure.

Much work remains to be done in the studies of tidal forcing. Its impact on preexisting spiral structure is perhaps one of the most interesting problems. Continuation of studies of the scenario of repeated encounters would be desirable.

5. HYPOTHESIS OF QUASI-STATIONARY SPIRAL STRUCTURE

Having addressed the issues on interacting galaxies, we now turn our attention to essentially isolated galaxies and to the issues raised earlier concerning a more precise statement of the working hypothesis of quasi-stationary structure. The physical and dynamical bases of this improved description is also discussed.

5.1. Physical Arguments and Observational Evidence for the QSSS

In view of the above discussions, we should give primary emphasis to Pop II objects in the statement of the QSSS hypothesis: *the Pop II objects would support a smooth regular standing wave pattern on a large scale*, and that *this pattern would evolve slowly*. This Pop II structure would also dominate the dynamics of Pop I objects on a large scale, and thus serves as a *backbone* for maintaining the overall structure (cf. [3, Section 10.6]), since the mass of the stellar disk generally is much larger than that of the interstellar medium, especially in the inner disk. Indeed, the large scale behavior of the Pop I objects plays an important role in the excitation and the maintenance of the resultant global structure. In contrast, the irregular structures in Pop I objects can evolve at faster rates. The dynamical basis for this difference in behavior will be described further in Subsection 5.2.

The Pop I objects, with dispersive velocity limited by dissipative turbulent motion of the interstellar medium, is very responsive to the spiral gravitational field. It is further reinforced by its own gravitational forcing.

The resultant standing pattern is supported by and composed of waves traveling in opposite directions along the radius vector and rotating in the angular direction at a uniform angular velocity. This suggests the formation of an interference pattern with amplitude modulation along the spiral arms. Indeed, such a pattern was revealed by the analysis of observational data in in neutral hydrogen in M81 [46]. (See patterns in FIG. 8, where good agreement is shown between the observational data and the mode computed by Lowe *et al.* [26]).

Modal calculation is a sensitive method for the determination of the basic state required to support the observed pattern. Good agreement has been found between the basic state required and that determined by Kent [12] from detailed studies of the

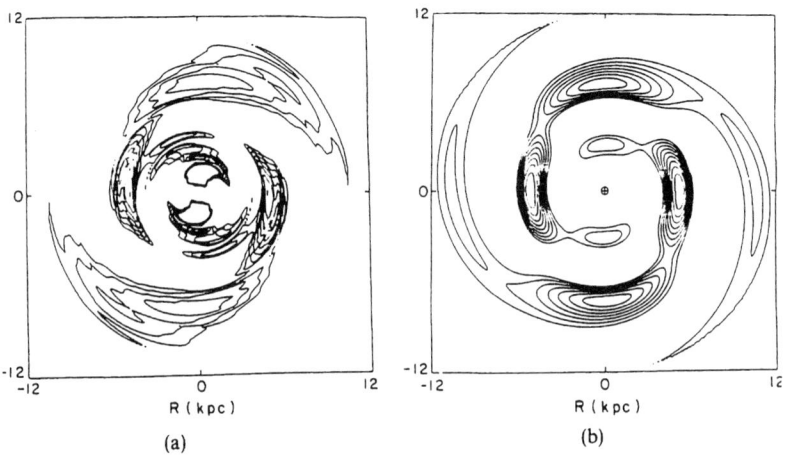

FIGURE 8. Observed amplitude distribution of neutral hydrogen in M81; the contours of the $m = 2$ component are shown rectified to a face-on aspect of the galaxy. (See [3, Sect. 5.2.2] for further discussions of morphological types observed.)

inner rotation curve. This is an important confirmation of the modal approach, since two very different ways are used to reach the same result for the basic state.

In contrast, N-particle simulation of the M81 system as a newly-generated spiral shows no amplitude modulation along the still evolving spiral arm when it is in temporary agreement with the observed pattern. Thus the evolving scenario also lacks observational support in this case.

Similar amplitude modulation has been found in a number of barred spirals. In NGC 1300; good agreement is again found between theoretical predictions and observational data (see [3, p. 44, FIG. 5.3]).

Continuation of work along this line should be pursued for a number of other galaxies, including both barred spirals and normal spirals. It would be interesting to find models for barred galaxies to match good systematic observation data at several wavelengths (cf. [29], [24]).

5.2. Dynamics of the Galactic Disk

The above discussions are based on the concept of the modal approach which is well established; indeed, the basic requirements for it to be dynamically viable has been addressed by Bertin and Lin [3, Section 10.6, p. 239]. It is pointed out there that the cooperation of gas and stars, with self-regulation of the interaction between the two components, is favorable to the excitation of modes and that the mechanism of wave absorption at inner Lindblad resonance is the key mechanism in the limitation of the number of modes excitable. Especially to be noted in this context is the very different behavior between stellar and fluid systems.

Since the conclusions of a semi-empirical theory needs the cumulative support of an ever increasing amount of empirical evidence and theoretical analysis, we wish to supplement that discussion by further addressing two points: (1) the dynamical basis for the coexistence of order and chaos in the galactic disk (as discussed above in the case of M51); and (2) the likelihood for the existence of a single spiral structure with a dominant two-armed mode in the galactic disk.

After these points are made, we would have a basis for supporting a more definite and *specific* statement to be added to the hypothesis of quasi-stationary spiral structure: There is a quasi-stationary spiral structure in the distribution of the Pop II objects that consists of a dominant single two-armed component, a secondary one-armed component, and other minor components of less significance. In contrast, the structure in the Pop I objects may have other significant structures which are evolving at faster rates (cf. first paragraph of Section 5.1).

The need for these discussions goes beyond these two objectives. Fluid dynamicists familiar with the modal and nonmodal approaches to the basic mechanisms of transition to turbulence are often skeptical of the modal approach. With the chaotic behavior of the Pop I component clearly in evidence, they question the justification of the modal approach to global spiral structure. Thus, we are brought back to one of the central points of our theoretical studies: What is the dynamical basis for the difference in behavior of the two components of the galactic disk? More specifically, why does the evolved stellar population form a *single* smooth and regular spiral pattern? Indeed, the structure of this pattern must be quite robust, since it appears to maintain its regularity despite the constant random disturbances from the turbulent gaseous component.

We shall see that these points can be understood if we note that, in a collisionless stellar system, the individual stars exercise epicyclic motions at a length scale not much smaller than that of their collective behavior (see [3, p.15]). In this perception, the global structure is actually quite compact, not extensive. It should therefore not be surprising if it is quite robust and devoid of substructures. We note that these considerations of collisionless systems are very similar to the simpler case of the classical studies of rarefied gas dynamics by Knudsen. In such studies, it is well known that turbulent motion is not observed.

Demonstration of the above statements will be presented in Subsections 5.2.1 and 5.2.2 in the context of nonbarred spirals. In Subsection 5.2.3, we shall present an explanation on a broader basis by noting the perennial "heating" of the stellar disk [3, p.10]. It will be shown that the process of evolution of the spiral pattern under the influence of intrinsic dynamical mechanisms would naturally prefer the eventual establishment of a single quasi-stationary global pattern, independently of its specific morphology.

5.2.1. Length Scales in the Disk

Let us begin with a discussion of the relevant length scales in the disk of nonbarred spiral galaxies, for we have more specific information here. Barred spirals will be considered below. *The length scale of basic state* is the exponential length scale h_*. Another important length scale is the typical diameter d of the epicycles. In general, we have

$$d < h_*.$$

When a global wave pattern is observed, it is likely that its characteristic length scale, the half wavelength $\lambda/2$, is somewhere in between:

$$d < \frac{\lambda}{2} < h_*.$$

Indeed, all macroscales are just a few times the microscale d. Specifically, for the model of M81 constructed by LRYBL [26] we have the following values for these parameters at the representative radius $r = 7$ kpc

$$h_* = 2.8 \text{ kpc}, \quad d = 1.6 \text{ kpc}, \quad \text{and} \quad \lambda = 4.5 \text{ kpc}.$$

The value of $\lambda/2$ lies almost midway between d and h_* in this case. Thus, there can be at most only one wavelength for the global structure. Indeed, if the epicyclic diameter were too high, no global spiral structure could be accommodated. The galaxy would be flocculent.

5.2.2. Length Scales Perpendicular to the Disk

Compared to the above length scales, the disk thickness is much smaller, with typical values $z_* \sim 300$ pc, $z_g \sim 100$ pc, respectively, for the stellar and the gaseous components. This would seem to suggest that the stellar disk is three-dimensional in nature while the gaseous disk is two-dimensional. Actually, the reverse is true for the consideration of the dynamical process, since the former has a larger "microscale" $d \sim 1.6$ kpc. In contrast, the gaseous component has a mean free path estimated to be of the order of 100 AU or less, which is much less than the thickness scale of 100 pc. Hence, the gaseous disk is in a three-dimensional distribution, and is therefore expected to be *turbulent* at such high speeds.

Discussions of turbulent (irregular) and laminar (regular) motions are usually done in terms of the Reynolds number. For the stellar systems, this concept is not proper. But we can define a ratio similar to it to assure the fluid dynamicists that there is nothing unusual for the stellar disk to have a smooth structure. If we represent the Reynolds number in terms of the thermal speed and the mean free path of the gas, the proper ratio should clearly be (within a factor of 3)

$$R = \frac{V_M}{V_m} \cdot \frac{\ell_M}{\ell_m},$$

where the subscripts M and m denote, respectively, macro- and microscales of velocity V and length ℓ. Applying this formula to the gas, one would indeed find a very large Reynolds number. Applying this formula to the stellar system, one finds a very low Reynolds number, since the ratio of the length scales is small. Regular smooth behavior is thus expected for the stellar disk. Physically speaking, this simply states that the cooperative behavior cannot have a scale smaller or comparable to the micro-scale (cf. rarefied gas dynamics studied by Knudsen as mentioned above).

5.2.3. A Process of Global Self-Regulation[2]

We continue our discussion of the observed existence of a *single* global spiral structure in the Pop II disk with little complication from structures on smaller scales.

In the above discussion of normal spirals, we pointed out that the stars are in epicyclic motion on scales comparable to that of the observed structures. One might expect that this is also true, in some sense, for barred spirals, although it is somewhat more difficult to quantify. We shall thus take a more general approach by emphasizing the tendency for the velocity dispersion of the stars to increase with time if they are under the influence of forces that are not strictly ordered. There is in general a perennial heating of the stars [3, p. 10].

The question is thus turned around: Why is there any global structure remaining despite this perennial trend. The answer is to be found by noting the following well-known fact. For an exact standing wave pattern rotating at a uniform angular velocity, the dynamical system is time independent in the framework rotating at that angular velocity. There is thus expected to be little further increase of velocity dispersion. In reality there would still be minor but perhaps unimportant increases, and the galactic disk would remain in a quasi-stationary state for a long time around the ideal solution. In other words, the natural evolution of the disk, under intrinsic mechanisms, would indeed *prefer* a single nonlinear mode. This is a process of *global* self-regulation between the mechanisms of the generation of global modes and the smoothing out of macroscopic structures by random stellar velocities. The nature of the mechanism can be better appreciated by contrasting it with that in the gaseous disk, where the resultant state is turbulent. The nonlinear process of having energy cascading down the scales of macroscopic motion is basically the same. The main

---

[2]This process of self-regulation is obviously distinct from that between gas and stars mentioned at the beginning of Section 5.2.

difference lies in the ratios of the length scales discussed above (5.2.1). It follows that there is a difference in the relative importance of the transfer of energy directly to the microscopic scale or first to smaller macroscopic scales. Eventually, the large size of the microscale of the stellar system makes the essential difference. On this last point, there is indeed similarity of the stellar problem with that in plasma physics.[3] In a number of experiments, it has been noted that the preferred wavenumber k may be estimated by the simple rule of thumb

$$k\rho \approx 1,$$

where ρ is the radius of gyration around the main magnetic field. In the galactic problem, the corresponding parameter (denoted by $\sqrt{x}$) appeared in the early first-studies of the dispersion relationship of density waves (see, e.g., [20]). It is generally of the order of unity.

5.3. N-Particle Simulation

Additional support for the existence of a single quasi-stationary wave pattern is provided by the N-particle simulation of the stellar system through the use of high speed computation [8], [43]. These authors found essentially standing wave patterns apparently in rigid-body rotation that can last as long as six revolutions. They have apparently succeeded in simulating smooth arm spirals without gas but with open pitch-angles similar to those found in many Sc spirals. Donner and Thomasson appeared to be surprised that the pitch angle cannot be correctly predicted by the original dispersion relation obtained by Lin and Shu (which holds only for small pitch angles) and Kalnajs (which was obtained for zero-pitch angle). The proper dispersion relationship to be used for comparison is the *cubic dispersion relation* for *finite*-pitch angles derived with an additional parameter J, which measures the surface density (see [3, p. 214] and the related discussions and references). According to this dispersion relationship, there are three branches of waves: the short, the long, and the open branches. It is the open branch (which is absent in the old dispersion relationship) that would represent spiral patterns at high pitch angles in the smooth arm-spirals. There then appears to be no discrepancy between analytical calculation and N-particle simulation.

6. SOME HIGHLIGHTS

As a conclusion to the brief review, we highlight a few of the general issues and their answers. These are (1) the need to formulate multiple mathematical models, each to be examined for its observational support; (2) the application of this approach to the working hypothesis of quasi-stationary spiral structure, and (3) the contrasting of the roles of intrinsic dynamical mechanisms and external tidal forcing, again in the perception of observational support.

---

[3]I wish to thank Y.Y. Lau for bringing this point to my attention.

6.1. Complexity and Simplicity

In a physical system as complicated as a galaxy, it is very difficult to embody the description of all aspects of observed phenomena in a single mathematical model, even though the relevant basic physical laws are well formulated mathematically. Different mathematical models may be needed to account for the phenomena observed in different physical contexts. We are challenged with the formulation of models with different initial conditions and boundary conditions appropriate to the physical context in question. The differences in physical contexts can be appreciated when we attempt to resolve the following issues. (A longer list may be found in [6, pp. 50–56]).

(1) How do barred spiral galaxies differ from normal spiral galaxies? Why have certain galaxies become barred, while others have not?
(2) Why are regular grand design spiral structures generally, but not always, two-armed?
(3) How can we explain the observed coexistence of morphologies within the same galaxy, especially in relation to observations in different wave bands?
(4) Is large-scale spiral structure in galaxies typically quasi-stationary or, rather, fast-evolving?
(5) If it is fast-evolving, is it a transient, occasional phenomenon or, rather, continually regenerated by internal mechanisms?
(6) How are spiral structures, of the various observed morphologies, excited in galaxy disks?
(7) How can we construct galaxy models that simulate categories of galaxies with a given spiral morphology?

Answers for a number of these issues have been given through the use of the modal approach in the Bertin-Lin monograph [3]. It should be noted that the fact we can reach clear understanding and even make some quantitative predictions is dependent on the fact that sufficient *simplicity* is already exhibited in the observational data. As pointed out above, the essence of this simplicity is contained in two early discoveries: the law and order in morphology discovered by Hubble and the coexistence of regular and irregular spiral structures in M51 discovered by Zwicky in different wave bands. The concept of a backbone of regular spiral structure in Pop II stars is thus strongly suggested and indeed well confirmed later by extensive data observed in infrared. On the other hand, Pop I objects exhibit only partial order, as shown through observational data obtained at optical and radio frequencies. Indeed, the high degree of irregularity observed on smaller scales suggests rather fast evolution in such structures.

6.2. The Working Hypothesis of Quasi-Stationary Spiral Structures (Empirical Support)

The theory is centered around the working hypothesis of quasi-stationary spiral structure whose dynamical basis is the modal approach. It is now clearly demonstrated, through both theoretical inferences and analysis of observational data. The idealized two-armed *mode* is established as a good first approximation to the spiral structure observed in the Pop II objects. Strong observational support is essential and available: there is a wide variety of empirical evidence in favor of the QSSS hypothesis. Some of them are the following.

(1) Interpretation of the Hubble sequence. FIG. 4.5, p. 115; FIG. 49, p. 122.
(2) Images of NGC 309 in blue and in infrared. FIG. 5.2, p. 146.
(3) Multiple wave band images of M51 (Rix and Rieke).
(4) Synchrotron emission in M51 with peak roughly coinciding with dust lane. FIG. 5.6, p. 151.
(5) Modelling M81. FIG. 5.2, p. 143; FIG. 5.9, p. 155; FIG. 5.10, p. 157.
(6) Modelling NGC 1300. FIG. 5.3, p. 144.
(7) Interpretation of the van den Bergh luminosity class (see [35, p. 296, FIG. 13.11]).

Some of these have been reexamined in this paper for further clarification of the issues and for further strengthening the observational support of the conclusions.

6.3. Intrinsic Dynamical Mechanisms and External Tidal Forcing

Let us now turn to a brief sketch of a few different physical contexts in which different roles are played by these two types of dynamical mechanisms.

In most cases of spiral galaxies, a nearby companion often appears to be absent even when the galaxy does have a well-developed spiral structure. As discussed above (Section 4), an examination of observational evidence leads to the conclusion that external excitation should be studied as an additional disturbance on a preexisting spiral structure, and that its effect is generally minor. (Of course, there are cases where the whole galaxy is greatly distorted by tidal forcing, but these are rare). Advocates of generation of spiral structure through tidal interaction should continue to identify cases where there is clear observational support. So far there are few cases, if any, where one can clearly demonstrate the absence of a a preexisting spiral structure prior to the most recent close encounter.

Steady external tidal forcing has been found to be important in the study of Saturn's rings and some other astrophysical phenomena. Shepherding satellites play an important role in the maintenance of such spiral structures. The role of bar driving has also been shown to be important in some cases, e.g., in the case of the 3-kpc arm in the Milky Way. (See papers by Shu, Yuan and their collaborators cited as references [28], [29], [32]–[34]). Block [6] suggested that the long and smooth dust-gas spirals in NGC 2841 is possibly a response to a bisymmetricpotential.

As mentioned above, barred structures should generally be regarded as a modal structure supported by intrinsic dynamical mechanisms. Specific cases of NGC 2589 and NGC 1300 are discussed in the Bertin-Lin monograph, where the morphology appears to be well consistent with the expectations of intrinsic modal dynamics. A broadened empirical support is always desirable in a semi-empirical theory. In this context, we are fortunate to have good systematic observational data at several wave lengths (e.g, [29]). Attempts to develop theoretical models for such galaxies would be desirable.

ACKNOWLEDGMENTS

I wish to thank George Contopoulos, Giuseppe Bertin, Louis N. Howard, Christopher Hunter, Y.Y. Lau, Donald Lynden-Bell, Ronald Probstein, and other persons at the conference for illuminating discussions. The exposition of this paper also benefited from the continual discussions over the years on the modelling of complex systems in fluid mechanics and galactic dynamics. These discussions culminated in the discussions with Lee Segel on the subject of transition to turbulence, where alternative scenarios have to be considered to get the total picture, as in many other subjects. These discussions led Christopher Hunter to suggest that I should present this paper to address the unsettled issues in galactic dynamics in the spirit described in Segel's paper. My attempts to resolve controversies in this approach to complexity and multiple modelling showed that one should fully recognize the desirability to divide the modelling process into several steps such as (1) formulation of alternative scenarios, (2) matching the models with proper physical contexts, and (3) consideration of the frequency of occurrence of these contexts in the real world. There is nothing unusual in this recognition; unfortunately, however, the last two steps are sometimes given insufficient attention and care. Controversies then arise.

REFERENCES

1. ADLER, D.S. & D.J. WESTPFAHL. 1996. Astron. J. **111**: 735–749.
2. BERTIN, G. 1996. Spiral structure in galaxies: Competition and cooperation of gas and stars. *In* New Galactic Perspectives in the New South Africa, D.L. Block & J.M. Greenberg, Eds.: 227–242. Kluwer Dordrecht. (Other valuable papers on infrared imaging may be found in this volume.)
3. BERTIN, G. & C.C. LIN. 1996. Spiral Structure in Galaxies: A density wave theory. MIT Press. Cambridge, MA.
4. BERTIN, G., C.C. LIN, S.A. LOWE & R.P. THURSTANS. 1989. Astrophys. J. **338**: 78.
5. BERTIN, G., C.C. LIN, S.A. LOWE & R.P. THURSTANS. 1989. Astrophys. J. **338**: 104.
6. BLOCK, D.L. 1996. Changing morphology and dust content in galaxies; in the same volume as Bertin. 1996.
7. BLOCK, D.L., G. BERTIN, A. STOCKTON, P. GROSBOL, A.F.M. MORWOOD & R.F. PELETIER. 1994. Astron. Astrophysics **288**: 365–382.
8. DONNER, K.J. & M. THOMASSON. 1994. Astron. Astrophys. **290**: 785–795.
9. HERNQUIST, L. 1990. In Dynamics of Interacting Galaxies, R. Widen, Ed: 108–117. Springer-Verlag. Heidelberg.
10. HUBBLE, E. 1936. The Realm of the Nebulae. Yale University Press. New Haven, CT.
11. JEANS, J.H. 1929. Astronomy and Cosmology. 332. Cambridge University Press. (Reprinted in 1961 by Dover, New York.)
12. KENT, S.M. 1987. Astron. J. **93**: 816, 1062.
13. LIN, C.C. 1997. Global Spiral Patterns in Galaxies. Paper presented at the 21 Century Chinese Astronomy Conference, University of Hong Kong, August 1–4, 1996.
14. LIN, C.C. 1966. SIAM Review **4**: 839.
15. LIN, C.C. 1971. IAU Gen. Assem. Proc. 88–121.
16. LIN, C.C. & G. BERTIN. 1995. Global wave patterns in galaxies: Their generation and maintenance. *In* Waves in Astrophysics. Annals of the New York Academy of Sciences. New York. Vol. 773, 125–144..
17. LIN, C.C. & Y.Y. LAU. 1979. Stud. Appl. Math. **60**: 97.
18. LIN, C.C. & F.H. SHU. 1964. Ap. J. **140**: 646.

19. LIN, C.C. & F.H. SHU. 1971. Brandeis University Summer Institute in Theoretical Physics, 1968: Astrophysics and General Relativity, Vol. 2: 239–329. Gorden & Breach. New York.
20. LIN, C.C. & F.H. SHU. 1966. Proc. Nat. Acad. Sci. **55:** 229.
21. LIN, C.C. & L.A. SEGEL. 1974. Mathematics Applied to Deterministic Problems. Macmillan Co. New York.
22. LIN, C.C., C. YUAN & F.H. SHU. 1969. Astrophys. J. **155:** 744.
23. LINDBLAD, B. 1963. Stockholm Observatory Annalen **22:** no.5, p.3.
24. LINDBLAD, P.A.B., P.O. LINDBLAD & E. ATHANASSOULA. 1996. Astron. Astrophys. **313:** 65–90.
25. LINDBLAD, P.O. 1960. Stockholm Obs. Ann. **21:** 3–73.
26. LOWE, S.A., W.W. ROBERTS, J. YANG, G. BERTIN & C.C. LIN. 1994. Astrophys. J. **427:** 184.
27. LYNDEN-BELL, D. 1997. Observatory **117:** 70.
28. OORT, J.H. 1969. IAU Symposium Proceedings No. **38:** 1.
29. PRIETO, M., S.T. GOTTESMAN, J.L. ANGUERRI & A. VARELA. 1997. Astron. J. **114:** 1413–1426.
30. RIX, H.W. & M.J. RIEKE. 1993. Astrophys. J. **418:** 123.
31. SALO, H. & E. LAURIKAMIAN. 1993. Astrophys. J. **410:** 586.
32. SANDAGE, A. 1961. The Hubble Atlas of Galaxies, Carnegie Institution of Washington, Washington, D.C.
33. SANDAGE, A., K.C. FREEMAN & N.R. STOKES. 1970. Ap. J. **160:** 831.
34. SEGEL, L. 1995. Complexity **1:** 18–25.
35. SHU, F.H. 1982. The Physical Universe: An Introduction to Astronomy. University Science Books. Mill Valley, California.
36. SHU, F.H., C. YUAN & J. LISSAUER. 1985. Astrophys. J. **291:** 356–376.
37. TOOMRE, A. 1981. In The Structure and Evolution of Normal Galaxies. S.M. Fall & D. Lynden–Bell, Eds: 111. Cambridge Univ. Press. Cambridge, England.
38. VISSER, H.C.D. 1980. Astron. Astrophys **88:** 149, 159.
39. YUAN, C. 1984. Ap. J. **281:** 600–613.
40. YUAN, C. & CASSEN. 1994. Ap. J. **437:** 338–350.
41. YUAN, C. & YE CHENG. 1991. Ap. J. **376:** 104–114.
42. YUN, M.S., P.T.P. HO & K.Y. LO. 1994. Nature **374:** 530–532.
43. ZHANG, X. 1996. Astrophys. J. **457:** 125.
44. ZWICKY, F. 1957. Morphological Astronomy. Springer-Verlag. Berlin.
45. TOOMRE, A. & TOOMRE. 1972. Ap. J. **178:** 623.
46. ELMGREN, B.G., D.M. ELMGREN & P.E. SEIDEN. 1989. Ap. J. **343:** 602–607.

Candidates for Abundance Gradients at Intermediate Red-Shift Clusters

RENATO A. DUPKE

Department of Physics and Astronomy, University of Alabama, Box 870324, Tuscaloosa, Alabama 35487-0324

ABSTRACT: In this work I present evidence for the possible existence of central abundance enhancements in two clusters of galaxies at intermediate redshifts: Abell 1413 ($z \approx 0.14$) and Abell 2390 ($z \approx 0.23$). The results are based on ASCA (Advanced Satellite for Cosmolgy and Astrophysics) data reduction of x-ray spatially resolved spectra. The Fe abundance for A1413 drops from 0.37 ± 0.08 (in the central 2 arcmin) to 0.20 ± 0.06 (in the outer 2–6 arcmin), while in A2390 the very central abundance is 0.77 ± 0.45 (central arcmin) and drops to 0.12 ± 0.12 in the outer regions.

1. INTRODUCTION

That heavy elements shed from stars in galaxies may contaminate intracluster gas was suggested [1] even before the discovery of iron lines in the x-ray spectra of clusters of galaxies [2], [3]. It became clear that an understanding of the mechanisms through which heavy elements were ejected from galaxies is of fundamental importance to the understanding of the origin and evolution of the intracluster medium (ICM) and provides strong clues to the evolutionary history of the clusters of galaxies themselves. Nevertheless the precise contribution of different enrichment mechanisms is not well known. Ram-pressure stripping of gas from galaxies and protogalactic winds are considered the most prominent mechanisms for enriching the inner and outer parts of the ICM, respectively. Resolving this issue is of major astrophysical and cosmological importance since it bears on theories of galaxy and cluster formation.

Protogalactic winds and ram pressure stripping can be discriminated by their correspondent supernova ejecta contribution. Protogalactic wind models assume an early-established wind-powered mainly by SN type II explosions [4]. This is due to the fact that SN II are typically lonely massive stars with a very short life span. The explosion in these stars occurs due to a core collapse of the massive progenitor (with plenty of hydrogen). After the SN II wind is established, star formation ceases and with it the formation of SNe II themselves, so that the wind tends to stall [5], [6].

SNe type Ia are believed to originate from deflagration or detonation of an accreting white dwarf. Consequently SN Ia require longer times to form than SN II (on the order of the evolution time for the binary system). In this way SN Ia may actually help maintain protogalactic winds for a longer time, provided their production rate declines less steeply than the stellar mass loss from the original wind [7]. Whether or not SN Ia contribute to prolong protogalactic winds they will continuously contribute to the enrichment of the interstellar medium in galaxies, so that, at later times during the cluster evolution, if the galaxy ejects its interstellar medium into the ICM, it will be enriched, considerably, by SN Ia material. Therefore, ram pressure stripping, acting constantly during cluster evolution (as long as galaxies have gas to be stripped) should enrich the ICM considerably with SN Ia material. Furthermore, since the condition for ram pressure to occur depends on the density of the external medium, this mass deposition by SN Ia should be significant in the central regions of clusters, where the gas density is higher.

The best way to determine the relative importance of these enrichment mechanisms and, therefore, to understand the evolution of the ICM (and of the galaxies involved in the process of its makeup) is through the analysis of spatially resolved x-ray spectra. In the last few years several clusters have been observed to show significant central abundance enhancements (gradients), allowing us to test the contribution of different metal injection mechanisms at different spatial scales in clusters. *ASCA* provided, for the first time, the possibility of discriminating between supernovae type enrichment for some nearby clusters, through the analysis of detailed radial abundance distributions and individual abundance ratios.

The results for nearby clusters suggests that, for well-behaved clusters that present abundance gradients (e.g., Abell 496), the central abundance (and the gradient itself) could be due to an excess of SN Ia ejecta if the bulk of the cluster is enriched mainly by SN II [8]. The overall abundance ratios measured for clusters [9]–[11], and in particular the O/Fe ratio (which can be 2 orders of magnitude higher for SN II than for SN Ia) suggests that the main contributor of metals in the outer parts of the cluster are SN II. The excess of SN Ia material in the central regions could be due to ram-pressure stripping, but could also be due to normal stellar mass loss from the central galaxy. The spatial resolution of *ASCA* does not allow such fine discrimination easily.

The direct observation of abundance and temperature distributions at intermediate-high redshifts is rather nontrivial due to the spatial resolution and calibration uncertainties of *ASCA* [12]. However, the physical state of the intracluster gas at those distances may provide very important clues of the evolution and energetics of the intracluster gas. For this reason we undertook a systematic search of public *ASCA* data for abundance and temperature distributions in intermediate red-shift clusters. We show below evidence for abundance gradients in two of these clusters: Abell 1413, Abell 2390 (FIGURES 1a and 1b). These clusters are well-behaved cooling flow clusters. We used standard *ASCA* data reduction techniques and specific details on the reduction will be described in a separate publication.

2. RESULTS AND DISCUSSION

2.1. ABELL 1413

Abell 1413 is a mostly regular cooling flow cluster ($dM/dt \approx 200$–$300\ M_\odot\ \mathrm{yr}^{-1}$) [13]. It is at a redshift of 0.1427. Previous x-ray analysis of this cluster with Ginga indicated a high average temperature (≈ 8.5 keV) [14]. The average x-ray temperature is of 7–8 keV and the average abundance is 0.25 (determined from simultaneous spectral fittings of GIS2 & 3 and SIS0). *ASCA* reduction of spatially resolved spectra for this cluster indicates a mild but significant central abundance enhancement (FIGURE 1a) varying from 0.37 ± 0.08 (in the central 2 arcmin) to 0.20 ± 0.06 (in the outer 2–6 arcmin). One arcmin for this cluster corresponds to $\approx 250\ h_{50}^{-1}$ kpc so that this abundance gradient, if confirmed, seems to be present at large spatial scales, similar to the case for AWM 7 [15].

Even though the cooling flow is not observed in *ASCA* temperature distribution within a 2 arcmin region, it was included in the models to test whether the abundance gradient described above is an artifact of our choice of spectral model. The cooling flow spectral model is characterized by a maximum and minimum temperature, an abundance, a slope parametrizing the temperature distribution of emission measures, and a normalization, which is the cooling accretion rate. Adopting a slope correspondent to an isobaric cooling flow we obtain a mass inflow of $\approx 320\ M_\odot$ yr^{-1}. There is no significant enhancement on the fitting statistics with the addition of the cooling flow component and the central abundance enhancement is still apparent (FIGURE 1a).

2.2. ABELL 2390

Abell 2390 is a rich lensing cluster with a strong cooling flow ($dM/dt \approx 1500\ M_\odot$ yr^{-1}). It is one of the brightest clusters known at high redshifts [16]. Lensing indicates a double mass clump, but only one bright galaxy is observed [17]. It is at a red-shift of 0.230. The average x-ray temperature is 8.9 keV and the average abundance is 0.27 (from simultaneous spectral fittings of both GISs and SISs). As in the case of Abell 1413, *ASCA* reduction of spatially resolved spectra also shows a significant abundance enhancement (FIGURE 1b) within the central arcmin ($\approx 400\ h_{50}^{-1}$ kpc), where the abundance is 0.77 ± 0.45 while in the outer 1–4 arcmin the abundance is 0.12 ± 0.12.

Abundance gradients in clusters of galaxies may provide a powerful tool in discriminating between different mechanisms of metal injection into the ICM. If ram-pressure stripping contributes considerably to the ICM enrichment, then one should expect a mass deposition in the central regions of clusters and this material would be significantly enriched by SN Ia. The analysis of other centrally abundance-enhanced clusters at low red-shift (e.g., A496, A2199 [8]) suggests that this scenario is consistent with observations.

The existence of abundance gradients in A1413 and A2390 is consistent with the overall properties of these clusters; they are both cooling flow clusters with smooth gas distributions and no sign of strong interactions (so that abundance inhomogeneities can be undisturbed for a long time).

Even though *ASCA* observations of intermediate red-shift clusters cannot constrain well the abundance of individual elemental abundances (other than Fe) and abundance ratios, they allow for the determination of the existence of such Fe central abundance enhancements. The analysis of individual abundance ratios will probably have to wait for AXAF observations. These two clusters were proposed as targets for

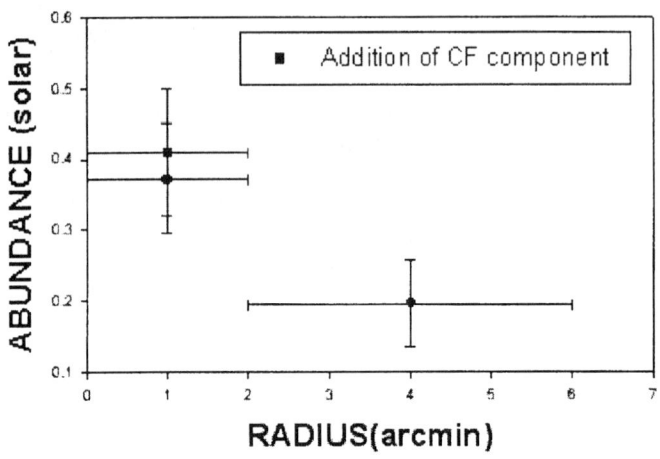

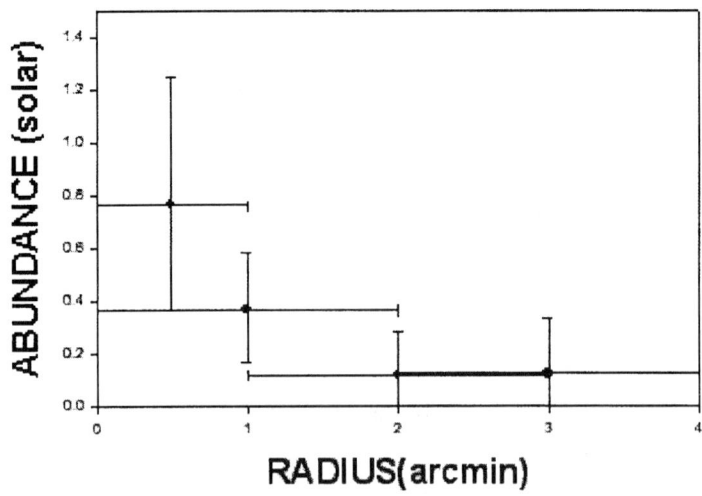

FIGURE 1. (a, top) and (b, bottom). Abundance distributions in A1413 (a) and A2390 (b). Joint spectral fittings of both GISs and SISs. The erors are 90% confidence level.

AXAF by Speybroec (Abell 1413) and Fabian (Abell 2390), each with requested exposures of 10 ksec.

ACKNOWLEDGMENTS

The author would like to thank Barbara Eckstein for helpful discussions and for carefully reviewing of the manuscript and Dr. Henry Kandrup for very interesting and stimulating discussions. The invitation to take part in the meeting is gratefully acknowledged.

REFERENCES

1. LARSON, R. B. & H.L. DINERSTEIN. 1975. Publ. Astron. Soc. Japan **87**: 911.
2. SERLEMITSOS P.J., B.W. SMITH, E.A. BOLDT, S.S. HOLT & J.H. SWANK. 1977. Astrophys. J. **11**: 63.
3. MITCHELL, R.J., J.L. CULHANE, P.J.N. DAVISON & J.C. IVES. 1976. Mon. Not. R. Astr. Soc. **175**: 29.
4. LARSON, R.B. 1974. Mon. Not. R. Astr. Soc. **169**: 229.
5. DAVID, L.P., W. FORMAN & C. JONES. 1990. Astrophys. J. **359**: 29.
6. RENZINI, A, L. CIOTTI, A. DERCOLE, S. PELLEGINI. 1993.Astrophys. J. **419**: 52.
7. DERCOLE, A., A. RENZINI, L. CIOTTI, S. PELLEGINI. 1989. Astrophys. J. **341**: 9.
8. DUPKE, R.A. & R.E. WHITE. 1998a,b. In preparation.
9. MUSHOTZKY, R., M. LOEWENSTEIN, K.A. ARNAUD, T. TAMURA, Y. FUKAZAWA, K. MATSUSHITA, K. KIKUCHI & I. HATSUKADE. 1996. Astrophys. J. **466**: 686.
10. LOEWENSTEIN, M. & R.F. MUSHOTZKY. 1996. Astrophys. J. **466**: 695.
11. WHITE, R.E. III. 1991. Astrophys. J. **367**: 69.
12. TAKAHASHI, T., Y. IKEBE, M. MARKEVITCH, Y. TAWARA, Y. FUKAZAWA, A. FUKAZAWA, H. HONDA, Y. ISHISAKI, K. KIKUCHI, K. MAKISHIMA, H. MATSUMOTO, H. MATUZAWA, T. OHASHI, M. TASHIRO, M. WATANABE & N. YAMASAKI. 1994. ASCA Internal Report, October.
13. ALLEN, S.W., A. FABIAN, A.C. EDGE, H. BOHRINGER & D.A. WHITE. 1995. Mon. Not. R. Astr. Soc. **175**:741.
14. DAVID, L.P., A. SLYZ, W. FORMAN, D. VRTILEK & K.A. ARNAUD. 1993. Astrophys. J. **412**: 479.
15. EZAWA, H., Y. FUKAZAWA, K. MAKISHIMA, T. OHASHI, F. TAKAHARA, H. XU & N.Y. YAMASAKI. 1997. Astrophys. J. **490**: 33.
16. EBELING, H., W. VOGES, H. BOHRINGER, A.C. EDGE, J.P. HUCHRA & U.G. BRIEL. 1996. Mon. Not. R. Astr. Soc. **252**: 428.
17. MIRALDA-ESCUDE, J. & A. BAUL. 1995. Astrophys. J. **449**: 18.

Scaling Regimes in the Distribution of Galaxies

G. MURANTE,[a] A. PROVENZALE,[a] E. A. SPIEGEL,[b] AND R. THIEBERGER[c]

[a]Istituto di Cosmogeofisica, Corso Fiume 4, 1-10133 Turin, Italy
[b]Department of Astronomy, Columbia University, New York, New York 10027
[c]Department of Physics, Ben Gurion University, Beer Sheva, Israel

ABSTRACT: If we treat the galaxies in published redshift catalogues as point sets, we may determine the generalized dimensions of these sets by standard means, outlined here. For galaxy separations up to about 5 Mpc, we find the dimensions of the CfA galaxy set to be about 1.2, with only a modest indication of multifractality. For larger scales, out to about 30 Mpc, there is also good scaling with a dimension of about 1.8. For even larger scales, the data seem too sparse to be conclusive, but we find that the dimension is climbing as the scales increase. We report simulations that suggest a rationalization of such measurements, namely that in the intermediate range the scaling behavior is dominated by flat structures (pancakes) and that the results on the smallest scales are a reflection of the formation of density singularities.

1. INTRODUCTION

An interesting question in modern cosmology is how to reconcile the very inhomogeneous distribution of galaxies with the very isotropic distribution of the cosmic background radiation. In the current conventional picture the marginally detectable anisotropies of the radiation field reveal the seeds of the impressive structures seen in the galaxy distribution. The transition from weak initial perturbation to strong clustering is supposed to be initiated by gravitational instability. This process, which is weak in an expanding medium, is nevertheless called upon to take the perturbations into a nonlinear collapse phase that produces the well-defined structures that are perceived in the lumpy distribution of galaxies.

In the present work, we try to make sense of the complicated galaxy distribution in terms of simple — perhaps overly simple — pictures. We are suggesting that the data on the galaxy distribution imply that there are two principal scales of structure formation, each with its own characteristics. The aim of our informal general discussion is to explain why we believe that we can detect these regimes in the analysis of publically available galaxy catalogues. The existence of a possible third regime at the largest scales may also be discernable. The final clarification of course awaits the coming of the tidal wave of data that should soon engulf us and that may perhaps wash away all our present conceptions.

Before we describe how the galaxy distribution is quantified and the regimes are detected, we briefly describe the simulations of gravitational clustering that have led to the preconceptions that we bring to the analysis. Then we outline one of the main methods of quantification of structure that has been used in this problem. Finally, we present our general interpretation of the observed structure.

2. DENSITY SINGULARITIES

Gravitational instability is usually assumed to be the primary mechanism that shapes the galaxy distribution. There may be other interactions that play a role in the process, depending on the era and on what matter and fields are present, but we ignore those here. Most of the models on the market are based on a strong preponderance of dark matter, and this is consistent with many lines of evidence. It is not clear what this matter is, but here we shall suppose that it is rather cool and describe a CDM (cold dark matter) scenario.

Early analytic studies have indicated that gravitational collapse may generate virialized mass concentrations, with approximately isothermal density profiles [11], [6], [1], [7] with details depending on the initial density perturbations and on the value of Ω. A study by Henriksen and Widrow [13] based on the Vlasov equation in expanding space produces local singular density distributions with density proportional to $r^{-\alpha}$ where $1.5 \leq \alpha \leq 3$ and r is the distance from the singularity.

The means for simulating structure formation are diverse. Some effort has gone into modeling with the Vlasov equations complemented by the Poisson equation. This latter would in principle be a good way to proceed if one could really carry it out. In fact, it is difficult to follow in detail the development of plasma turbulence through the collective effects that it contains. Such collective effects perhaps make the gravitational problem more like a fluid dynamical problem than a particle dynamics problem.

More typically, numerical simulation of the structure formation process has been pursued by N-body techniques, where point masses, representing fluid elements, move under the influence of their own gravitational interaction in an expanding background. On the scales of galaxies and galaxy clusters, such simulations generally suggest the presence of strong density concentrations with a local density growing algebraically, at least in a certain range of scales [19], [20].

Many simulations have been performed and we may mention the recent report of large scale calculations of this kind by Zurek and Warren [29]. Our own simulations, which we describe in this section, were performed with an adaptation of Couchman's [5] public AP3M code for present purposes. These simulations have $N_p = 128^3$ massive particles on a Cartesian grid of $N_g = 128^3$. For the initial conditions, we introduce, as is usual, a small perturbation superposed onto a homogeneous density field that is a solution to the Friedmann-Lemaitre equations. The average density has been taken to be equal to the critical density ρ_{cr} so that $\Omega = \rho_{av}/\rho_{cr} = 1$; this makes it easier for structures to form than do the observationally suggested values which are noticeably less than unity. The initial conditions have Gaussian, scale-free density perturbation spectra with random Fourier phases and density power spectrum

$$P(k) \propto k^n. \tag{2.1}$$

During the computed evolution, strong density concentrations form. For each particle cluster, we have located the center of mass and determined the radial density profile. Each profile has been normalized to the half-mass radius (the radius within which half of the mass in a cluster is containted) and the average density profile has been calculated.

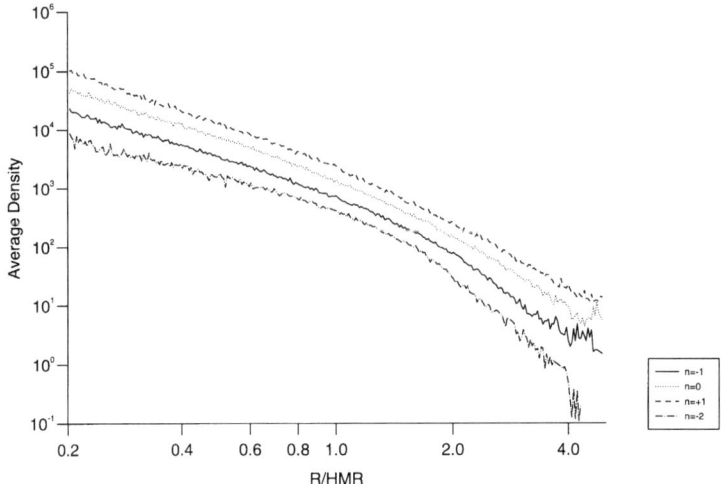

FIGURE 1. Average density profiles of the dark halos for the four scale-free initial conditions considered in the text, at the time when the density variance is 1 at 1/16th of the simulation box.

FIGURE 1 shows the averaged dark halo profiles for the four initial conditions considered at the time when the variance of the density field, coarse-grained on a scale of 1/16 of the domain size, is unity. At scales smaller than the half-mass radii, the dark halos have power-law profiles for all the four initial conditions considered. These profiles are slightly steeper for larger n, showing also a tendency to display a larger range of power-law behavior for larger values of n.

What is the reason for the differences between the four average profiles, and what is their time evolution? Are they evolving in time or do they have a logarithmic slope which is constant in time?

FIGURE 2 shows the average profiles of collapsed objects, obtained from the three initial conditions with $n = 1, 0, -1$, at different times. This figure shows that, independently of the initial conditions, the slope of the profile is a function of the central density of the collapsed object, the profile being steeper for larger central density. The average profiles of objects with similar central density, produced by simulations with different initial conditions, are the same. For the three spectral indices shown in FIGURE 3, the profile steepens, and comes closer to a power-law, as time increases.

A different behavior is observed for $n = -2$. In this case, shown in FIGURE 3, the profiles at different times do not evolve; rather, they keep the same shape and logarithmic slope. The initial conditions for the CDM scenario are often assumed to have a power spectrum with $n \approx -2$ on small scales. This suggests that dark halos for CDM initial conditions do not evolve in time, once the transients have decayed.

We turn now to a discussion of the observed structures. In doing this we leave several questions unanswered: Why do the $n = -2$ initial conditions behave so differently from the other three cases considered? For $n = 1, 0, -1$, do the dark halo profiles evolve in time forever, with an ever-increasing steepness, or do they reach a limiting

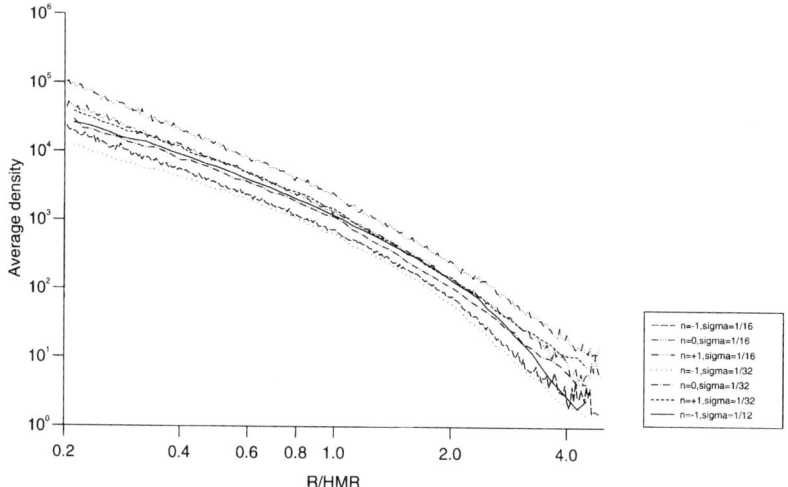

FIGURE 2. Average profiles of the dark halos produced by cosmological N-body simulations for three scale-free initial conditions with spectral indices $n = 1, 0, -1$, at different evolutive times.

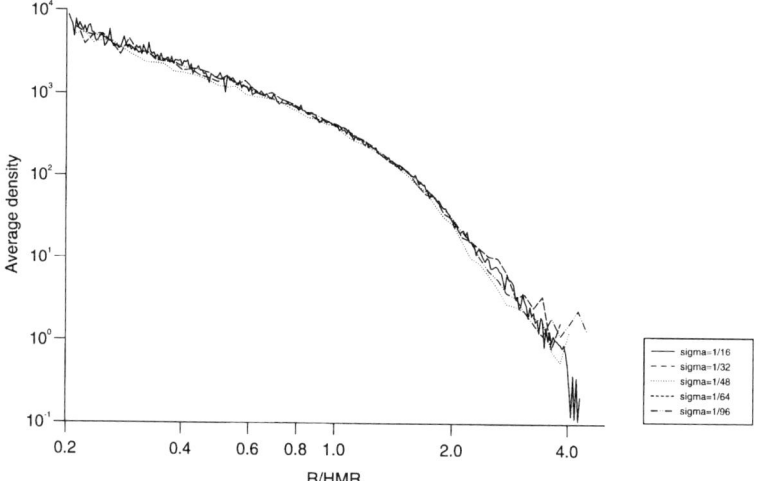

FIGURE 3. Average profiles of the dark halos produced by a cosmological N-body simulation with scale-free initial conditions and spectral index $n = -2$, at different evolutive times.

slope after which they do not evolve further? To answer these questions, one should properly understand the behavior of gravitational collapse at small scales. On more general grounds, we recall that the results reported here (and in several other numerical studies) refer to cold initial conditions. Does the addition of random dispersion velocities significantly modify the picture? And, finally, what happens when one includes hydrodynamical effects, such that the collapse is not purely gravitational any longer?

3. QUANTIFYING COSMIC SCALING

Early analyses of the galaxy distribution were based upon the the correlation function [10], [22], but, in recent years, the correlation integral [9], [21], [2] has been increasingly used for such studies. If the galaxy distribution is a fractal, the latter is advantageous since the definition of the correlation involves the density of galaxies. Though the notion of density is intuitively reasonable, it is not well-defined for a fractal [3], [4] and this makes the use of correlations problematic. At any rate, in the study of analytically understood point sets, the correlation integral has been found more reliable [27], so it seems worthwhile to briefly describe it here. We treat the galaxies as a point set for this purpose.

Let $\mathcal{N}_i(r)$, be the number of galaxies of the set lying within a distance r of the typical (the ith) galaxy. This number is then averaged over the galaxies in the set to give us the number of galaxies within a distance r of a typical galaxy. If there are a lot of galaxies around, we may replace the sum in the calculation of $\mathcal{N}$ by an integral and write

$$\mathcal{N}(r) = 4\pi \int_0^r n(s)s^2 ds . \tag{3.1}$$

If the galaxies in the set are found in a finite volume, we may divide the total number of them by this volume and call this n_0. The correlation used in the study of the galaxy distribution may then be written as $\xi = n/n_0 - 1$. (The choice of n_0 is not crucial and it may be defined otherwise than here.) For a uniform Poisson distribution, $\xi = 0$, the galaxies are uncorrelated and $n = n_0$. If ξ is positive, the galaxy positions are correlated.

When r goes to zero, so must $\mathcal{N}(r)$ and the simplest way to achieve this is to let n go as a power, say, $n = n_0(r/r_0)^{-\gamma}$, where r_0 is a conveniently chosen length. For small r/r_0, we see that $\xi \propto r^{-\gamma}$, so there seems little to choose between the two approaches. But when the data are few, the tendency is to let r/r_0 get close to unity, and differences appear in the results. Enough has been written about these differences so that we have simply chosen one approach, which we now outline.

The correlation integral [9] can be defined as

$$C_2(r) = \frac{1}{N(N-1)} \sum_i \sum_{j \neq i} \Theta(r - |\mathbf{X}_i - \mathbf{X}_j|) , \tag{3.2}$$

where Θ is the Heaviside function, N is the number of galaxies in the set, and their coordinates are $\mathbf{X}_j$. (Modifications of this formula to allow for the effects of the finiteness of the sample have been used in dealing with the data. We omit these technicalities here, but refer the reader to Ref. [18].) We see from the formula that $C_2(r) = \mathcal{N}(r)/N$. The quantity $C_2(r)$ is thus proportional to the volume integral of $n(r)$. As r gets small, C_2 must go to zero and for general distributions, in the limit

$r \to 0$, we have $C_2 \propto r^{D_2}$. The exponent D_2 is called the correlation dimension and we see that $D_2 = 3 - \gamma$ [23].

As with the correlations, it is possible to study higher-order statistics. To generalize the correlation integral formalism [9] one introduces correlation integrals defined as

$$C_q(r) = \left(\frac{1}{N}\sum_i \left[\frac{1}{N-1}\sum_{j \neq i} \Theta(r - |\mathbf{X}_i - \mathbf{X}_j|)\right]^{q-1}\right)^{\frac{1}{q-1}} \quad (3.3)$$

where q is a parameter that defines the order of the moment. For $q = 2$ a two-point probability (second-order moment) is evaluated and the standard correlation integral is recovered. More generally, for any integer value of q, $C_q(r)$ is the fraction of q-tuples in the set whose members lie within a distance r of one another. For sufficiently small r, C_q will go to zero for $q > 1$ and, for a typically well-behaved set, it will vanish like r^{D_q}. The index D_q is called a generalized or Renyi dimension [25], [12]. A fractal for which the dimensions are all the same (D_q independent of q) is called a homogeneous fractal, or a monofractal. The more general cases with D_q depending on q are called multifractals. Some authors reserve the general term fractal for the special case of a monofractal, but here we retain the general sense of the term fractal, with the multifractal as a particular case.

4. LACUNARITY

Another feature of the scaling analysis may ultimately prove interesting for the study of the galaxy distribution once the data are sufficiently abundant. As yet we have only some preliminary results on this, but we find them intriguing enough to mention here.

For the typical case, the correlation integrals go to zero like a power of the separation. This behavior may be considered as the first term in an asymptotic expansion of $\log C_q$ in $\log r$. More generally, we may seek an expansion of the form

$$\log C_q(\log r) = D_q \log r + E_q + \frac{F_q}{\log r} + \ldots . \quad (4.1)$$

Higher terms are hard to detect when there are few data, but it may be possible to get an estimate of the second term. When only two terms are kept in the expansion, we have the representation

$$C_q = \Lambda_q r^{D_q}. \quad (4.2)$$

Mandelbrot [15] has named Λ the lacunarity.

A weak dependence of E_q on r is compatible with this representation. This is easily seen for the case of the homogeneous fractal, for which the generalized dimensions, D_q, are all the same, $D_q = D$, for example. We may expect in that case that any statistical moment $C(r)$ satisfies the scaling law

$$C(r) = aC(br) \tag{4.3}$$

with constant a and b. This functional equation has solutions of the form (4.2) with

$$D_q \equiv D = \frac{\log a}{\log b}. \tag{4.4}$$

As we see, (4.2) and (4.4) satisfy the functional equation (4.3) even when Λ depends on $\log r$ provided that $\Lambda(\log r - \log b) = \Lambda(\log r)$. In this case, we call Λ the lacunarity function, or LF, and we may identify $\log \Lambda$ with E_q in the asymptotic development, so long as E_q remains of order unity. The appearance of a periodic dependence of E_q on $\log r$ is typically seen in monofractals, when there are enough data. Solis and Tao [26] have seen corresponding oscillations in theoretical multifractals, though more weakly. In the case of the theoretical fractals, the period of the LF is a remnant of the decimation procedure that produced the fractal.

5. COSMIC SCALING REGIMES

We concentrate here on the analysis of redshift catalogues, so we shall not enter into the issues that the study of position catalogues entail, such as the effect of projection of the spatial distribution onto the celestial sphere. A full discussion of this issue would involve the basic notions developed in the study of stellar statistics and the analysis of radio source counts. In the latter, a dependence on r was often overlooked and anomalies in the data were rationalized by assuming a dependence on time. We shall not go that far back into the history of the subject.

The first analyses of the distribution of the galaxies on the celestial sphere to gain wide acceptance were based on the correlation function and they revealed an approximate scaling regime on scales smaller than about $5\ h^{-1}$ Mpc, where h is the Hubble constant in units of 100 km/sec Mpc^{-1} [10], [22]. Out to these scales, the two-point galaxy correlation function has an approximate power-law shape with $\xi(r) \propto r^{-\gamma}$ where $\gamma \approx 1.8$ [22]. In the fractal vernacular, this would mean that the correlation dimension of the galaxy distribution is $D_2 \approx 1.2$ out to an outer scale of $5\ h^{-1}$ Mpc.

When redshift data were later analyzed using correlation integrals, the result $D_2 \approx 1.8$ was found [27], [16], [2]. The explanation for this discrepancy with the earlier work, at least the one that we adopt, is that the scaling law on scales under 5 Mpc is different from those at larger scales [8], [17]. We have $0.8 < D_q < 1.4$ for $q > 2$ at scales below $5\ h^{-1}$ Mpc, whereas, on intermediate scales, the value of D_2 is just under 2; we find 1.8 [18]. On the very largest scales, above about 30 Mpc, we find that D_2 has begun another rise.

To understand the meaning of $D_2 \approx 1.2$ for scales less than 5 Mpc, we recall the results of the N-body simulations. Once strong condensation has begun, pancakes form. These break up into clusters and the simulations reported in Section 2 suggest that the collapse into clusters continues into the formation of local structures with approximate power-law density profiles around their centers. Within the singular concentrations, there is no fractal structure in the sense that this term is now normally used. When there are such local smooth structures, the usual notion of density may

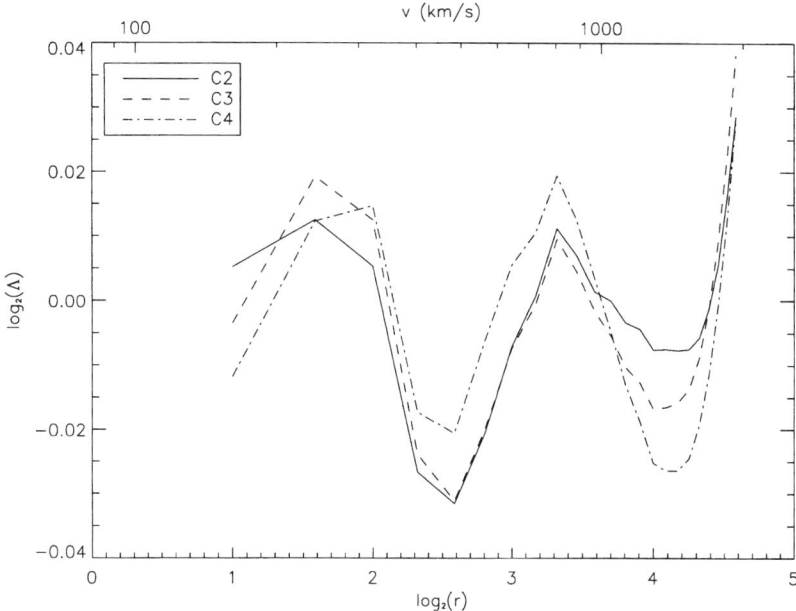

FIGURE 4. Lacunarity function for $q = 2, 4, 6$ for the CfA galaxies in the Northern galactic emisphere.

be used, even if its definition is still not so clear. As mentioned in the discussion on dynamics, roughly spherical structures form, having densities varying like $r^{-\alpha}$, with r being the distance from the singularity center. The value $\alpha = 2$, which is appropriate for the isothermal sphere, is just the one that produces flat rotation curves in a disk around a singularity center.

To see the connection to the analysis of the data, we have computed the dimension of a random distribution of density singularities with power-law density distributions. We have found that this simple representation of the structure with $\alpha \approx 2$ leads to global scaling properties like those observed on scales seen for sizes under about 5 Mpc, for a density exponent, α, of order 2. On the other hand, since the simulations show that the first structures to form are highly flattened, as anticipated by Zeldovich [28] it makes sense that we find fractal dimensions of about 2 on the intermediate scales. These two separate regimes correspond to the two observed correlation dimensions.

As we go to larger scales, D_2 seems to be increasing, but the data are too sparse as yet for a definite conclusion. Many believe that there has not been enough time to have allowed the formation of well-defined structures on scales of hundreds of megaparsecs. On the other hand, Pietronero and collaborators defend the de Vaucouleurs idea that there is structure at all scales in terms of these same data. That vision may call for a "spooky action at a distance" in Einstein's famous phrase and so test our notions of causality.

We conclude that the considerations we have outlined here rationalize the observed exponents and the presence of the two principle regimes. We are trying to go further with the data currently available to us by looking at finer details. To do this, we have made an attempt to detect the lacunarity function of the galaxy distribution. For this purpose, we have used a sample of 30,000 galaxies from the CfA-ZCAT galaxy catalogue [14]. The reductions are described elsewhere [24]. A plot of

$$\log \Lambda = \log C_q - D_q \log r \tag{5.1}$$

versus $\log r$ for $q = 2, 3, 4$ is shown for the northern galaxies (see FIGURE 4). We see perhaps one period of the anticipated oscillation.

Since the amplitude of the oscillation in the LF is so small, we must of course be wary in accepting it as real. Nevertheless, there are some features that do make it seem worth pursuing this matter further. First, there is the apparent constancy of what we may optimistically call the period for the three values of q that we have examined. This is suggestive of monofractality. Second, although the LF for the southern hemisphere is not nearly so well defined as for the northern hemisphere, its "period" is the same as in the northern hemisphere. If this result holds up, it is striking evidence of statistical homogeneity of the galaxy distribution on large scales. We do not see this kind of oscillatory LF for the smallest scales and this may be in keeping with our interpretation of the dimensions on those scales.

REFERENCES

1. BERTSCHINGER, E. 1985. Astrophys. J. Supp. **58:** 39.
2. BORGANI S. 1995. Phys. Rep. **251:** 1.
3. COLEMAN P.H., L. PIETRONERO & R.H. SANDERS. 1988. Astron. Astrophys. **200:** L32.
4. COLEMAN P.H. & L. PIETRONERO.1992. Phys. Rep. **231:** 311.
5. COUCHMAN, H.M.P. 1991. Astrophys. J. Lett. **386:** L23–L26.
6. FILLMORE J.A. & P. GOLDREICH.1984. Astrophys. J. **281:** 1.
7. GUREVICH A.V. & P. ZYBIN. 1988. Sov. Phys. JETP **67:** 1957.
8. GUZZO L., A. IOVINO, G. CHINCARINI, R. GIOVANELLI & M.P. HAYNES. 1991. Astrophys. J. Lett. **382:** L5.
9. GRASSBERGER, P. & I. PROCACCIA. 1983. Phys. Rev. Lett. **50:** 346.
10. GROTH, E.J. & P.J.E. PEEBLES. 1977. Asrophys. J. **217:** 385.
11. GUNN, J. & J. R. GOTT. 1972. Astrophys. J. **281:** 1.
12. HALSEY T.C., M.H. JENSEN, L.P. KADANOFF, I. PROCACCIA & B.I. SHRAIMAN. 1986, Phys. Rev. A **331:** 1141.
13. HENRIKSEN, R.N. & L.M. WIDROW. 1995. Mon. Not. R. Astron. Soc. **276:** 679.
14. HUCHRA, J.P., M.J. GELLER, C.M. CLEMENS, S.P. TOKARZ & A. MICHEL. 1993. CfA Redshift Catalogue; 10-3-93 version available at ftp://cdsarc.u-strasbg.fr/pub/cats/VII/164.
15. MANDELBROT, B.B. 1982. The Fracial Geometry of Nature. Freeman. San Francisco, CA.
16. MARTINEZ, V.J. & B.J.T. JONES. 1990. Mon. Not. R. Astron. Soc. **242:** 517.
17. MURANTE G., A. PROVENZALE, S. BORGANI, A. CAMPOS & G. YEPES. 1996. Astroparticle Phys. **5:** 53.
18. MURANTE, G., A. PROVENZALE, E.A. SPIEGEL & R. THIEBERGER. 1997. Mon. Not. R. Astron. Soc. **291:** 585.
19. NAVARRO, J.F., C.S. FRENK & S.D.M. WHITE. 1996. Astrophys. J. **462:** 563.
20. NAVARRO, J.F., C.S. FRENK & S.D.M. WHITE. 1997. Astrophys. J. **490:** 493.
21. PALADIN, G. & A. VULPIANI. 1984. Nuovo Cimento Lett. **41:**82.

22. PEEBLES, P.J.E. 1980. The Large Scale Structure of the Universe. Princeton Univ. Press. Princeton, NJ.
23. PROVENZALE, A. 1991. *In* Applying Fractals in Astronomy, A. Heck & J.M. Perdang, Eds. Springer. Berlin.
24. PROVENZALE, A., E.A. SPIEGEL & R. THIEBERGER. 1997. CHAOS **7:** 82.
25. RENYI, A. 1970. Probability Theory. North-Holland. Amsterdam.
26. SOLIS F.J. & L. TAO. 1997. Phys. Lett. A **228:**351.
27. THIEBERGER, R., E.A. SPIEGEL & L.A. SMITH. 1990. *In* The Ubiquity of Chaos, S. Krasner, Ed. AAAS Press.
28. ZELDOVICH, YA.B. 1970. Astrofizika(A) **6:** 319.
29. ZUREK W.H. & M.S. WARREN. 1994. Los Alamos Science, No. 22: 58.

Recent Progress in the Study of One-Dimensional Gravitating Systems

BRUCE N. MILLER,[a] KENNETH YAWN, AND PAIGE YOUNGKINS

Physics Department, Box 298840, Texas Christian University, Fort Worth, Texas 76129

ABSTRACT: We review recent developments in the study of two different interacting gravitational systems: the system of parallel planar mass sheets and the system of concentric spherical mass shells. The approach to equilibrium of a system of parallel planar mass sheets is investigated. Parallels with three-dimensional systems are described. Mass segregation and kinetic energy equipartition in a two-component system of planar mass sheets is demonstrated via numerical simulation. The existence of two distinct phases is demonstrated in the system of spherical mass shells. The nature of the transition in the microcanonical, canonical, and grand canonical ensembles is studied both theoretically in terms of mean-field theory and via dynamical simulation.

I. INTRODUCTION

One-dimensional models have provided a testing ground for astrophysical theories of gravitational evolution for several decades. The systems studied to date fall into two classes, parallel mass sheets of infinite spatial extent, and concentric spherical mass shells. It has been conjectured that these idealized systems can be identified with processes occurring in nature: The parallel sheet system with the motion of stars perpendicular to the plane of a highly flattened galaxy [1], and the shell system with the dynamics of a spherical globular cluster [2]. However strained these connections may appear, the central motivation for studying these systems is the ease and accuracy with which their dynamical evolution may be simulated with the computer. In contrast with the evolution of three-dimensional point masses, for these systems the slow, stepwise, numerical integration of the orbits can be replaced by the direct computation of successive sheet (or shell) crossings, permitting accurate orbit computations over many dynamical time scales. Another reason to study these systems is the realization that they are the simplest N-body gravitational systems available, and therefore worthy of consideration in their own right. It is difficult to claim understanding of the more complex three-dimensional systems when there is still much we don't know about their lower-dimensional cousins.

A nagging problem associated with the planar sheet model is its week ergodic properties. Following the initial "violent" relaxation phase [3], the system "hangs up" in a slowly varying state resembling one of the (infinite in number) stationary solutions of the Vlasov (collisionless Boltzmann) equation [4]. As a consequence, for decades, the average system properties, revealed in dynamical simulations of all

---

[a]Address for telecommunication: phone, 817-257-7123; fax, 817-257-7742; e-mail, B.Miller@tcu.edu

but the smallest systems, depended strongly on the specific initial condition in phase space, as this targets the ensuing stationary Vlasov state, and hence the associated system properties observed in the simulations. The simple fact is that until very recently, due to improvements in processor speed, it was not possible to observe evolution away from the first metastable state visited by the system. Thus, for example, attempts to observe the equipartition of planar systems with two species of different masses consistently failed [5], [6]. Moreover, systems initiated by directly sampling a stationary distribution failed to evolve to a different state in the observed time [7].

At the present time three groups, including ours, have studied the system for dynamical times of sufficient length to observe the system evolution through a succession of metastable "Vlasov" states. For populations on the order of 100 sheets in a one species system, it has been shown by Tsuchiya *et al.* that the time average of the energy of each sheet approaches its predicted equilibrium value in about 10^7 system crossing times (hereafter T_c) [8]. Milanovic *et al.* have shown that the single particle position and velocity distribution functions also approach their equilibrium values and the Lyapunov exponents converge to unique values on similar time scales [9]. These results have been known for smaller systems since 1984 [10], [11]. Recently, Yawn and Miller demonstrated that a planar system consisting of two mass species does, in fact, exhibit both equipartition of kinetic energy and mass segregation [12]. In Section II we describe this work and the dynamical features of the single component system which are responsible for the slow convergence of time averages.

It has been known for some time that the thermodynamics of systems of particles interacting via gravitational forces differs strongly from typical "chemical" systems, where the interactions have finite range and are repulsive at short distances [3]. The classic distinction is that a gravitational system of fixed total energy can support a negative heat capacity! Some time ago Antonov showed that, due to the singularity in the two-body interaction potential, in the mean-field (Vlasov) approximation a confined spherical system of gravitating point masses lacks a global, and for sufficiently low energy a local, entropy maximum [13]. Thus it is not subject to the usual thermodynamic analysis and may undergo a spontaneous collapse, presently referred to as the gravothermal catastrophe following Lynden-Bell and Wood who also conjectured that, if the singularity is screened by a hard sphere interaction, the collapse may be replaced by a phase transition to a more centrally condensed state [14]. (Such a system has been investigated theoretically by both Stahl and Kiessling [15] and Podmonabhan [16] by introducing a local pressure due to the short-range repulsion.) Later Hertel and Thirring showed that the thermodynamics of a model system interacting via purely attractive forces of finite range and potential depth can undergo a phase transition to a more condensed state [17]. Lynden-Bell may be the first to have demonstrated a phase transition in a particular model *gravitating* system in which all of the particles are restricted to lie on the same spherical surface [18]. More recently, Kiessling has rigorously proven that (1) a truly isothermal, three-dimensional, Newtonian, gravitating system will condense to a droplet of zero size and that (2) a phase transition is possible in the three-dimensional gravitating gas of fixed energy restricted to a spherical box with a *regularized* two-body interaction potential [19]. However, regardless of all of these predictions, until now the occurrence of a gravitational phase transition has not been demonstrated either experimentally (granted this would be a challenge) or by dynamical simulation.

Here we consider a model system of uniform, concentric, mass shells. The shells undergo radial motion, i.e., they expand and contract, under the influence of their mutual gravitational forces. Other investigators have used concentric mass shells to model the evolution of globular clusters [2], [20], [21]. In contrast with these studies, our shells have no angular momentum or rotational energy and, in addition, are confined between two reflecting barriers with inner and outer radii (a, b). In contrast with the system of planar sheets, our shells appear to have strong ergodic properties and spread out rapidly in their μ-space [22], [23]. We have studied the system both theoretically and via dynamical simulation under three different conditions: microcanonical (constant energy), canonical (constant temperature), and grand canonical (constant temperature and chemical potential). For all ensembles we have shown that stationary mean-field (Vlasov) theory predicts the existence of two possible phases when the inner barrier radius is less than a critical value, a_c. Each phase has a smooth density profile in (a, b), but one of them has a higher central concentration of mass. For each $a < a_c$ mean-field theory predicts that the more concentrated phase is favored thermodynamically for sufficiently low energy (temperature) in the constant energy (temperature) ensemble. However, for the open system, the more uniform phase is always more stable. As we shall show in Section III, the dynamical simulations support the predictions of mean-field theory in all cases. As far as we are aware, this is the first dynamical demonstration of a phase transition in a gravitational system.

II. RELAXATION OF PLANAR MASS SHEETS

The dynamics of the planar sheet system is simple: each sheet experiences a constant acceleration proportional to the difference in mass of the sheets on either side. Between crossings, therefore, the equations of motion of each sheet can be integrated analytically. Dynamical simulation of the system proceeds by determining the correct sequence of crossings of each adjacent pair of particles (sheets). The subtleties of constructing an efficient algorithm, and there are some, have been discussed in detail elsewhere and will not be repeated here [24].

Until about 1980 it was believed that the planar sheet system came to equilibrium in a time scale proportional to $N^2 T_c$, where N is the system population and T_c is a typical crossing time [24], [25]. However, in 1982 it was shown that systems prepared with an initial virial ratio of 10 did not thermalize in twice that time [4]. Early dynamical studies of small systems ($N < 10$) showed that stable regions of finite extent exist in the phase space so that the system is not ergodic. These studies also showed that regions having diverging orbits increase in size [11] and that the exact, microcanonical, velocity and position distribution functions derived by Rybicki for a fixed population [26], are more closely approached [10] as the population is increased. More recently it was shown that both stable and unstable fixed points exist in the phase space and that the size of the stable regions surrounding the former decrease with increasing population [27]. These studies leave open the possibility that, for a given large N, the phase space has a single, large ergodic component while numerous, coexisting, small, stable regions persist.

For $N > 10$ the number of independent phase-space dimensions ($2N - 3$ considering energy and momentum conservation and the possibility of locking the center of mass) is too large to easily study the phase-space geometry. Dynamical simulations of systems where N is of the order of 10^2 have demonstrated that (1) the system quickly relaxes to a quasi-stationary "dynamical" equilibrium which depends strongly on the initial conditions [28], (2) correlation functions have extremely long time tails, at least on the order of $10^3 \, T_c$ [29], and (3) a system prepared by sampling a stable, stationary, Vlasov distribution in μ-space shows no tendency to evolve to a different state for times on the order of $10^3 \, T_c$, i.e., the full duration of the simulations [7].

In the last few years it has been possible to examine larger systems for runs of much longer duration, on the order of $10^7 \, T_c$. This has finally allowed the observation of the approach of time averages to their predicted equilibrium values. In the first such study Tsuchiya et al. [8] showed that $\Delta(t)$, the variance in the time-averaged single particle energy ε_j,

$$\Delta(t) = \frac{1}{N}\sum_1^N [\bar{\varepsilon}_j(t) - \bar{\varepsilon}]^2, \quad \bar{\varepsilon}_j(t) = \frac{1}{t}\int_0^t \varepsilon_j(t')dt', \tag{1}$$

decays to zero on this time scale for $N = 64$. However, the decay is not uniform and there are long, erratic, periods where $\Delta(t)$ increases before resuming its downward trend.

Milanovic et al. recently carried out simulations on systems with $N \leq 40$. They have extended the earlier conclusions of Wright and Miller by showing that the time-averaged position and velocity distributions closely agree with Rybicki's exact derivations. They observed small, systematic, deviations for $N = 16$, which become hard to discern at $N = 40$. They also computed the largest Lyapunov exponents of the planar system, extending the work on smaller systems of Benettin et al. [11] also to $N = 40$. While their recent simulations show that the Lyapunov exponents converge, the time required is extremely long, with periods during which the exponents appear to have values far from their global average.

Further insight is gained by considering $\Delta_2(t, \tau)$, the mean square deviation of each particle's energy from the average during the finite time window τ:

$$\Delta_2(t) = \frac{1}{N}\sum_1^N [\varepsilon_j^*(t, \tau) - \bar{\varepsilon}]^2, \quad \varepsilon_j^*(t, \tau) = \frac{1}{\tau} \int_t^{t+\tau} \varepsilon_j(t')dt'. \tag{2}$$

In FIGURE 1 we plot $\Delta_2(t, \tau)$ as a function of t with fixed $\tau = 10^4 \, T_c$. From the figure it is apparent that there are long periods during which $\Delta_2(t, \tau)$ remains small, followed by random bursts where it becomes large, suggesting periods where the system is captured in a small region of phase space for a substantial period $\gg \tau$.

The equipartition theorem of statistical mechanics [30] tells us that if a system is ergodic, each dimension contributes $kT/2$ to the mean kinetic energy per particle. Thus in a two component system consisting of equal populations of "heavy" and "light" particles, the ratio of the average kinetic energy of the heavy component to that of the light component kinetic energy, say R, is unity. Typically in simulations

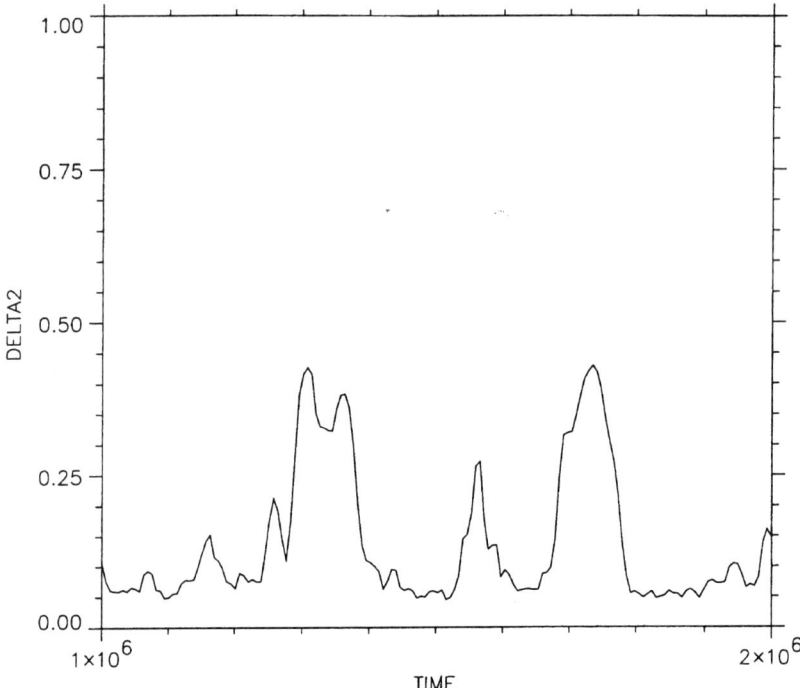

FIGURE 1. Time windowed mean energy variance.

the initial conditions are chosen by uniformly sampling a rectangular region in the μ space. The initial mean square velocity of a particle in such a sample is independent of its mass. Thus, initially, $R \approx m_H/m_L$ where $m_H(m_L)$ is the mass of a star in the heavy (light) component. Clearly, in a simulation, we should observe the approach of $R(t)$ to unity if the system is ergodic. In early attempts to observe equipartition, there was no significant change in $R(t)$ from its initial value throughout the duration of the runs [5], [6].

In FIGURE 2 we plot $\overline{R}(t)$, the cumulative time average of R, versus time for a two-component system with a particle mass ratio $m_H/m_L = 3.0$. Following the initial violent relaxation we observe that $\overline{R}(t)$ remains nearly constant for a long period, exhibiting the fact that the system is stalled in the stationary Vlasov state to which it initially relaxed. Finally, the initial state is forgotten and $\overline{R}(t)$ appears to slowly approach unity as time progresses (note the logarithmic time scale), confirming equipartition. By averaging backward in time from the end of the run, we eliminated the influence of the long lasting transient associated with the initial, far from equilibrium, state and found that $R(t)$ sits right on unity for most of the simulation [12].

Another consequence of equipartition is mass segregation, the sinking of the heavy particles toward the system center while the light particles spread out in a halo in the μ-space. In FIGURE 3 we plot the cumulative time average of the mean distance from the system center of particles in each component. The separation from a com-

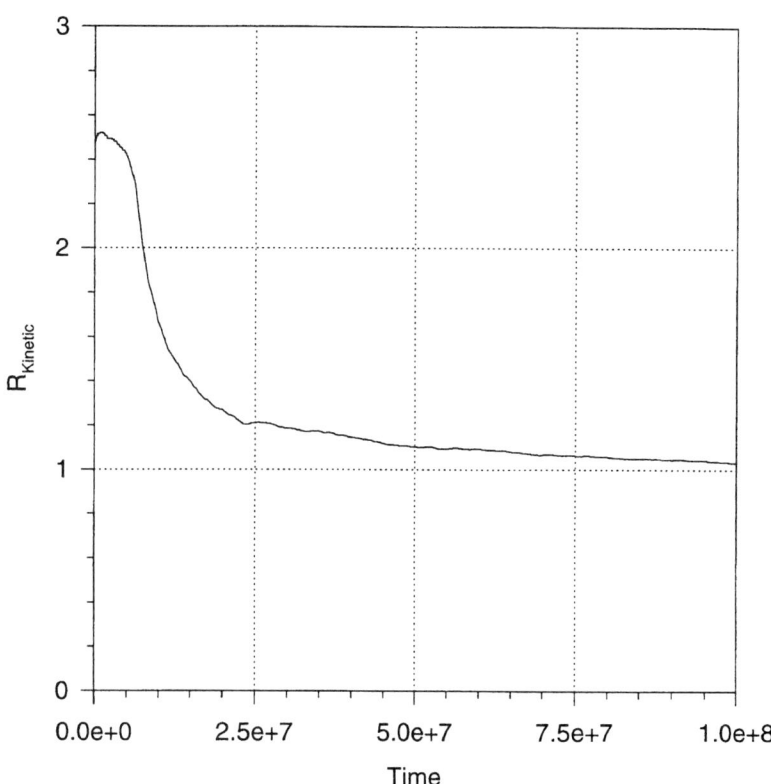

FIGURE 2. Cumulative time average of the kinetic energy ratio of the two species.

mon initial value is apparent, confirming mass segregation. Again we see the rapid change accompanying violent relaxation, then a long period with little change, followed by slow convergence (again note the scale) to the equilibrium values.

To further explore the unusually protracted memory effects in the system dynamics, we computed the average of $R_2(t, \tau)$ for a sequence of successive windows of fixed duration and starting times $t = n\tau$, $n = 1, 2, \ldots$ (see the previous definition of $\Delta_2(t, \tau)$). We then computed the autocorrelation of the $R_2(t, \tau)$ as a function of n, starting backward in time from the end of the run to avoid the long-lived initial transient. The results are displayed in FIGURE 4 for a window size of $\tau = 6 \times 10^4$ time units (about 10^4 crossing times). It is clear from the graph that memory effects persist on time scales on the order of 10^6 crossing times! In recent related work we have investigated equipartition for a wide range of mass ratios and system populations and developed the equilibrium Vlasov solutions for the two component system. The sim-

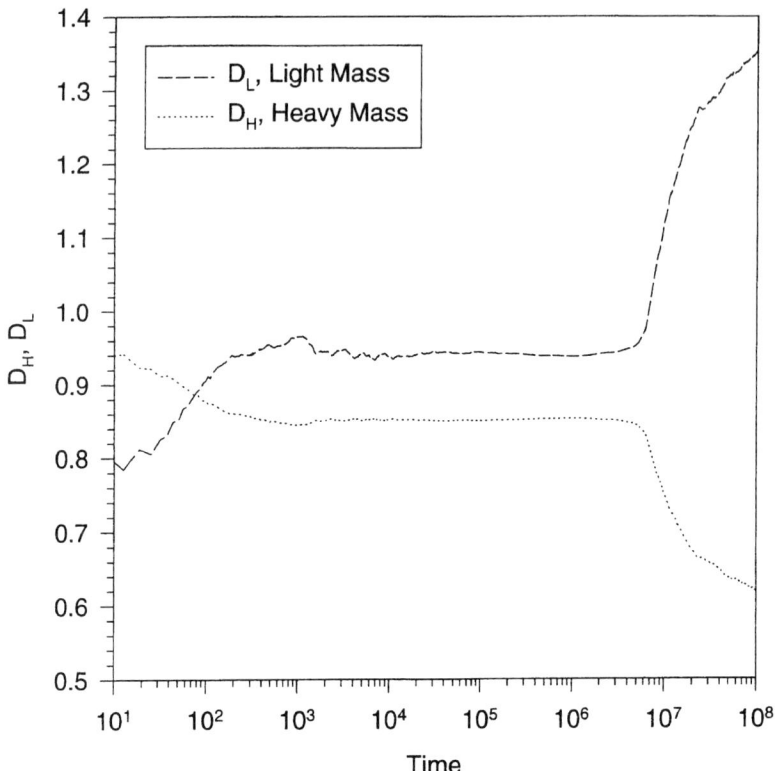

FIGURE 3. Cumulative average distance from the system center versus time for each species.

ulations generally confirm a slower approach to equilibrium with increasing population.

At the present time we lack a real understanding of the connection between the geometry of phase-space orbits and the macroscopic behavior of the system. Is there a direct link between the stable periodic orbits in the phase space and in the incredibly long memory effects observed in simulations? Perhaps the system is trapped for long times near the last tori surrounding the remaining, stable, periodic orbits. Certainly all of the recent research is consistent with this view. Since at the present time we cannot identify these orbits for $N > 10$, we can only conjecture that this is the true scenario. A related question is the dependence of the relaxation time scale on the system population. By making the strongest possible stochastic assumption concerning the distribution of crossing events, it is possible to derive kinetic equations which predict that relaxation of single particle properties scales directly with the popula-

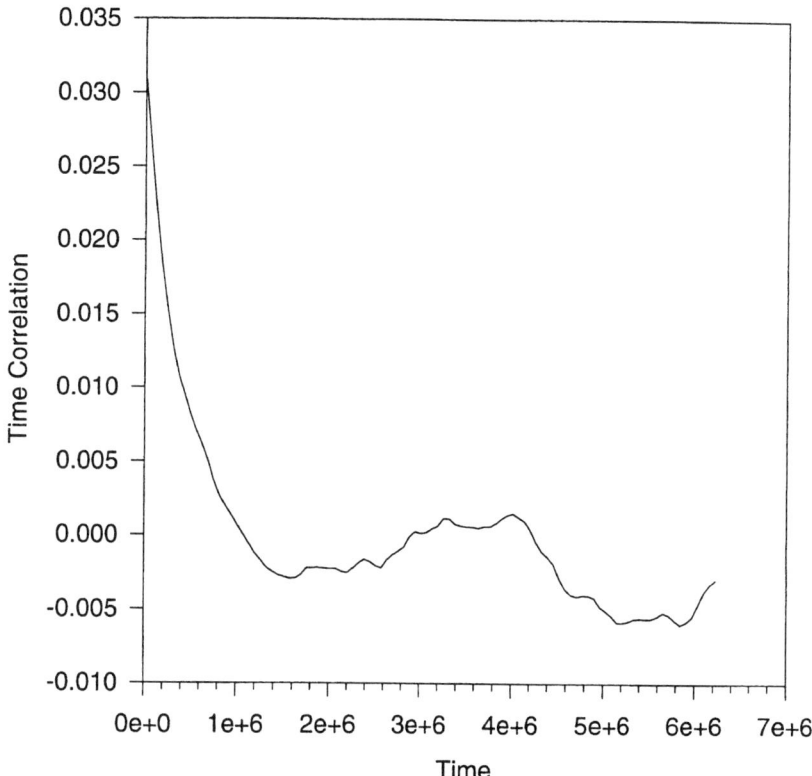

FIGURE 4. Time correlation of windowed kinetic energy ratio: backward average.

tion [31]. Future work needs to compare the predictions of this kinetic theory with actual simulations.

III. PHASE TRANSITIONS IN A SYSTEM OF CONCENTRIC SHELLS

Consider a system of N concentric, uniform, irrotational, Newtonian, shells of mass m ($= M/N$). Each shell has a single coordinate, its radius r, so that, as in the planar sheet model, the system phase space has $2N$ dimensions. It is easy to show that the acceleration of a shell with radius r is $-G[M(r) + m/2]/r^2$ where, as usual, G is the gravitational constant and $M(r)$ is the mass of the interior shells, i.e., each shell accelerates as if the interior mass as well as half of its own mass were concentrated at the system center. To both prevent escape and shield the singularity at the

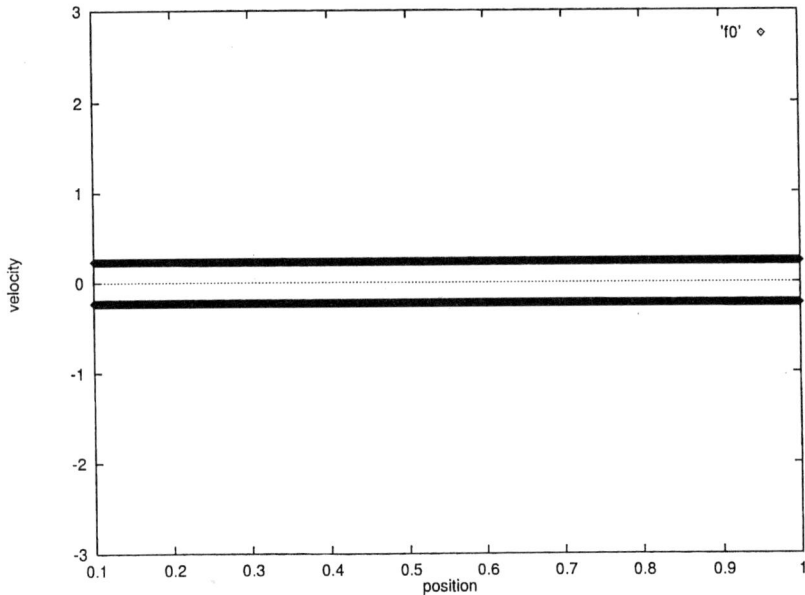

FIGURE 5. Initial condition in μ-space for a 3200 shell system.

system center, the shells are confined to move between two likewise spherical reflecting barriers located at $r = (a, b)$. For convenience, and without loss of generality, we define units of distance, time, and mass for which $b = G = M = 1$.

Numerical simulation of the shell dynamics is facilitated by the fact that, between events (shell crossings, turning points, and barrier collisions) the equations of motion can be integrated analytically to yield the time as a function of position. In two earlier studies we showed that the shell system has stronger ergodic properties than the parallel sheets [22] and relaxes on a much shorter time scale [23]. This can be clearly seen in the following example of a system of 3200 shells prepared by uniformly distributing them in position and alternately assigning velocities of equal size and opposite sign (see FIGURE 5). After only about fourteen characteristic infall times, we see that the system is well mixed in the μ-space (see FIGURE 6).

In the Vlasov limit, i.e., letting $N \to \infty$ while both the total mass and energy are held constant, the system is represented by a fluid in the $m(r, v)$-space with mass density $f(r, v, t)$. Intuitively, it seems natural to assume that our system obeys the identical Vlasov (collisionless Boltzmann) equation for $f(r, v, t)$ as for a spherically symmetric three-dimensional system. In particular, one anticipates that equilibrium solutions, where they exist, are identical to Emdens' isothermal spheres [3], [32]. However, since here the dynamical system is one-dimensional, this is not the case. The stationary maximum entropy solution of the CBE is still of the form $f(r, v) = MC \exp[-\beta(v^2/2 + \varphi(r)]$ where C is a normalization constant and $\varphi(r)$ is the gravitational potential, but here v is the radial velocity of a shell, and the system density,

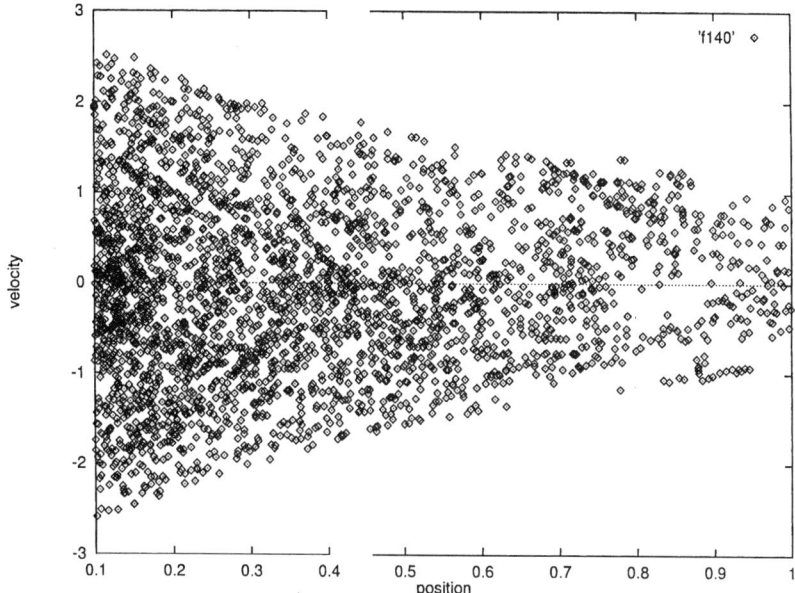

FIGURE 6. Final condition in μ-space for a 3200 shell system.

$\rho(r)$, is the mass per unit length. It is straightforward to show that the normalized, *linear*, mass density, $g(r) = \rho(r)/M$, now obeys

$$\frac{d}{dr}r^2\frac{d}{dr}\ln g(r) = \beta M G g(r)$$

subject to

$$\left.\frac{d}{dr}g(r)\right|_{r=a} = 0 \text{ and } \int_a^b g(r)dr = 1.$$

(3)

The solutions of Eq. (3) depend on the single parameter βMG (= β in our units). By direct numerical integration we find that for sufficiently large $a < b = 1$ there is only one solution of (3). However, as the value of a is reduced, we reach a critical value, say a_c, such that, for $a < a_c$, it is possible to construct three distinct solutions of (3) for $\beta \in \Delta\beta(a)$, a particular interval of $\Re+$. Outside of this interval the solution is unique. The solutions for large β, i.e., to the right of $\Delta\beta(a)$, are more centrally concentrated than those to the left.

It is possible to construct a complete formulation of the equilibrium thermodynamics of this system in the mean-field (Vlasov) limit. All quantities of interest can be expressed in terms of the solutions of (3). It is natural to regard a, the inner barrier radius, as the generalized thermodynamic coordinate of the system and $\theta \equiv 1/\beta$ as the generalized temperature [30]. In order to track the phase transitions which occur in the microcanonical and canonical ensemble, it is convenient to choose the virial ratio, i.e., the ratio of kinetic to potential energy, as the system order parameter.

With a little work the explicit dependence of the energy, entropy, Helmholtz free energy, and grand potential on $g(r)$ can be determined. They are single valued functions of β except when $a < a_c$ and $\beta \in \Delta\beta(a)$, where they can take on three values. This is the parameter region where phase transitions can occur.

A. Microcanonical Ensemble

The system is isolated and the energy is conserved. The thermodynamic state is determined by the total energy, E, and inner barrier radius, a. Equilibrium states are states of maximum entropy $S = S(E, a)$. To explore the system properties in the context of mean-field theory we need to determine the solution of (3) corresponding to a given (E, a) which, in turn, requires finding $\beta = \beta(E, a)$. When $a < a_c$ and $\beta \in \Delta\beta(a)$ multiple phases are possible; the stable phase corresponds to the branch with maximum entropy. From FIGURE 7 we see that, corresponding to $\beta \in \Delta\beta(a)$, there is an energy interval in which both phases can exist, but one is more stable. As usual, the third solution tying the stable phases together is thermodynamically unstable [27]. As the energy is lowered, there is a sharp transition to the more centrally concentrated phase.

At the transition point (the intersection in FIGURE 7) each phase is equally stable. Thus we can construct a "coexistence" curve in the (E, a) plane along which the transition occurs. Here, however, we must use the term coexistence guardedly. In contrast with chemical systems, our phases are associated with a particular density

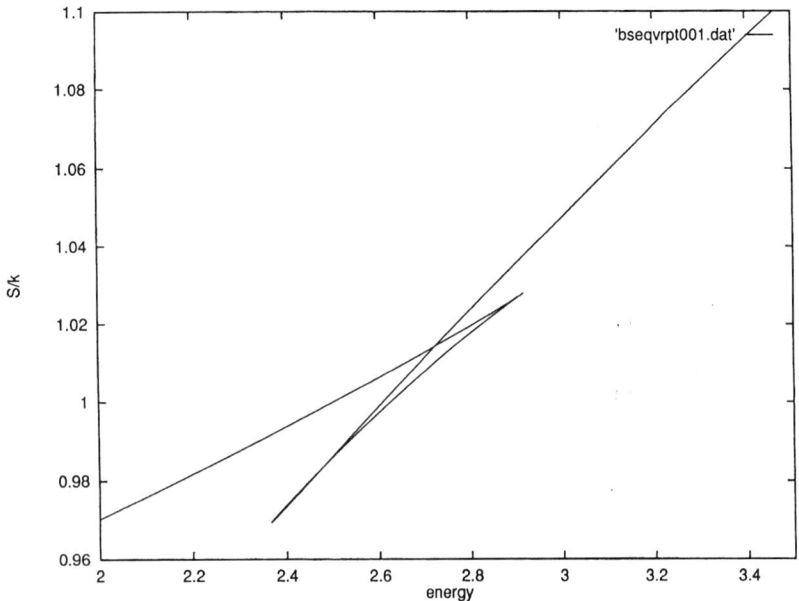

FIGURE 7. Entropy versus energy for isolated system in the two-phase region.

profile in (a, b) and cannot simultaneously exist in the same volume. Note that there is a jump in temperature across the transition: the temperature decreases suddenly as the system becomes less concentrated. This is a consequence of the fact that, in contrast with a chemical system, above the transition $(a < a_c)$ the isolated gravitational system can support a negative heat capacity [15].

In earlier work we compared dynamical simulations of systems with $N = 16$, 32, and 64 shells with the Vlasov equilibrium predictions for $a < a_c$ [23]. We found good agreement with increasing N. At the largest value, the time-averaged density profile agreed within a few percent with the mean-field prediction obtained from the numerical solution of (3) over the complete interval (a, b). Here we have extended the simulations into the transition region. Away from the transition energy we also find close agreement with the mean-field density profile. However, as the transition is approached, differences occur. This is expected: Finite size scaling theory [33] predicts that the transition is only sharp in the limit $N \to \infty$. For finite N the transition energy is shifted as $N^{-\lambda}$ and broadened as $N^{-\gamma}$. By comparing the time-averaged virial ratio in simulations with the predicted mean-field values, we found that the results accurately conformed to the scaling predictions.

B. Canonical Ensemble

The system is no longer isolated and may now exchange energy with a reservoir at constant temperature $\theta \equiv 1/\beta$. For a given population the thermodynamic state is determined by a and θ. Equilibrium states are states of minimum Helmholtz free energy, $F(a, \theta)$ [30]. In common with the entropy, for $a < a_c$, F is a multivalued function of θ for $1/\theta \in \Delta\beta(a)$ (see FIGURE 8) and the thermodynamically stable phase

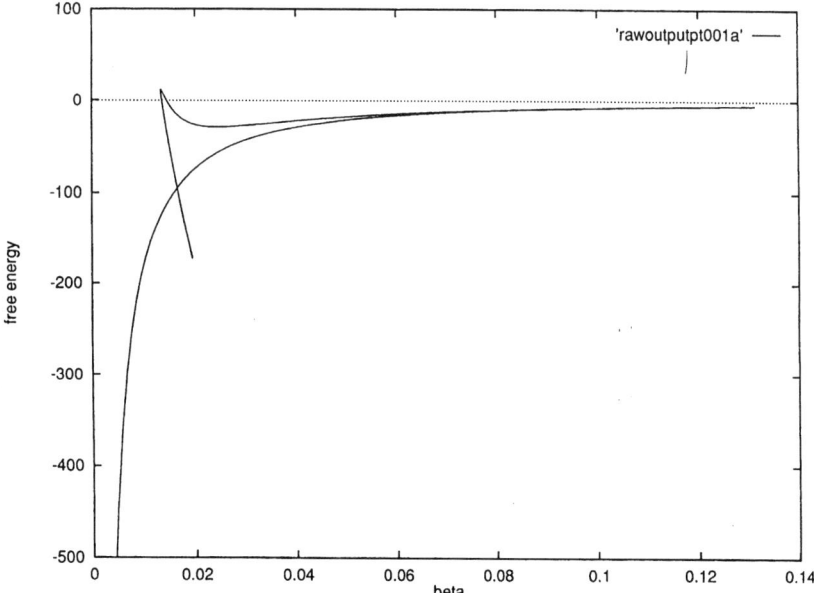

FIGURE 8. Free energy versus β for thermostated system in the two-phase region.

corresponds to the lower curve. As before, mean-field theory predicts a sharp transition where the free energies of the two phases intersect. Note that although the canonical transition also depends of the solutions of (3) for $g(r; \beta)$, the features of the transition are different than in the microcanonical case. Here there is a jump in the value of E while β remains constant. This is just one consequence of the fact that the thermodynamic behavior of a gravitational system depends on the specific ensemble. In the canonical ensemble negative heat capacities cannot be supported in equilibrium, thus excluding some of the states which are permitted in the microcanonical ensemble [15]. The "coexistence" curve now bisects a portion of the (a, θ) plane.

Dynamical simulation of the canonical ensemble was accomplished by converting the reflecting barrier at $r = b$ into an isothermal wall for every hundredth collision. In such an event, detailed balance is respected by returning the shell to the system with kinetic energy determined by randomly sampling the exponential distribution with mean θ. By thermostating only once every hundred collisions, we ensured that the effect of the "reservoir" was simply a perturbation on the system dynamics, a requirement of the thermodynamics of interacting systems [30]. In common with the microcanonical ensemble, in these isothermal simulations good agreement was found between the time-averaged density profile and that predicted by mean-field theory away from the transition temperature. Near the transition the results were again consistent with finite size scaling.

An interesting feature of the simulations was the robustness of the concentrated phase above the transition temperature, where it is thermodynamically less stable than the more uniform phase. For $N > 16$, if the system was prepared in the former, we were never able to observe a dynamical phase transition to the more stable phase during the typical run times of the numerical experiments. Another interesting feature was the lack of critical slowing down. The point on the coexistence curve where $a = a_c$ is a thermodynamic critical point of the system. In chemical systems, relaxation times and correlation lengths diverge at this point [30]. In our dynamical simulations, measurements of both temporal and positional correlation functions were carried out for the thermostated system. While differences in temporal and positional relaxation were found in each phase, no hint of divergence of either was indicated near the critical point.

C. Grand Canonical Ensemble

The system can now exchange both energy and particles (shells) with a reservoir. An equilibrium thermodynamic state is defined by θ and μ, the temperature and chemical potential of the reservoir, as well as the inner barrier radius a. For the sake of comparison, for each a and θ the value of μ was fixed by requiring that the average system mass is unity, i.e., the same as for each of the ensembles described above. In equilibrium, the grand potential has a global minimum for the stable state [30]. Plots of Φ versus θ do not exhibit the kink structure found for the other thermodynamic potentials, even for $a < a_c$. In contradistinction with both the microcanonical and canonical ensembles, here mean-field theory predicts that the more uniform phase is more stable over the complete (a, θ) plane! Thus a transition to a more a centrally concentrated density profile is not predicted.

Dynamical simulation of the open system (grand canonical ensemble) was accomplished by randomly introducing new shells at the outer barrier with a carefully chosen mean creation rate. This was selected by imagining the effect of a virtual reservoir surrounding the system with the desired temperature and chemical potential. To weaken the interaction between the system and reservoir, the boundary was assumed to be only "semipermeable" [30]. The creation rate was chosen to be only one hundredth of the virtual external particle flux striking the outer barrier. The kinetic energy of a new shell was chosen by employing the same thermostat described above for the canonical ensemble. To balance the flux, every hundredth system shell striking the outer barrier from within was removed from the system.

As above, the results of the dynamical simulations were in complete accord with the mean-field predictions. The system was simulated with a variety of initial conditions. In each case the time-averaged density profile agreed with that predicted for the more uniform phase. No sign of a transition was ever observed, even when the initial state was highly concentrated near the inner barrier.

IV. CONCLUSIONS

The one-dimensional systems described above are the simplest gravitating systems we can construct, lacking the complicating features associated with escape and binary formation, yet much is not known about their behavior. Because simulating their evolution is similar to iterating a multidimensional map, they are ideal testing grounds for dynamical theories.

With regard to the system of planar sheets, the situation is in much better shape than it was as recently as five years ago in that we can now observe the convergence of time averages to their predicted means. For both the parallel sheets and concentric shells, mean-field theory seems to be adequate for predicting mean values of single particle functions. Of course, it is incapable of predicting the system evolution or the behavior of statistical correlations between the particles.

In common with the planar system, we have demonstrated that dynamical simulation of the shell system completely conforms to the mean-field predictions with increasing system population in all three ensembles considered. Although relaxation in the shell system proceeds at a much higher rate, the reason for this is not known. The presence of the reflecting barriers may play an important role. Relaxation in either system is still not understood. A diffusion model of relaxation based on a strongest possible stochastic assumption has been introduced [31] which may explain the gross features but, to date, it has only been marginally tested for the sheet system [34]. It predicts a relaxation time scale proportional to the population, and has yet to be applied to the shell system. Since the shell system relaxes much more rapidly than the sheet system a separation of microscopic and macroscopic relaxation time scales may not occur.

A final issue is whether the shell system behavior has relevance for astrophysics. Certainly the observations of galaxies and globular clusters reveal a wide variety of structures. Some have very dense cores which may support black holes, while others do not. Is there a connection between the observed core-halo structures and the centrally concentrated phase described in the preceding pages? It could be argued that both globular clusters and galaxies may be closest in essence to a grand canonical ensemble since each can exchange both energy and mass with its surroundings, i.e., a globular cluster with its parent galaxy, and a galaxy with its parent galactic cluster, and therefore a transition may not be possible. Future work will have to consider both the importance of the rate of energy and mass exchange, the role of binary formation in the case of the globular cluster, and the role of dimension for both.

REFERENCES

1. OORT, J.H. 1932. Bull. Ast. Inst. Neth. **6:** 289; CAMM, G. 1950. Mon. Not. R. Ast. Soc. **110:** 305.
2. HENON, M. MEM. 1967. Soc. R. Sci. Leige (V) **15:** 243.
3. BINNEY, J & S. TREMAINE. 1987. Galactic Dynamics. Princeton University Press. Princeton, NJ.
4. WRIGHT, H.L., B.N. MILLER & W.E. STEIN. 1982. Astrophys. Space Sci. **84:** 421–429.
5. SEVERNE, G., M. LUWEL & P.J. ROUSSEEUW. 1984. Astron. Astrophys. **138:** 365.
6. HOHL F. & J. CAMPBELL. 1968. Astron. J. **73:** 611.
7. LEWEL, M. & G. SEVERNE. 1985. Astron. Astrophys. **152:** 305.
8. TSUCHIYA, T., T. KONISHI & N. GOUDA. 1994. Phys. Rev. E **50:** 2607–2615; **53:** 2210–2216 (1996).
9. MILANOVIC, L., H.A. POSCH & W. THIRRING. 1998. Phys. Rev. E **57:** 2763–2775.
10. WRIGHT H. & B.N. MILLER. 1984. Phys. Rev. A **29:** 1411.
11. BENETTIN, G., C. FROESCHLE & J.P. SCHEIDECKER.1979. Phys. Rev. A **19:** 2454; FROESCHLE, C. & J.P. SCHEIDECKER. 1975. Phys. Rev. A **12:** 2137–2143.
12. YAWN, K.R. AND B.N. MILLER. 1997. Phys. Rev. Lett. **79:** 3561.
13. ANTONOV, V.A. 1962. Vest. Leningr. Gos. Univ. **7:** 135.
14. D. LYNDEN-BELL & R. WOOD. 1968. Mon. Not. R. Ast. Soc. **138:** 495.
15. STAHL, B., M. KIESSLING & K. SCHINDLER. 1995. Planetary Space Sci. **43:** 271–282.
16. PODMANABHAN, T. 1990. Phys. Rep. **188:** 285–362.
17. HERTEL, P. & W. THIRRING 1971. Commun. Math. Phys. **24:** 22; **28:** 159 (1972).
18. LYNDEN-BELL, D. & R.M. LYNDEN-BELL. 1977. Mon. Not. R. Ast. Soc. **181:** 495.
19. KIESSLING, M. 1989. J. Stat. Phys. **55:** 203–257.
20. YANGURAZOVA, L.R. & G.S. BISNOVATYI-KOGAN. 1984. Astrophys. Space Sci. **100:** 319.
21. HENRIKSEN, R. N. & L.M. WIDROW. 1997. Phys. Rev. Lett. **78:** 3426.
22. MILLER, B.N. & P. YOUNGKINS. 1997. CHAOS **7:** 187.
23. YOUNGKINS, P. & B.N. MILLER. 1997. Phys. Rev. E **56:** R4963.
24. HOHL, F. 1967. Ph.D. dissertation, College of William and Mary; SEVERNE, G. & M. LEWEL. 1980. Astrophys. Space Sci. **72:** 293.
25. RYBICKI, G. 1971. Astrophys. Space Sci. **14:** 56–72.
26. REIDL, C.J. JR. & B.N. MILLER. 1993. Phys. Rev. E **48:** 4250; **51:** 884 (1995).
27. REIDL, C.J. JR. & B.N. MILLER. 1987. Ap. J. **318:** 248–260.
28. MILLER, B.N. & C.J. REIDL JR.1990. Ap. J. **348,** 203–211.
29. REICHL, L.E. 1998. A Modern Course in Statistical Physics, 2nd edit. John Wiley. New York.
30. MILLER, B.N. 1996. Phys. Rev. E **53:** R4279–4282.
31. CHANDRASEKHAR, S. 1939. Introduction to the Study of the Stellar Structure. Dover. New York.
32. BINDER, K. 1987. Ferroelectrics **73:** 43–67.
33. YAWN, K.R., B.N. MILLER, & W. MAIER. 1995. Phys. Rev. E **52:** 3390.

Modeling the Time Variability of Black Hole Candidates

DEMOSTHENES KAZANAS AND XIN-MIN HUA[a]

Laboratory of High Energy Astrophysics, NASA/GSFC Code 661, Greenbelt, Maryland 20771

ABSTRACT: We present model light curves for accreting black hole candidates (BHC) based on a recently developed model of these sources. According to this model, the observed light curves and aperiodic variability of BHC are due to a series of soft photon injections at random (Poisson) intervals and the stochastic nature of the Comptonization process in converting these soft photons to the observed high energy radiation. The additional assumption of our model is that the Comptonization process takes place in an extended but non-uniform hot plasma corona surrounding the compact object. We compute the corresponding power spectral densities (PSD), autocorrelation functions, time skewness of the light curves, and time lags between the light curves of the sources at different photon energies and compare our results to observation. Our model reproduces the observed light curves well, in that it provides good fits to their overall morphology (as manifest by the autocorrelation and time skewness) and also to their PSDs and time lags, by producing most of the variability power at time scales $\gtrsim$ a few seconds, while at the same time allowing for shots of a few msec in duration, in accordance with observation. We suggest that refinement of this type of model along with spectral and phase lag information can be used to probe the structure of this class of high energy sources.

1. INTRODUCTION

The study of the physics of accretion onto compact objects (neutron stars and black holes) whether in galactic (X-ray binaries) or extragalactic systems (Active Galactic Nuclei) involves length scales much too small to be resolved by current technology or that of the foreseeable future. As such, this study is conducted mainly through the theoretical interpretation of spectral and temporal observations of these systems, much in the way that the study of spectroscopic binaries has been used to deduce the properties of the binary system members and the elements of their orbit. In this endeavor, the first line of attack in unfolding their physical properties consists of the analysis of their spectra. For this class of objects and in particular the black hole candidate (BHC) sources, a multitude of observations have indicated that their energy spectra can be fitted very well by Comptonization of soft photons by hot electrons; the latter are "naturally" expected to be present in these sources, a result of the dissipation of the accretion kinetic energy within an accretion disk. It is thus generally agreed upon that Comptonization is the process by which the high energy ($\gtrsim$ 2–100 keV) spectra of these sources are formed, with the study of this process

---

[a]Universities Space Research Association.

hence receiving great attention over the past couple of decades (see, e.g., [23], [24], [7]). Thus, while the issue of the detailed dynamics of accretion onto the compact object is still not resolved, the assumption of the presence of a thermal distribution of hot electrons in the vicinity of the compact object has proven sufficient to produce models which successfully fit the spectra of the emerging high energy radiation. Spectral fitting has subsequently been employed as a way to constrain or even determine the dynamics of accretion onto the compact object.

It is well known, however, that the Comptonization spectra cannot provide, in and of themselves, any information about the size of the scattering plasma, because they depend (for optically thin plasmas) on the product of the electron temperature and the plasma Thomson depth. Therefore, they cannot provide any clues about the dynamics of accretion of the hot gas onto the compact object, which require the knowledge of the density and velocity as a function radius. To determine the dynamics of accretion, one needs, in addition to the spectra, time variability information. It is thought, however, that such information may not be terribly relevant, because it is generally accepted that the X-ray emission originates at the smallest radii of the accreting flow and as such, time variability would simply reflect the dynamical or scattering time scales of the emission region, of order of msec for galactic accreting sources and 10^5–10^7 times longer for AGN.

Recent RXTE (Focke, private communication) as well as older HEAO-1 observations [13] of the BHC Cyg X-1, which resolved X-ray flares of duration approximately a few milliseconds, appear to provide a validation of our simplest expectations. At the same time, however, the X-ray fluctuation power spectral densities (PSD) of accreting compact sources generally contain most of their power at frequencies $\omega \lesssim 1$ Hz, far removed from the kHz frequencies expected on the basis of the arguments given above. Flares of a few msec in duration, while present in the X-ray light curves of Cyg X-1, contribute but a very small fraction to its overall variability as manifest in its PSD, which exhibits very little power at frequencies $\gtrsim 30$ Hz [3]. Interestingly, this form of the PSD has been reported in addition to Cyg X-1 also in the BHC GX 339-4 and Nova Muscae GINGA data (see, e.g., [15]).

This discrepancy, between the observed and the expected distribution of the variability power of BHC sources, hints that one may have to revise the notion that the entire hard X-ray emission of these sources derives from a region a few Schwarzschild radii in size and indicates the need of more detailed models of the timing properties of these systems. In this respect, models of the timing properties of accreting compact sources have been rather limited (with the exception of models of the quasiperiodic oscillations), the reason being (on the theoretical side), the largely aperiodic character of their light curves and (on the experimental side) the lack of sufficiently large area detectors, in conjunction with high telemetry rates which would provide high timing resolution data. As a consequence, the study and modeling of these class of sources has concentrated mainly on their spectra, whose S/N ratios can be improved by longer exposure times. Thus, while there exists a large body of literature associated with the spectra of BHC sources, the literature and analysis associated with their timing properties is comparatively sparce.

In the present work we will briefly outline earlier attempts to uncover the underlying dynamics of accretion from timing observations as well as attempts to model these dynamics in terms of simple dynamical models. Then, we will focus our attention in the constraints imposed by the recent spectro-temporal observations of lags between different energy bands and we will present a kinematic model which is capable of accounting for these observations; we argue that the same model is also consistent with the observed PSD features and finally we will present model light curves computed on the basis of the earlier discussion.

2. FROM LIGHT CURVES TO DYNAMICS

One of the major obstacles in modeling the time variability of BHC has been their aperiodic character and the notion that most of the information relevant to the dynamics of accretion could in fact be obtained from their energy spectra. Nonetheless, closer look into the BHC light curves did generate a fair amount of interest in the high energy astrophysics community. This interest focused mainly on two issues: The shape of the variability power spectral densities (PSD) which were shown to exhibit the ubiquitous $1/f$ Fourier frequency dependence, and the possibility that the observed aperiodic variability is the result of the underlying chaotic dynamics of accretion, namely dynamics with a small number of degrees of freedom but with associated strange attractors in the respective Hamiltonians.

Concerning the second of the above issues, the most thorough attempt of uncovering the underlying dynamics from the observed light curves has been that of Lochner [11] (also presented in reference [12]). These authors followed techniques developed in the field of nonlinear dynamics for recovering the properties of the Hamiltonian of a given dynamical system from a time series resulting from its equation of motion. The general scheme followed in this technique consists generally of the following steps:

- There exist a time series represented by the function $f(t_i)$ which indicates in the present case either the X-ray flux of the source as a function of time, or simply, in the case of low photon fluxes, a series of delta functions at the arrival times of the X-ray photons.

- Creation of the n-dimensional vectors V composed of the value of the flux

$$V_n = \{f(t), f(t + \tau), \ldots, f(t + [n-1]\tau)\} \quad (1)$$

- Formation of the correlation integral $C(r)$ defined by

$$C(r) \equiv \frac{1}{M^2} \sum_{i,j}^{M} \theta(r - |V_i - V_j|) \quad (2)$$

as a function of the norm r of the vectors V, where M is the total number of points in the time series.

- Search for a value of the *embedding dimension D* defined by

$$D = \frac{d \log C(r)}{d \log r} \quad (3)$$

above which the correlation integral slope does not change.

- The embedding dimension D is in general noninteger and represents the dimensionality of the underlying (strange) dynamical attractor of the corresponding dynamical Hamiltonian.

Application of this method by Lochner, Swank, and Szymkoviak [12] to the data of the source with the best data, namely the BHC source Cygnus X-1 has indicated that the dimension of the underlying attractor must be $D \gtrsim 14$, suggesting that the dynamics giving rise to the observed variability are truly stochastic.

In a number of different approaches, attempts have been made to produce light curves with the requisite timing properties. Thus, Abramowicz et al. [1] produced model light curves with PSDs similar to those observed, resulting from a large number of bright spots rotating with different angular velocities in an accretion disk. Clearly, such models are always possible as they contain an arbitrary number of free parameters, generally larger than the number of parameters imposed by observations.

More recently, a number of dynamical models have appeared in the literature. Thus, Chen and Taam [2] produce time variability as a result of hydrodynamic instabilities of an accretion disk, while Takeuchi, Mineshige, and Negoro [21] use a model of self-organized criticality to simulate accretion onto the compact object. Both these models provide reasonable fits to the observed PSD shapes (however, not necessarily to their normalizations) by producing a modulation of the accretion rate onto the compact object. However, their associated light curves are very much different than those observed, testimony to the fact that the power spectrum erases all the phase information available in the signal and that very different light curves can in fact have identical PSDs.

3. THE LAG OBSERVATIONS

The discrepancy between the expected and observed variability of BHC discussed above, as gauged by their PSDs, is generally attributed to a modulation of the accretion rate onto the compact object, effected by processes far away from the compact object. This suggests that one may have to search for a different measure of variability for this class of sources. Indeed, the Comptonization process, thought to be responsible for the high energy emission, offers such an alternative measure of variability, which is, actually, more refined than the PSD (see, e.g., [25]): Because, in the inverse-Compton scattering of soft photons, it takes longer to up-scatter them to higher energies, the energy of the emergent photons increases with their residence time in the scattering medium. As a result, the hard photon light curves lag with respect to those of softer photons by amounts which depend on the photon scattering time. Thus, observations of these time lags provide a measure of the local electron density, a quantity inaccessible to analysis restricted to spectral fits alone. Therefore, if the high energy radiation is produced in the vicinity of the compact object and the observed PSDs are due to a modulation of the accretion rate by a hitherto unknown process, these lags should be roughly of the order ~msec or shorter, the electron scattering time near the compact object, independent of the modulation of overall fluctuations of the light curve.

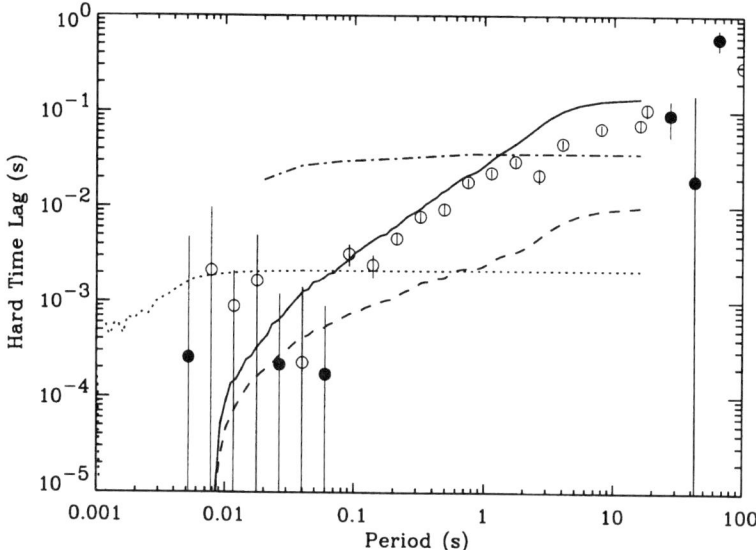

FIGURE 1. The time lags of hard X-rays (15.8–24.4 keV) with respect to soft ones (1.2–5.7 keV) as a function of Fourier period, obtained from the analysis of Ginga data of Cyg X-1 [14]. The circles and dots represent the measured positive and negative lags, respectively. The dash-dotted curve represents the lag predicted based on analytical Comptonization model given in Miyamoto et al. [14] and assumed corona electron density 10^{16} cm^{-3}. The three other curves are calculation results for Comptonization in the model coronae that produce the energy spectra shown in FIGURE 5.

Observations and analyses of the lags have given results drastically different from our expectations. The first such analysis was published by Miyamoto [14] for the light curves of the black hole candidate Cyg X-1 obtained by Ginga. The Fourier analysis of the lags between photons in the energy band 15.8–24.4 with respect to those in the 1.2–5.75 keV band has revealed that these lags increase linearly with Fourier period $P = 1/\nu$ to $\delta t \simeq 0.1$ sec for $\nu \simeq 0.1$ Hz (FIGURE 1). Similar results were obtained by the analysis of the higher quality RXTE data from the same source [3] (see FIGURE 5). Analysis of the light curves of the source GX 339-4 obtained by Ginga [15] and the source GRS 1758-258 obtained by RXTE [20, Fig. 2] has given a similar dependence of the lags on the Fourier period. In the latter case, the time lags of hard X-rays (6–28 keV) with respect to soft ones (2–6 keV) as a function of Fourier period extend to $\delta t \simeq 1$ sec for $\nu \sim 0.02$ Hz! Finally, the high energy transient GRO J0422+32 was observed by CGRO/OSSE during outburst and analysis of the lags between the hard X-rays (75–175 keV) and the softer ones (35–60 keV) indicated a similar behavior [4, Fig. 3]. These results suggest that the frequency-dependent hard X-ray time lags may be a common property of these sources, at least during particular states of their emission.

The magnitude and ν-dependence of these lags are very hard to understand in the context of models which assume that the X-rays are produced by Comptonization in the vicinity of a compact object. In such models, they should be roughly constant and reflect the photon scattering mean free time of the region, or $\delta t \sim$ msec. Miyamoto et al. [14] found this dependence sufficiently disturbing to question the process of Comptonization as the one responsible for the production of the high energy spectra of these sources. Thus, observations indicate that both the PSDs and the hard X-ray lags of the aperiodic variability of galactic BHC is in serious disagreement with our expectations based on the generic accretion disk-Comptonizing corona models, even though the associated spectral fits by Comptonization models are generally satisfactory. We cannot overemphasize the v-dependence of the lags, because, unlike the PSD, it is independent of modulations in the accretion rate and thus excludes from the outset the possibility that its form can be understood in terms of this process.

4. THE EXTENDED, NONUNIFORM CORONA MODEL

Kazanas, Hua, and Titarchuk [9] proposed that the discrepancies between the observed and the expected variability properties of X-ray binaries can be resolved if the corona of Comptonizing electrons is: (a) nonuniform and (b) extends over several orders of magnitude in radius. Specifically, they proposed the following density profile for the Comptonizing medium:

$$n(r) = \begin{cases} n_i & \text{for } r \le r_1 \\ n_1 (r_1/r)^p & \text{for } r_2 > r > r_1 \end{cases} \quad (4)$$

where the power index p is a free parameter; r is the radial distance from the center of the spherical corona; r_1 and r_2 are its radii of the inner and outer edges, respectively. The density profile of Eq. (4) allows scattering to take place over a wide range of densities, thereby introducing time lags over a similar range of time, leading to v-dependent lags as well as injecting fluctuation power to lower frequencies. Hua et al. [5] further showed that these inhomogeneous models, due to the linearity of Compton scattering, yield highly coherent light curves with coherence functions consistent with those obtained in the timing observations of X-ray binaries by Ginga [15], [26], and also the more recent observations by RXTE [3].

Compared to the standard, homogeneous Comptonization models, the models based on electron densities given by Eq. (4) contain additional adjustable parameters, most notably the index of the radial density dependence p; before these models can be used to interpret the lag data, the question arises as to how the corresponding energy spectra depend on and are affected by these extra parameters.

4.1. The Effects of Density Profile on the Energy Spectrum

Once the choice in favor of a nonuniform Comptonizing corona is made, the question that is immediately raised is that of the resulting spectra, since the earlier uniform density Comptonization models have produced very good fits to observations. A moment's thought can in fact convince the reader that the quality of the spectra will not be compromised: The photon spectra in the Comptonization process are the result of the convolution of the spatial photon transport (diffusion) through the hot plasma cloud with that of their energy gain in the scattering process with the electrons. As discussed in Sunyaev and Titarchuk [23], the index of the (generally power law) photon spectra is the product of the electron temperature with the lowest eigenvalue of the diffusion operator associated with the specific problem. Going from a uniform to a nonuniform source will, in general, change the numerical value of the corresponding eigenvalue, however, by judicious change of the total Thomson depth of the corresponding corona, we can end up with the same numerical figure of this eigenvalue and hence spectrum of the same quality as in a uniform case.

This point can be directly exhibited by fitting observations with spectra resulting from several different nonuniform coronae. In FIGURE 4, we use three photon spectra resulting from coronae with the same temperature but different optical depths and density profiles to fit the set of observational data by BATSE aboard the CGRO space craft taken from Ling *et al.* (1997), which tabulates the spectrum of the black hole candidate Cyg X-1 at its γ_0 state observed in early 1994. In the same figure we also show the PCA (circles) and HEXTE (dots) data covering an energy range 2–200 keV from RXTE observations of Cyg X-1 in 1996 [3], while the source was in a spectral state very similar to that of the Ling *et al.* [27] observations of 1994.

Along with the data we also present Monte Carlo fits corresponding to coronae with different density profiles and total Thomson depths and the same electron temperature $T_e = 100$ keV and soft photon injection from a black body source of temperature $T_0 = 0.2$ keV. Thus, the dotted line corresponds to $n_1 = 10^{16}$ cm^{-3}, and the total Thomson depth $\tau_0 = 0.5$; the solid line to a corona with $p = 1$, $n_1 = 4.35 \times 10^{16}$ cm^{-3} and uniform density inner core of radius $r_1 = 10^{-4}$ light second and Thomson depth $\tau_1 = 0.2$, while its total Thomson depth is $\tau_0 = 1$; the dashed line corresponds to a corona with $p = 3/2$, $n_1 = 1.594 \times 10^{17}$ cm^{-3}, $r_1 = 10^{-4}$ light second but a Thomson depth $\tau_1 = 0.07$ (the uniform core has density less than n_1) and total Thomson depth of the corona $\tau_0 = 0.7$.

4.2. The Hard X-Ray Time Lags of Nonuniform Coronae

As indicated by specific models outlined above, spectral fits alone are incapable to distinguish amongst configurations of Comptonizing clouds of as diverse structures as those used in these models. It is our contention however, that the particulars of the associated spatial structures can be revealed through the study of their timing properties. In fact, it was precisely these properties which motivated Kazanas *et al.* [9] to consider inhomogeneous models. In the following we will apply such models to provide fits of specific data sets associated with accreting black hole candidate sources.

In FIGURE 5 we plot the lags corresponding to the three different corona models used in producing the spectra which fit the observations of Cyg X-1 shown in FIGURE 4. These were obtained from the Monte Carlo calculations by collecting the escaping photons according to their arrival time to the observer as well as their energy. The photons were collected in the two energy bands 13–60 and 2–6.5 keV in order to be directly compared to the observational data of [3]. In each energy band, the photons were collected into 4096 bins over 16 seconds, each 1/256 seconds in length. The light curves so obtained were then used to calculate the time lags of the emission in the higher energy band with respect to that in the lower one.

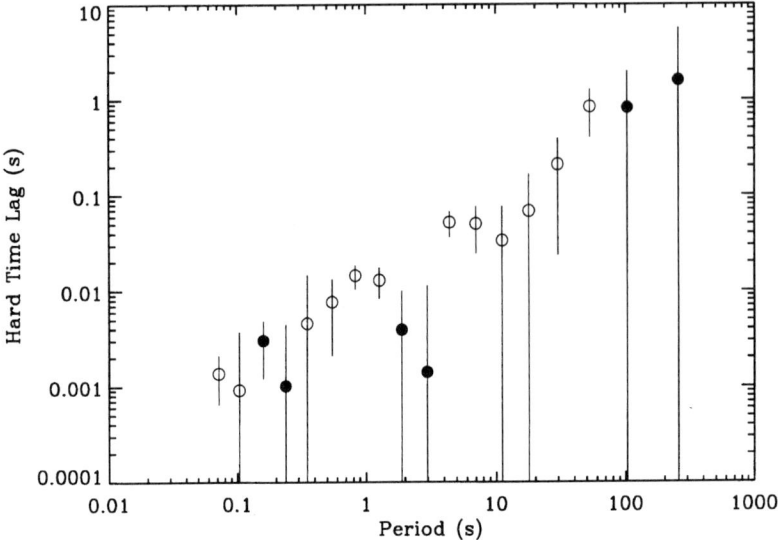

FIGURE 2. The time lags of hard X-rays (6–28 keV) with respect to soft ones (2–6 keV) as a function of Fourier period for the source GRS 1758-258. This is converted from the phase-lag-versus-frequency relation obtained by Smith et al. [20]. Dots are negative lags.

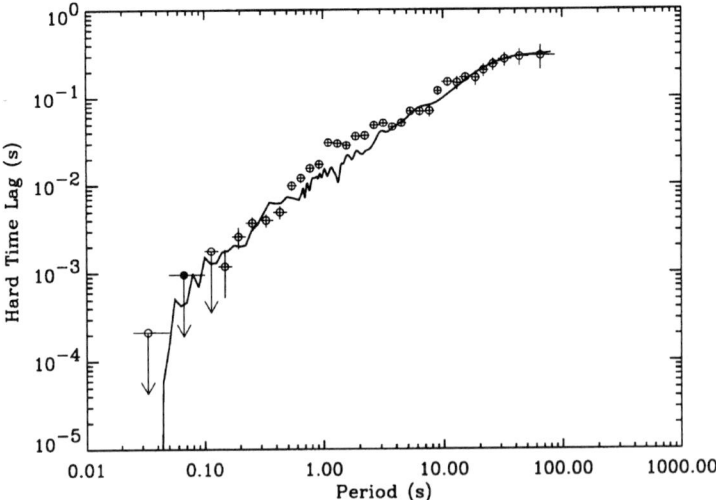

FIGURE 3. The time lags of hard X-rays (75–175 keV) with respect to soft ones (35–60 keV) as a function of Fourier period for the source GRO J0422+32, obtained by Grove et al. [4] from OSSE observation. The solid curve is obtained from a calculation based on Comptonization in a model corona with $p = 1$, $r_2 = 1.9 \times 10^{11}$ cm, $kT_e = 100$ keV and $\tau_0 = 0.25$ (see next sections for the model description).

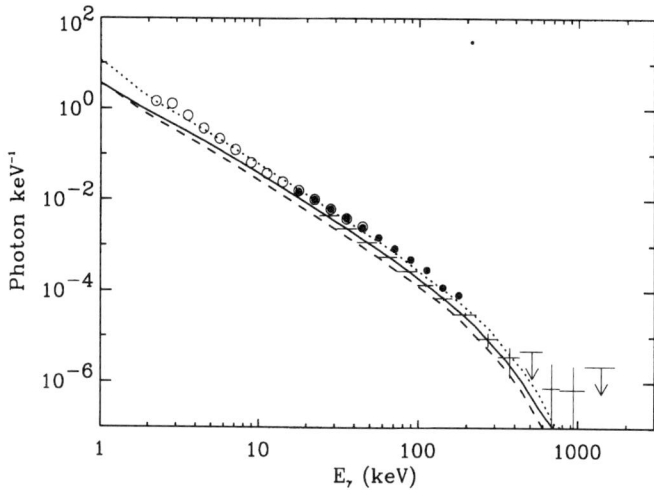

FIGURE 4. Three calculated energy spectra which fit equally well the Cyg X-1 data (crosses) in its γ_0 state observed by CGRO/BATSE in 1994 [12]. These spectra result from Comptonization in coronae with the same temperature but different optical depths and density profiles: dotted, $p = 0$, $\tau_0 = 0.5$; solid, $p = 1$, $\tau_0 = 1.0$; and dashed, $p = 3/2$, $\tau_0 = 0.7$. The dotted and dashed curves are slightly displaced to separate the otherwise nearly identical curves. Also plotted are RXTE/PCA (circles) and HEXTE (dots) data from the same source observed during its high state in 1996 [3].

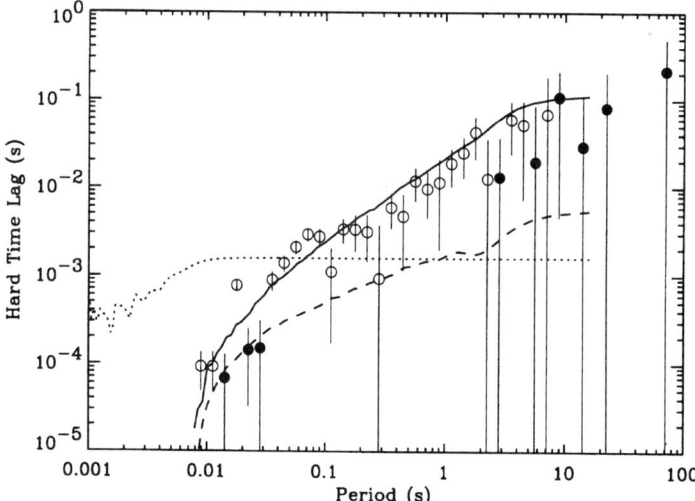

FIGURE 5. The time lags of hard X-rays (13–60 keV) with respect to soft ones (2–6.5 keV) resulting from Comptonization in the same coronae that produce the energy spectra shown in FIGURE 2. The dotted, solid, and dashed curves, as in FIGURE 2, represent the density profiles $p = 0$, 1, and 3/2, respectively. The time lag between the same energy bands based on RXTE data from Cyg X-1 [3] are also plotted in the figure.

The same parameters were applied in fitting the lag data obtained by Miyamoto *et al.* [14] which are shown in FIGURE 1 with the addition of the lags as computed by the analytic formula of Payne [28] for a uniform cloud of density $n = 10^{16}$ cm^{-3}. In FIGURE 2 we present the lags of the source GRS 1758-258 without any fits, while in FIGURE 3 those corresponding to the X-ray nova GRO J0422+32 obtained by OSSE [4] with the corresponding parameters given in the figure caption.

5. LIGHT CURVES, PSD, AND LAGS OF EXTENDED CORONAE

The linearity of the Comptonization process suggests that most of the relevant information is contained in the Green's function of the corresponding problem, which in the present case is the response of the extended corona discussed above to a δ-function in both time and energy input of soft photons; these soft photons are generally considered to be input at the center of the corona though different modes of injection are also possible. Thus the differences between coronae of various extents and density profiles can be assessed by looking into these response functions. Schematically, their form can be approximated by the gamma function

$$g(t) = \begin{cases} t^{\alpha-1} e^{-t/\beta}, & \text{if } t \geq 0; \\ 0 & \text{otherwise,} \end{cases} \quad (5)$$

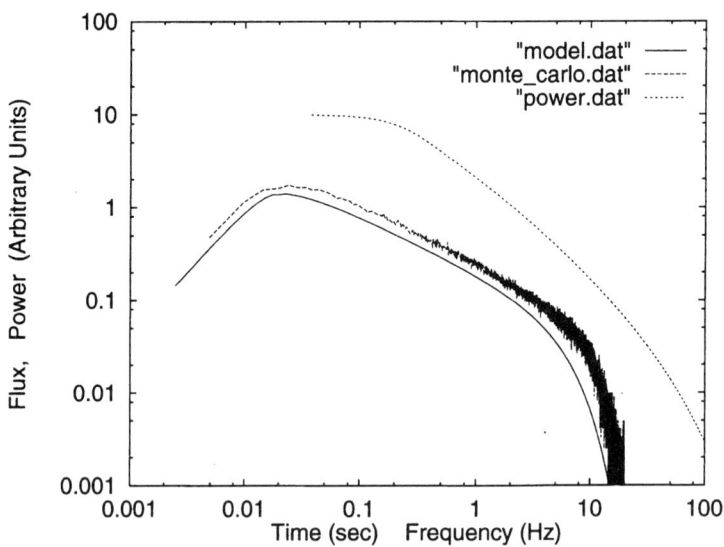

FIGURE 6. The response function is a corona with $p = 1$ and $r_1 = 6.35 \times 10^{-3}$ light sec $= 1.9 \times 10^8$ cm, $r_2 = 6.24$ light sec $= 1.87 \times 10^{11}$ cm, $T_e = 100$ keV, $\tau_0 = 1$ and $n_1 = 10^{15}$ cm^3 (long dashed curve) along with its fit by Eq. (4) with $\alpha = 0.4$, $\beta = 10$ sec, $t_0 = 0.02$ ' tsec, $b = 1$ (solid curve), and the corresponding power spectrum (short dashed curve).

where t is time; $\alpha > 0$ and $\beta > 0$ are parameters determining the shape of the light curves. The parameter β is associated with the outer edge of the corona [9] while α is related to the density profile (i.e., the parameter p) and the total Thomson depth of it τ_0. While the above form is well suited for analytical work, the exact form of the response function has to be obtained numerically (Monte Carlo simulation). An example of this is given in FIGURE 6 along with an analytic fit to the data of a form which deviates from the gamma function form of Eq. (5) at the short time scales, indicating that the rise of the response function is not infinitely sharp. The rising part is indicative of the size of the uniform core of the considered corona. The parameters of the corona associated with this response function are $p = 1$ and $r_1 = 6.35 \times 10^{-3}$ light sec $= 1.9 \times 10^8$ cm, $r_2 = 6.24$ light sec $= 1.87 \times 10^{11}$ cm, $T_e = 100$ keV, $\tau_0 = 1$ and $n_1 = 10^{15}$ cm^{-3} (long dashed curve).

5.1. Computing Light Curves and PSD

With the response function at hand it is an easy task to compute the resulting light curves, simply by convolving it with the soft photon injection function. However, since the latter is unknown, care must be taken so that the answer is not introduced by hand into the problem. The prescription we follow, therefore, is the following: The observed light curves consist of the incoherent superposition of elementary events, of the form given by Eq. (5) each triggered by the injection of a soft photon pulse of zero duration into the extended corona. The resulting light curve should then be of the form

$$F(t) = \sum_{i=1}^{N} Q_i g(t - \tau_i) \qquad (6)$$

the variable τ_i in the above equation is a random variable indicating the injection times of the individual shots, while Q_i is their normalization; to avoid introducing additional information we assume the Q_i's to be constant while the values of τ_i's to be Poisson distributed with a given constant rate. As such, the τ_i's are chosen using the relation $\tau_i = -f \cdot t_0 \cdot \log R_i$, where R_i is a random number uniformly distributed between 0 and 1 and f a real number, indicating the mean time between shots in terms of their rise time t_0.

In FIGURE 7 we present such a model light curve. The parameters of the shots used in constructing these curves were the same as those fitting the response function of FIGURE 6 (solid line). The value of the parameter f, i.e., the parameter which indicates the mean arrival time between shots was chosen in this case to be $f = 10$, while the value of the time scale t_0 at which the response function achieves its maximum value is, in this case, $t_0 \simeq 10^2$ sec.

Due to the Poisson nature of the injection of soft photon shots, the PSD of the light curve given in FIGURE 7 is the same at that of the PSD of a given individual shot, as given in FIGURE 6. The PSD associated with the analytic approximation of the response function of the corona depicted in FIGURE 6 is also shown in the same figure (dashed line). It is apparent there that the PSD has a white noise spectrum at frequencies $\nu \lesssim 1/\beta$, a power law component $\propto \nu^{-2\alpha}$, determined by the power-law like ($\propto t^{\alpha-1}$) section of the response function, and an additional steepening at $\nu \gtrsim 1/t_0$ due to the finite rise time

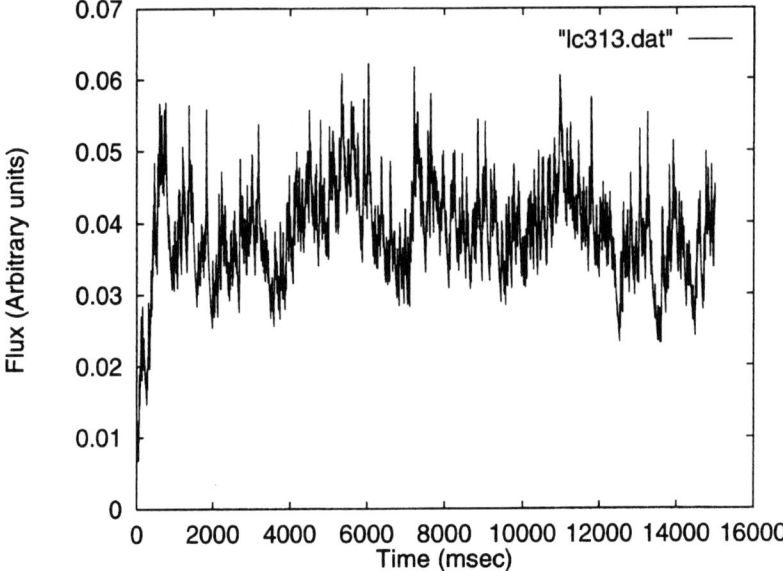

FIGURE 7. Model light curve constructed using the prescription of Eq. (6). The following values for the relevant parameters have been used: $\alpha = 0.5$, $\beta = 1.5$ sec, $t_0 = 0.001$ sec, $b = 1$, while the parameter f takes the value $f = 10$.

of the response function. Our model is thus capable of reproducing the observed PSD's of BHC, but at the same time attributes a direct physical significance to all its features, setting it apart from most alternative models of BHC variability.

5.2. Computing The Phase Lags

As is well known, the PSD by being the square of the Fourier transform, in effect, throws away all the information contained in the phases. While the absolute phases of the light curves of BHC appear to be totally random, the *relative* phases are well correlated, since the energy of the photon is detemined by its time of residence in the hot scattering plasma. As discussed in Sections 3 and 4.2, the observed lags do not conform with our notions of emission of most of the power associated with BHC from a region near a compact source and that they are indeed consistent with emission from the extended coronae discussed in Section 4.2. To make these arguments more transparent, we present in this section an analytic demonstration of this fact, based on the general form of the response function given above (Eq. (5)). The Fourier transformation of $g(t)$ (Eq. (5)) is

$$G(\omega) = \frac{\Gamma(\alpha)\beta^\alpha}{\sqrt{2\pi}}(1 + \beta^2\omega^2)^{-\alpha/2} e^{i\alpha\theta}, \tag{7}$$

where $\Gamma(x)$ is the Gamma function; $\omega = 2\pi\nu = 2\pi/P$ is the Fourier circular frequency and θ is the phase angle that we are interested in and

$$\tan\theta = \beta\omega. \tag{8}$$

If we consider two light curves in two energy bands distinguished by the different values of α and β (see FIGURE 8a), say α_1, β_1 and α_2, β_2, the time lag between them will be given by their phase lag θ divided by the corresponding frequency ω, i.e.,

$$\delta t = \frac{1}{\omega}[\alpha_2 \arctan(\beta_2\omega) - \alpha_1 \arctan(\beta_1\omega)]. \tag{9}$$

For large periods P (small ν) or $\beta_1\omega$, $\beta_2\omega \ll 1$, δt approaches the constant $\alpha_2\beta_2 - \alpha_1\beta_1$. On the other hand, for small P, $\delta t \simeq (\alpha_2 - \alpha_1)P/4$. The transition from the latter to the former occurs at

$$\beta\omega \sim 1 \quad \text{or} \quad P_c \sim 2\pi\beta.$$

It becomes apparent now that the leveling off of the time lags comes about at a frequency determined by the outer edge of the corona, r_2, as is the corresponding break in the PSD. In our model with the $p = 1$ density profile, β_1 and $\beta_2 \sim 1$ second. This explains what we see in FIGURES 1, 3, and 6; namely, for $P \lesssim$ a few seconds, $\delta t \propto P$ and the curve levels off for large P. On the other hand, for the case of uniform corona, $\alpha_1 = \alpha_2 = 1$; β_1 and $\beta_2 \sim 1$ msec. As a result, for large periods P, δt approaches the constant $\beta_2 - \beta_1$, which is of the order 1 millisecond. For small P, $\delta t \propto (1/\beta_1 - 1/\beta_2)P^2$. The transition occurs at $P_c \sim 2\pi\beta \simeq 0.006$ second.

In FIGURE 8a, we plot two pairs of light curves, one with $\alpha_1 = 0.1$ $\alpha_2 = 0.2$, $\beta_1 = 1$ and $\beta_2 = 1.25$, the other with $\alpha_1 = \alpha_2 = 1.0$, $\beta_1 = 0.001$ and $\beta_2 = 0.0025$. The former represents the light curves resulting from the corona with $p = 1$ density profile while the latter from a uniform one. The time lags between these two pairs of light curves are presented in FIGURE 8b. It is seen that the light curve parameters α and β determine the shape of the time lag curve: For $\alpha = 1$, or pure exponential light curves, the time lag is proportional to P^2 for small Fourier period P and turns to constant for large P. For $0 < \alpha < 1$, the time lag is proportional to P for small P and becomes constant for large P. In both cases, the level-off point is $P_c \simeq 2\pi\beta$. However, the time lag resulting from the exponential light curves has no portion linear in P. Thus the existence of a linear portion in the time lag curves obtained from observations clearly favors the power law light curve. From [9], we know that power law light curve is a signature of the nonuniform density distribution of the Comptonizing corona and the values of α and β are closely related to the physical size and density distribution of the source corona.

As indicated in the earlier sections, the numrically computed time lag curves corresponding to $p = 3/2$ are also distinctly different from those corresponding to $p = 1$ density profile. One can understand that by bearing in mind that the lags are a measure of the size of the scattering corona multiplied by the probability of scattering. Since the latter is constant for $p = 1$, while it decays with radius like $\propto r^{-1/2} \propto P^{-1/2}$ for $p = 3/2$, the results obtained in the numerical calculations given in FIGURES 1 and 5 become apparent.

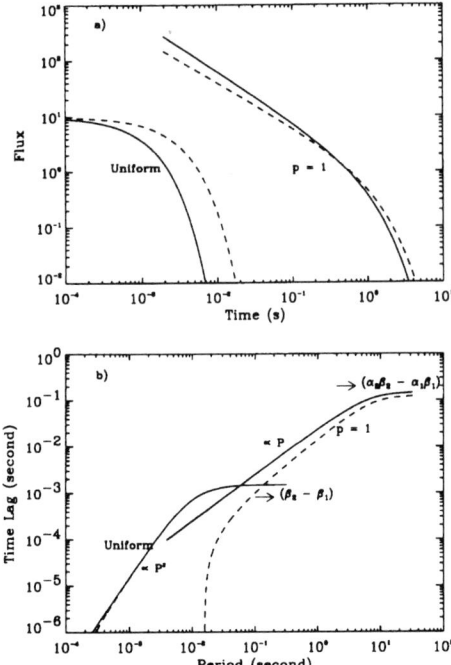

FIGURE 8. (a) Two pairs of light curves corresponding to two sets of values of α and β in Eq. (2). The pair with $p = 1$ represents the light curves from a cloud with $1/r$ density profile, the other pair from a uniform one. The solid curves correspond to the light curves in the lower energy bands. (b) The time lags based on Eq. (7) between the two curves in each pair in FIGURE 8. The dashed curves are times lags obtained by numerical Fourier transformation with finite time resolution.

Thus, under the assumptions of the present calculations (i.e., Comptonization as the main process of high energy emission, uniform temperature, nonuniform density) the time dependence of the photon flux, or light curves, at various energies can be used to map the radial density distribution of the hot electron Comptonizing corona. This fact provides the possibility of deconvolution of the density structure of these corona through a combined spectral-timing analysis of their light curves.

6. CONCLUSIONS

The main results of the present investigation are the following:

1. Spectral information is in itself incapable of resolving the issue of the dynamics of accretion onto compact objects. Additional timing information is necessary.

2. The available timing information, as determined by the PSD, is at odds with our notion that the X-ray emission in BHC sources is emitted from a region of only

several gravitational radii. Most dynamical models attempt to account for the observed PSD's as due to modulation of the accretion rate.

3. The phase, or equivalently time lag information in the light curves of different energies, under the assumption that the observed X-ray radiation is the result of the Comptonization process, demand that the extent of the X-ray emitting region be $\sim 10^3$ Schwarzschild radii!

4. The dependence of the time lags on the Fourier period P, can serve as a means of probing the density structure of the X-ray emitting plasma in these sources.

5. One can easily produce models of the X-ray light curves based on models of very extended coronae (i.e., $\sim 10^3$ Schwarzschild radii) which are consistent with all observed costraints, i.e., energy spectra, PSD, and time lags, suggesting that their existence should be seriously considered.

REFERENCES

1. ABRAMOWICZ, M.A., G. BAO, A. LANZA & N.-H. ZHANG. 1991. Astron. & Astrophys. **245**: 454.
2. CHEN, X. & R.E. TAAM. 1995. Astrophys. J. **441**: 354.
3. CUI, W. et al. 1997. Astrophys. J. **484**: 383.
4. GROVE, J.E. et al. 1997. Proceedings of the 4th Compton Symposium. AIP Conf. Proc. 410. Dermer, Strickman & Kurfess, Eds.: 122.
5. HUA, X.-M., D. KAZANAS & L. TITARCHUK. 1997. Astrophys. J. (Lett.) **482**: L57.
6. HUA, X.-M., D. KAZANAS & W. CUI. 1997. Astro-ph/9710184.
7. HUA, X.-M. & L. TITARCHUK. 1995. Astrophys. J. **449**: 188.
8. KAZANAS, D. & D.C. ELLISON. 1986. Astrophys. J. **304**: 178.
9. KAZANAS, D., X.-M. HUA & L. TITARCHUK. 1997. Astrophys. J. **480**: 735.
10. KROLIK, J., C. DONE & G. MADEJSKI. 1993. Astrophys. J. **402**: 432.
11. LOCHNER, J.C. 1989. Ph.D. Thesis. University of Maryland.
12. LOCHNER, J.C, J.H. SWANK & A.E. SZYMKOWIAK. 1991. Astrophys. J. **376**: 295.
13. MEEKINS, J.F. et al. 1984, Astrophys. J. **278**: 288.
14. MIYAMOTO, S. et al. 1988. Nature. **336**: 450.
15. MIYAMOTO, S. et al. 1991. Astrophys. J. **383**: 784.
16. MIYAMOTO, S. et al. 1992. Astrophys. J. (Lett.) **391**: L21.
17. NARAYAN, R. & I. YI. 1994. Astrophys. J. (Lett.) **428**: L13.
18. PRIEDHORSKY, W. et al. 1979. Astrophys. J.(Lett.) **233**: 350.
19. SHAPIRO, S. L., A.P. LIGHTMAN & D. M. EARDLEY. 1976. Astrophys. J. **204**: 187.
20. SMITH, D.M., et al. 1997. Astrophys. J. (Lett.) **489**: L51.
21. TAKEUCHI, M., S. MINESHIGE & H. NEGORO. 1995. PASJ, 47, 617
22. WILMS, J., et al. 1997, Proc. 4th Compton Symp., AIP Conf. Proc. 410, Dermer, Strickman & Kurfess, Eds., p. 849.
23. SUNYAEV, R.A. & L.C. TITARCHUK. 1980. Astron. & Astrophys. **86**: 121.
24. L.C. TITARCHUK. 1994. Astrophys. J. **434**: 570.
25. VAN DER KLIS et al. 1987. Astrophys. J. (Lett.) **319**: L13.
26. VAUGHAN, B.A. & M.A. NOWAK. 1997. Astrophys. J. (Lett.) **474**: L43.
27. LING, J.C., et al. 1997. Astrophys. J. **484**: 375.
28. PAYNE. 1980. Astrophys. J. **237**: 951.

Stellar Oscillons

O.M. UMURHAN, L. TAO, AND E.A. SPIEGEL

Department of Astronomy, Columbia University, New York, New York 10027

ABSTRACT: We study the weakly nonlinear evolution of acoustic instability of a plane-parallel polytrope with thermal dissipation in the form of Newton's law of cooling. The most unstable horizontal wavenumbers form a band around zero and this permits the development of a nonlinear pattern theory leading to a complex Ginzburg-Landau equation (CGLE). Numerical solutions for a subcritical, quintic CGLE produce vertically oscillating, localized structures that resemble the oscillons observed in recent experiments of vibrated granular material.

1. INTRODUCTION

The excitation of sound waves in the envelopes of stars has been extensively studied for its diagnostic importance as well as for its intrinsic physical interest. The most familiar mechanism of stellar acoustic instability is the Eddington valve mechanism or kappa mechanism, so-called because it relies on the dependence of opacity, κ, on physical conditions. This mechanism, which is basically thermal, resembles the phenomenon of negative differential resistivity [1] familiar in condensed matter physics. However, sound waves can become unstable even when there is no kappa mechanism operating, and here we discuss a simple version of such instability.

The case of optically thin perturbations to a fluid layer stratified under gravity is one where the kappa mechanism cannot operate and yet it does show instabilities of sound waves under suitable conditions [2]. The space available for this work is not sufficient for a discussion of the conditions under which such instabilities can occur (but see Umurhan [3]). Moreover, this paper is a contribution to a symposium on nonlinear astrophysics, so our aim is to describe some nonlinear aspects of these acoustic instabilities, with pauses along the way for only a few indispensable remarks about the linear theory in Section 3, following an introduction to the basic equations in Section 2. In Section 4 we outline the nonlinear procedures and conclude in Section 5.

2. EQUATIONS AND EQUILIBRIA

We consider the dynamics of a plane-parallel fluid subject to some form of radiative heat exchange but not to viscosity. The equations of motion for this system are,

$$\partial_t \rho + \nabla \cdot (\rho \mathbf{u}) = 0 \tag{1}$$

$$\rho(\partial_t + \mathbf{u} \cdot \nabla)\mathbf{u} = -\nabla p + \rho g \hat{\mathbf{z}} \tag{2}$$

$$C_v\rho(\partial_t + \mathbf{u} \cdot \nabla) T + p\nabla \cdot \mathbf{u} = Q(T, \rho) \tag{3}$$

$$p = \mathcal{R}\rho T \tag{4}$$

where the variables ρ, $\mathbf{u}$, T, and p are, respectively, density, vector velocity, temperature, and pressure; $Q(T, \rho)$ represents the thermal sources and sinks of the medium due to radiation and possibly mechanical effects. The vertical coordinate z is measured positively downward and $\hat{\mathbf{z}}$ is a downward-pointing unit vector.

In equilibrium, $\mathbf{u} = 0$ and the state variables depend only on z. The governing equations are the hydrostatic equation, the equation of state, and the thermal equilibrium condition, $Q(T_0, \rho_0) = 0$, where the subscript naught denotes equilibrium values. Rather than go into the details of the transfer problem in the equilibrium state, we simply postulate that there is an equilibrium in which $T_0(z) = \beta z$, where β is a constant; this is in fact the state one obtains in the diffusion limit with no heating and with a fairly general form for the opacity [4]. With the hydrostatic condition and the equation of state, we then find that $\rho_0(z) = \rho_*(z/z_*)^m$ and $p_0(z) = p_*(z/z_*)^{m+1}$ where ρ_* and p_* are constants and $m = g/(\mathcal{R}\beta) - 1$ is an atmospheric analogue of the polytropic index. We let $T_* = \beta z_*$, and choose $p_* = p_0(z_*)$, $\rho_* = \rho_0(z_*)$, so we have $p_* = \mathcal{R}\rho_* T_*$.

We introduce natural units so that there remain only nondimensional equations in evidence. We let z_* be the unit of length, the speed of sound ($c_a = \sqrt{\gamma \mathcal{R} T_*}$) be the unit of speed, ρ_* be the unit of density and so on. The equilibrium temperature is then $T_0 = z$ and similarly for the other quantities.

We assume that the atmosphere is truncated so that the equilibrium thermodynamic quantities are nowhere zero. The fluid is confined between $z = 1$ below and $z = z_0 > 0$ above, where $(1 - z_0)$ is the nondimensional layer thickness. We require that the vertical velocity vanishes on $z = z_0$ and $z = 1$.

3. LINEAR THEORY

We now introduce small perturbations about the static basic state and linearize the resulting equations to study the stability of the equilibrium state. We shall consider a very simple version of the stability theory here since our aim is to bring out features of the nonlinear aspects.

The most complicated physical issue is the treatment of the transfer problem in the general case. For weak, optically thin perturbations to the equilibrium, we have that

$$Q(T, \rho) = Q(T_0, \rho_0) + Q_T(T_0, \rho_0)(T - T_0) + Q_\rho(T_0, \rho_0)(\rho - \rho_0) + \ldots \tag{5}$$

where the subscripts represent differentiation. In Newton's law of cooling, $Q_\rho = 0$ and the higher-order terms are neglected. We adopt that form here and write $Q_T = -\rho_0 C_v q$, where q is a characteristic inverse time.

For a grey, optically thin medium, we may express q in terms of the absorption coefficient and the state variables [5]. If we assume that this coefficient is proportional to a power of density times a power of temperature, the linear theory is

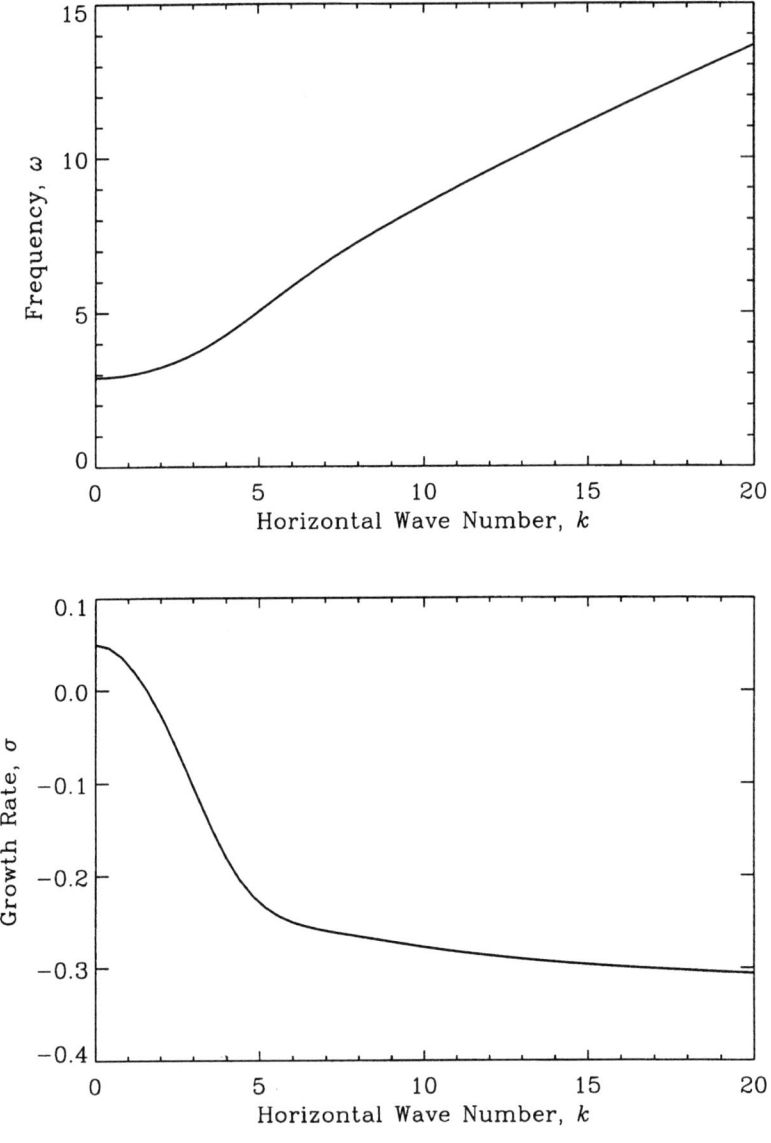

FIGURE 1. Dispersion relation of the fundamental acoustic mode ($\gamma = 1.28$, $m = 1.5$, $z_0 = 0.1$ and $q = 3$). The curves are frequency and growth rate as a function of the horizontal wavenumber.

straightforward and acoustic instabilities occur in several parameter regimes [2], [6], [7]. In particular, the case of constant q is very simple and we shall adopt it here and write for the rest of this discussion that $Q(T, \rho) = -q\rho_0 C_v(T - T_0)$ with q constant.

If we expand about the equilibrium solution and linearize, we obtain a set of equations that is tractable both analytically and numerically. These linear equations are separable and the general perturbation is written in the form

$$f(x, y, z, t) = F(z)\exp[\sigma t + i(\omega t + \mathbf{k} \cdot \mathbf{x})] \tag{6}$$

where $\mathbf{k}$ is the horizontal wavevector and σ and ω are real. This leads to a confluent hypergeometric equation if suitable dependent variables are introduced [2].

As usual, we find gravity (or convective) modes and acoustic modes, but we focus only on the fundamental acoustic mode for the purposes of this discussion. In FIGURE 1 we plot ω and σ versus $k = |\mathbf{k}|$ for the case of $\gamma = 1.28$, $m = 1.5$, $z_0 = 0.1$, and $q = 3$. We see that there is instability in a band of wavenumbers around zero. From this band, we then construct a nonlinear wave packet in the next section.

4. ACOUSTIC PATTERN EQUATIONS

The linear problem is characterized by a band of overstable wavenumbers around $k = 0$ and we have what is called a Hopf bifurcation in nonlinear stability theory. In this circumstance, the generic equation governing the nonlinear spatio-temporal evolution of a wave packet is a complex Ginzburg-Landau equation. To derive this equation, we use a multiple-scale analysis based on the availability of a small parameter, here the degree of instability of the layer.

The linear theory provides a condition on the parameters q, m, γ, and z_0 for the onset of acoustic instability. We may fix three of these and treat the fourth, say γ, as the control parameter governing the degree of instability. Thus, for fixed m, z_0, and q we find that instability begins as γ passes below the critical value γ_c and we examine values

$$\gamma = \gamma_c - \epsilon^2 \mu, \tag{7}$$

where μ is simply a fiducial quantity. It is ϵ that measures the degree of instability and we take it to be small here.

The band of significantly unstable wavenumbers around zero has a width of order ϵ and so we use this parameter to rescale the horizontal coordinate in the usual manner of nonlinear instability theory [8]. Similarly, we scale the time and we then look for solutions in terms of the scaled variables. In particular, the deviations from equilibrium temperature, density, and vertical and horizontal velocity are of the forms

$$\theta \sim \epsilon[A(\epsilon x, \epsilon y, \epsilon^2 t)\Theta(z)e^{i\omega t} + \text{c.c.}] + \ldots$$

$$\rho \sim \epsilon[A(\epsilon x, \epsilon y, \epsilon^2 t)P(z)e^{i\omega t} + \text{c.c.}] + \ldots$$

$$w \sim \epsilon[A(\epsilon x, \epsilon y, \epsilon^2 t)W(z)e^{i\omega t} + \text{c.c.}] + \ldots$$

$$u \sim \epsilon^2 [A(\epsilon x, \epsilon y, \epsilon^2 t)U(z)e^{i\omega t} + \text{c.c.}] + \ldots, \tag{8}$$

where A is the (generally complex) envelope function that describes the pattern of the instability. In linear theory, A is an arbitrary constant; in nonlinear theory it is a slowly varying function that is the focus of interest. The functions of z, on substitution and suitable asymptotic development, turn out to be the z-dependent parts of the linear eigenfunctions.

The asymptotic developments based on these scalings leads to an equation for A. On general grounds, an overstability of the kind we have here is known to lead to an equation of the form [8]

$$\partial_t A = \mu A + \alpha \Delta A + \mathcal{F}(|A|^2)A \qquad (9)$$

where the Laplacian operates in the two-dimensional space (ϵx, ϵy). The linear part of this equation describes the linear stability theory, while the quantity $\mathcal{F}(|A|^2)$ represents the renormalization of the linear growth rate by nonlinear effects, with $\mathcal{F}(0) = 0$.

Determination of $\mathcal{F}$ is not possible in general, so it is represented by Taylor series and the asymptotic development allows the computation of the coefficients in this development at each order in ϵ. Typically, only the leading order is needed for weak instability. Then Eq. (9) becomes the (cubic) complex Ginzburg-Landau equation (CGLE), which generally describes the patterns resulting from overstable systems with a continuous spectrum of horizontal wavenumbers. The cubic term is able to saturate the linear growth when $\text{Re}(\beta) < 0$. However, if the leading nonlinearity, the cubic term, has a coefficient that does not allow for nonlinear saturation of the instability, higher order terms must be sought. This may require a modification of the scaling.

For situations without a strong symmetry in the vertical, nonlinear saturation by the cubic term often does not occur and we have what is called subcritical bifurcation. This resembles a phase transition of the first kind and it is what we see at the larger values of q. In that case, the acoustic pattern equation is of the form

$$\partial_t A = \mu A + \alpha \Delta A + \beta |A|^2 A + \eta |A|^4 A. \qquad (10)$$

The complex constants α, β, and η are functions of the physical parameters of our model atmosphere. Calculation of the coefficients requires a working out of the nonlinear perturbation theory and we indicate some results of such work for selected values in the table below.

m	z_0	q	γ_c	α_r	α_i	β_r	β_i	Comments
1.5	0.1	3.0	1.38	+0.0173	−0.0661	+1.107	+4.754	subcritical
6.0	0.1	1.0	1.238	+0.0003	−0.041	−166.1	+35.94	supercritical

Both subcritical and supercritical instabilities may occur and produce differing behavior. For the case of one-dimensional patterns, spatio-temporal disorder is the rule in the supercritical case [2], [9]. But in the subcritical state, stable isolated structures may be expected, as argued by Thual and Fauve [10], [11]. In the two-dimensional case of subcritical bifurcation, we find oscillating, stable, localized structures

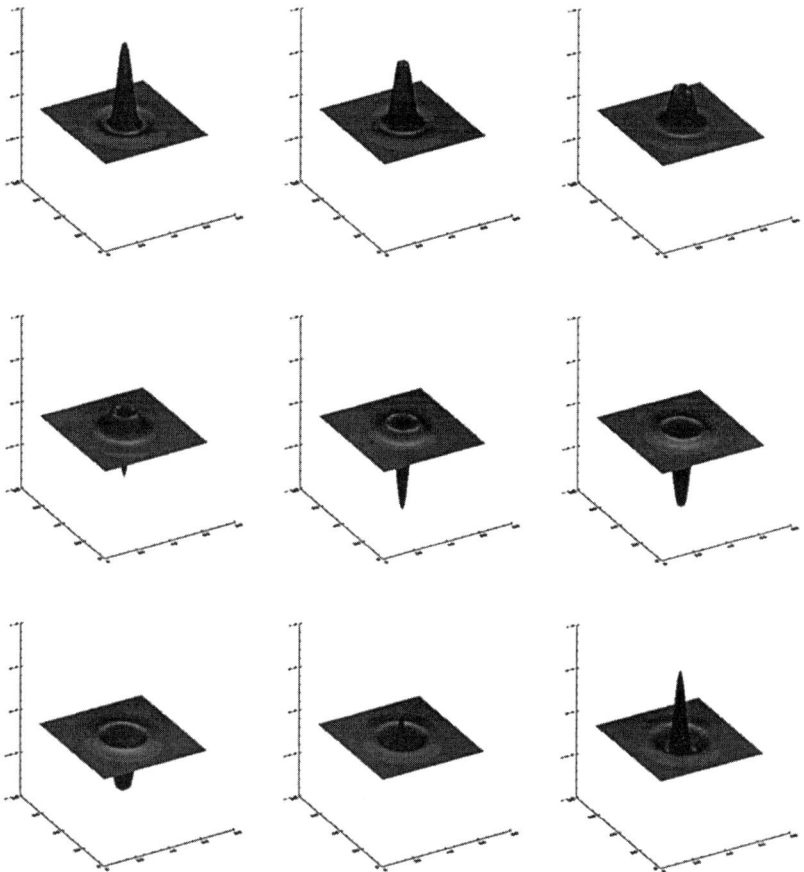

FIGURE 2. Evolution of an oscillon (CGLE parameters: $\alpha = 1.0$, $\beta = 3.0 + i$, and $\eta = -2.75 + i$). The domain is periodic in both spatial dimensions with a period of 10π. The shaded surfaces are the real parts of the amplitude. The oscillation starts from the top left panel and continues to the right.

whose time dependence is shown in FIGURE 2 in a series of snapshots. This structure is robust and we have seen it with a large range of values of the CGLE coefficients, emerging from a wide variety of initial conditions.

Such oscillating pulses resemble the oscillons observed in recent experiments of vertically-shaken layers of granular material [12]–[14], and they also have emerged from other pattern equations [15]–[18]. We have here used the term oscillon to characterize the similar object found with the subcritical CGLE. The "-on" ending normally smacks of integrability and it may be that, in the astronomical context, the word spicule would be more apt. This is a matter for later discussion.

5. DISCUSSION

Stratified layers with thermal dissipation frequently suffer acoustic instabilities for various reasons that have not all been clarified as yet. We have considered one of the simplest of these instabilities to show what they may lead to. In fact, it is well known that overstabilities in thin layers where the spectrum of horizontal wavenumbers is so dense as to be regarded as continuous have an amplitude function, or envelope, that satisfies an equation of the CGLE form when the amplitudes are not too large. As we shall discuss elsewhere, the inclusion of the effects of magnetic fields or rotation may not require much qualification of these remarks. So we are inclined to seek applications of these considerations to thin layers such as stellar chromospheres or slabs and disks. Global modes such as are familiar in helioseismology, being discrete, would require simpler treatments and call for ODEs for their nonlinear description. At any rate, we have been able to extract from this simple theory solitary structures of a kind that have been attributed to magnetic effects in discussions of solar atmospheric dynamics.

It is not even necessary that the acoustic waves be unstable for a description by the CGLE to be appropriate, as mechanical forcing could also be included in the theory. However, if the degree of instability or the amplitude of the forcing becomes too large, one may question the use of weakly nonlinear theory. The recently observed excitation by a solar flare of an expanding wave on the solar surface [19] is a case in point and it remains to be seen whether simple pattern equations can be appropriate in such situations where the initial amplitudes are quite large. Still, it is also true that the wave amplitude decays quickly when the general background is stable, and even then the weakly nonlinear theory may be brought to bear.

ACKNOWLEDGMENTS

L.Tao thanks National Science Foundation for a Postdoctoral Fellowship and Pierre Coullet for a helpful conversation.

REFERENCES

1. CONWELL, E.M. 1970. Physics Today. June: 35–41.
2. SPIEGEL, E.A. 1964. Astrophys. J. **139:** 959–974.
3. UMURHAN, OM. 1998. Ph.D. Thesis. Dept. Astron. Columbia University.
4. SPIEGEL, E.A. 1965. Astrophys. J. **141:** 1068–1090.
5. SPIEGEL, E.A. 1957. Astrophys. J. **126:** 202–207.
6. SYROVATSKII, S.I. & YU.D. ZHUGZHDA. 1968. Sov. Phys. Astron. **11:** 945–952.
7. MACDONALD, J. & D. MULLAN. 1997. Astrophys. J. **481:** 963–972.
8. MANNEVILLE, P. 1990. Dissipative Structures and Weak Turbulence. Academic Press. New York, USA.
9. BRETHERTON, C.S. & E.A. SPIEGEL. 1983. Phys. Lett. **A96:** 152–156.
10. THUAL, O. & S. FAUVE. 1988. J. Physique (Paris) **49:** 1829–1833.
11. FAUVE, S. & O. THUAL. 1990. Phys. Rev. Lett. **64:** 282–284.
12. MELO, F., P. UMBANHOWAR & H.L. SWINNEY. 1994. Phys. Rev. Lett. **72:** 172–175.
13. MELO, F., P. UMBANHOWAR & H.L. SWINNEY. 1995. Phys. Rev. Lett. **75:** 3838–3841.
14. UMBANHOWAR, P, F. MELO & H.L. SWINNEY. 1996. Nature (London) **382:** 793.

15. TSIMRING, L.S. & I.S. ARONSON. 1997. Phys. Rev. Lett. **79:** 213–216.
16. CERDA, E., F. MELO & S. RICA. 1997. Phys. Rev. Lett. **79:** 4570–4573.
17. ROTHMAN, D.H. 1998. Phys. Rev. **E57:** R1239–R1242.
18. CRAWFORD, C. &H. RIECKE. 1998. LANL preprint: patt-sol/9804005.
19. KOSOVICHEV, A.G. & V.V. ZHARKOVA. 1998. Nature (London) **393:** 317.

Chaos in Cosmological Hamiltonians[a]

HENRY E. KANDRUP[b] AND JOHN DRURY

Department of Astronomy, University of Florida, Gainesville, Florida 32611

ABSTRACT: This paper summarizes a numerical investigation which aimed to identify and characterize regular and chaotic behavior in time-dependent Hamiltonians $H(\mathbf{r}, \mathbf{p}, t) = \mathbf{p}^2/2 + V(\mathbf{r}, t)$, with $V = R(t)V_0(\mathbf{r})$ or $V = V_0[R(t)\mathbf{r}]$, where V_0 is a polynomial in x, y, and/or z and $R(t) \propto t^p$ is a time-dependent scale factor. When p is not too negative, one can distinguish between regular and chaotic behavior by determining whether an orbit segment exhibits a sensitive dependence on initial conditions. However, chaotic segments in these potentials differ from chaotic segments in time-independent potentials in that a small initial perturbation will usually exhibit a sub- or superexponential growth in time. Although not periodic, regular segments typically exhibit simpler shapes, topologies, and Fourier spectra than do chaotic segments. This distinction between regular and chaotic behavior is not absolute since a single orbit segment can seemingly change from regular to chaotic and visa versa. All these observed phenomena can be understood in terms of a simple theoretical model.

INTRODUCTION

The past several decades have witnessed a growing recognition that chaotic behavior is seemingly ubiquitous in Nature, and that chaos could play an important role in many problems of astronomical interest, extending from stellar pulsations to galactic dynamics (see. e.g., Ref. [1]). It thus seems natural to consider the possibility that chaos could play an important role in problems related to cosmology, large scale structure, and quantum-field theory in the early Universe. However, cosmological problems lead to an important new ingredient, namely the expansion of the Universe, which implies, e.g., that, even if the system of interest is symplectic, the Hamiltonian $H(t)$ generating the evolution will usually have an explicit systematic time dependence. The obvious question then is: what effects, if any, will this time dependence have on the possibility of chaotic behavior?

One example of some interest is the gravitational N-body problem, as formulated for a large system of objects of comparable mass. In the context of isolated systems like individual galaxies, where the expansion of the Universe plays no role, the N-body problem is known to be chaotic in the sense that small initial perturbations in the locations of individual "particles" grow exponentially (see. e.g., Refs. [2]–[4] and references contained therein). However, despite some preliminary investigations

---

[a]H.E. Kandrup was supported in part by National Science Foundation Grant No. PHY92-03333 and by Los Alamos National Laboratory through the Institute of Geophysics and Planetary Physics. Some of the computations were facilitated by computer time made available through the Research Computing Initiative at the Northeast Regional Data Center (Florida) by arrangement with IBM.

[b]Also with the Department of Physics and Institute for Fundamental Theory, University of Florida, Gainesville, Florida 32611.

[5] it is not yet clear whether this exponential instability persists for the cosmological N-body problem, as formulated in a comoving frame that expands with the Universe. Another cosmological example involving a time-dependent Hamiltonian is the problem of particle creation in the early Universe, e.g., in the context of the phenomenon of preheating, which leads to what Kofman, Linde, and Starobinsky [6] have termed a "stochastic resonance."

But why might one expect that chaos will manifest itself differently in a cosmological context than for systems characterized by a time-independent H? For time-independent Hamiltonian systems, an orbit is usually said to be chaotic if and only if it has one or more positive Lyapunov exponents (see, e.g., Ref. [7]). However, these Lyapunov exponents can be defined in terms of the average properties of the stability matrix associated with an orbit, as evaluated along that orbit in an asymptotic $t \to \infty$ limit. Assuming canonical variables, the problem of stability, and hence the possibility of chaos, thus hinges on the properties of solutions to

$$\frac{d\delta Z^i}{dt} = J^{ij} \frac{\partial^2 H}{\partial Z^j \partial Z^k}\bigg|_{Z_0} \delta Z^k \equiv \Lambda^i_k(t) \delta Z^k, \qquad (1)$$

where δZ^i denotes a perturbed phase space coordinate, J^{ij} is the cosymplectic form [8], and $\Lambda^i_k(t)$ is a function of t because of its dependence on the unperturbed trajectory $Z_0^i(t)$. In particular, for a Hamiltonian

$$H = \tfrac{1}{2}\mathbf{p}^2 + V(\mathbf{r}), \qquad (2)$$

the configuration space perturbation satisfies

$$\frac{d\delta r^a}{dt^2} = -\frac{\partial^2 V}{\partial r^a \partial r^b}\bigg|_{r_0(t)} \delta r^b. \qquad (3)$$

If, e.g., the second derivative matrix $\partial^2 V/\partial r^a \partial r^b$ is constant and has at least one negative eigenvalue, there exist solutions that grow exponentially in time. If, however, that matrix acquires a secular time dependence this is no longer guaranteed to be true. For example, the one-dimensional equation

$$\frac{d^2 \delta r}{dt^2} = \Omega^2(t)\delta r = \frac{\Omega_0^2}{t^2}\delta r \qquad (4)$$

admits solutions which exhibit a power law growth. Just as an expanding Universe can convert an exponential Jeans instability into a milder power-law instability [9], it might be expected to make chaotic orbits "less chaotic."

For time-independent Hamiltonian systems, sensitive dependence on initial conditions and the existence of one or more positive Lyapunov exponents is not the only way in which chaotic orbits differ from regular orbits. Regular and chaotic orbits also have very different Fourier spectra. Because regular orbits are multiply periodic, they will have computed spectra where (at least if one integrates long enough) most of the power is concentrated at or near a relatively small number of frequencies, whereas the spectra for chaotic orbits should exhibit substantially broader band power. (Strictly speaking, not every flow admitting one or more positive Lyapunov exponents must have nonzero power for a continuous range of frequencies, but one anticipates that, as a practical matter, positive Lyapunov exponents and broad band power go hand in hand [10].)

The situation is very different for Hamiltonian systems which manifest a systematic secular time-dependence. In this case, one anticipates generically that no orbit can be truly periodic, so that even spectra which one might wish to interpret as corresponding to "regular" orbit segments should have Fourier spectra with broad band power. This does not necessarily mean that an examination of the Fourier spectra cannot be used to distinguish between regular and chaotic behavior. However, it *does* imply that any satisfactory discriminant based on an inspection of Fourier spectra must be more subtle than determining whether power is concentrated at or near a small number of frequencies.

The next section suggests a simple theoretical model based on a generalized Matthieu equation which can be used to make substantive predictions regarding the existence and manifestations of chaos in time-dependent Hamiltonian systems. This is followed by two sections which describe in detail a collection of experiments which were performed to test the predictions based on this model and the results of those experiments. A final section concludes by denumerating the principal conclusions, and then speculating on possible implications.

THEORETICAL EXPECTATIONS

To make reasonable predictions regarding the behavior of orbits in time-dependent Hamiltonian systems, including possible sensitive dependence on initial conditions, it is useful to understand precisely why chaotic behavior can arise in time-*in*dependent Hamiltonian systems. For such systems, an exponentially sensitive dependence on initial conditions, as manifested by the existence of one or more positive Lyapunov exponents, is related to solutions to the linearized evolution equation (3) satisfied by a small initial δr^a, which can be interpreted as a time-dependent oscillator equation of the form

$$\frac{d^2 \delta r^a}{dt^2} = \Omega^2_{ab}(t) \delta r^b, \qquad (5)$$

If the matrix Ω^2_{ab} is constant in time and all its eigenvalues are nonnegative, solutions to this equation involve stable oscillations, so that a small initial perturbation cannot grow exponentially. Alternatively, if Ω^2_{ab} is constant but has one or more negative eigenvalues, there *do* exist small perturbations that grow exponentially, which implies a sensitive dependence on initial conditions. However, this latter possibility does not seem very realistic: if there is always at least one negative eigenvalue, the potential V is not bounded from below!

The important point, therefore, is that instability is not necessarily associated simply with the fact that the second derivative matrix $\partial^2 V/\partial r^a \partial r^b$ has a negative eigenvalue. Indeed, many nonintegrable potentials that admit large amounts of chaos, including all the finite-order truncations of the Toda potential [11], yield a second derivative matrix that is everywhere nonnegative. Rather, as has been discussed elsewhere [12]–[14] in the context of Maupertuis' principle, where the flow associated with a time-independent H is reinterpreted as a geodesic flow on a curved manifold, chaos in time-independent Hamiltonian systems can be understood as resulting from a parametric instability.

The idea is very simple. Given a knowledge of the unperturbed orbit, $\mathbf{r}_0(t)$, one could diagonalize Eq. (5) and then express the second derivative matrix in terms of its Fourier transform to conclude that each eigenvector δr^A satisfies an equation of the form

$$\frac{d^2 \delta r^A}{dt^2} = -\left[C_0^A + \sum_\alpha C_\alpha^A \cos(\omega_\alpha t + \varphi_\alpha) \right] \delta r^A, \tag{6}$$

where, of course, the sum must be interpreted as a Stiltjes integral. The obvious point, then, is that, even if the coefficients C_α^A are sufficiently small that the term in brackets, and hence the eigenvalues of the stability matrix, are always nonnegative, there is the possibility of resonant behavior leading to solutions that grow exponentially in time. One simple example, corresponding to the case where there is only one nonzero frequency ω_α, is the Matthieu equation [15], which can be written in the form

$$\frac{d^2 \xi}{dt^2} = -(A + B\cos 2t)\xi. \tag{7}$$

As is well known, a study of solutions to Eq. (7) as a function of A and B reveals that the A–B plane divides naturally into distinct, well-defined regions corresponding to stable and unstable motions. In the stable regions, solutions to Eq. (7) are purely oscillatory; in the unstable regions they exhibit a systematic exponential growth, i.e., $|\xi(t)| \sim \exp(\chi t)$. The precise value of χ depends on A and B, so that, e.g., unstable values of A and B especially close to stable regions correspond to especially small (but still positive) values of χ. However, the fact that $\ln|\xi|$ grows linearly in time is robust. Allowing for generalizations of Eq. (7), which incorporate one or more additional frequencies does not change the basic picture. In some cases, ξ is bounded but, for other choices of parameters, ξ grows in such fashion that $\ln|\xi|$ is reasonably well fit by a linear growth law.

The obvious question then is: how do things change if the Hamiltonian H acquires an explicit time dependence? Consider, e.g., the simplest possible time-dependence, where the potential is multiplied by an overall time-dependent factor, so that

$$H = \tfrac{1}{2}\mathbf{p}^2 + R(t)V(\mathbf{r}), \tag{8}$$

with $R(t)$ a specified function of time. In this case, the natural analogue of Eq. (7) becomes

$$\frac{d^2\xi}{dt^2} = -R(t)(A + B\cos 2t)\xi, \qquad (9)$$

(or, perhaps, a generalization thereof with $\cos 2t$ replaced by $\cos 2\tau(t)$). Even if $R(t)$ evidence a systematic secular time-dependence, one can often make reasonable distinctions between solutions to Eq. (9) that do and do not grow rapidly in time. However, in general the rapidly growing "unstable" solutions will not exhibit a purely exponential growth.

Consider, e.g., the case where $R(t) = R_0 t^p$, with p a real constant. Here trivial numerical computations reveal that, at least for values of p somewhat larger than $p = -2$, the evolution of ξ can be well understood in an adiabatic approximation. The factor $R \propto t^p$ in the potential implies that the instantaneous "natural" frequencies ω with which ξ grows or oscillates should scale as $R^{1/2}(t) \propto t^{p/2}$; but, in the adiabatic approximation this leads to a time dependence

$$\int dt \omega(t) \sim \int dt R^{1/2}(t) \sim t^{1+p/2}. \qquad (10)$$

It follows that, for $p > 0$, unstable solutions correspond to superexponential growth, so that $\ln|\xi| \sim a + bt^q$, with $q = 1 + \frac{p}{2} > 1$. Alternatively, for $-2 < p < 0$, unstable solutions correspond to subexponential growth with $q = 1 + \frac{p}{2} < 1$. The adiabatic approximation fails for values of p that are too small, the special case $p = -2$ corresponding instead to solutions that exhibit a (possibly oscillatory) power law time dependence.

The other obvious point is that a single initial condition evolved with Eq. (9) can exhibit transitions from stable to unstable motions and visa versa, this corresponding to transitions between regular and chaotic behavior. Solutions to the ordinary time-independent Matthieu equation involve either stable or unstable motion, depending on the values of A and B, which do not change in time. However, incorporating a time dependence as in Eq. (9) involves allowing for time-dependent "dressed" quantities $\hat{A} = t^{p/2}A$ and $\hat{B} = t^{p/2}B$. In the adiabatic approximation, the time dependence involves $\hat{A}$ and $\hat{B}$ evolving through a sequence of values corresponding to a line in the $A-B$ plane. This line will in general intersect both stable and unstable regions, corresponding to intervals of both regular and chaotic motions.

The basic inference is that, for Hamiltonian systems of the form given by Eq. (8) with $R \propto t^p$, power laws $p > 0$ yield small perturbations of "chaotic" orbit segments that exhibit superexponential growth, whereas power laws $p < 0$ yield small perturbations that exhibit subexponential or power law growth. For more complicated Hamiltonians, e.g., $H = \frac{1}{2}\mathbf{p}^2 + V[\mathbf{r}/R(t)]$, the simple scaling that leads to Eq. (10) no longer holds. However, by analogy with the preceeding, one would anticipate that if the characteristic size of the second derivative matrix $\partial^2 V(t)/\partial r^a \partial r^b$ is increasing systematically in time, small perturbations of chaotic orbits should grow faster than exponentially, whereas the growth should be slower than exponential if this matrix is decreasing systematically. The examples described in the following sections corroborate this physical expectation.

NUMERICAL EXPERIMENTS PERFORMED

The numerical experiments described here were performed for time-dependent extensions of the time-independent potential

$$V_0(x, y, z) = -(x^2 + y^2 + z^2) + \tfrac{1}{4}(x^2 + y^2 + z^2)^2 - \tfrac{1}{4}(ay^2z^2 + bz^2x^2 + cx^2y^2) \quad (11)$$

which is itself an obvious three-dimensional generalization of the two-dimensional dihedral potential of Armbruster, Guckenheimer, and Kim [16] for specific choices of parameter values. The simplest extension, most easily compared with theory, involved introducing an overall multiplicative factor, setting

$$V(x, y, z, t) = R(t)V_0(x, y, z), \quad (12)$$

with $R(t) = t^p$. Another alternative involved mimicking the effects of comoving coordinates by setting

$$V(x, y, z, t) = V_0[R(t)x, R(t)y, R(t)z], \quad (13)$$

again with $R(t) = t^p$. Some computations focused on fully three-dimensional orbits. Others focused on two-dimensional orbits with $z = p_z = 0$. It was found that, at least in terms of their sensitive dependence on initial conditions, two- and three-dimensional orbits behaved very similarly, but that, in terms of possible shapes, three-dimensional orbits exhibited a richer phenomenology.

Ensembles of ~1000 initial conditions for use in two-dimensional simulations were generated by freezing the energy of the time-independent H at a fixed value E, setting $x = 0$, uniformly sampling the energetically allowed regions of the $y - p_y$ plane, and then solving for $p_x(x, y, p_y, E) > 0$. Initial conditions for fully three-dimensional simulations were generated by freezing the energy at E, setting $x = z = 0$, uniformly sampling the allowed regions of the $y - p_y - p_z$ cube, and solving for $p_x(x, y, z, p_y, p_z, E) > 0$. Each ensemble was evolved into the future for a time $t = 256$ or longer, with the initial time t_0 chosen to vary between $t_0 = 1.0$ and $t_0 = 100$. A reasonably broad range of exponents p was considered. The simulations with the potential (12) allowed for $-1.5 < p < 1.5$. Those with the potential (13) allowed for $-1 < p < 1$.

The evolution equations were integrated using a fourth-order Runge-Kutta algorithm with fixed time step δt ranging between 10^{-3} and 10^{-4}. The integrator solved simultaneously for the evolution of a small, linearized perturbation, renormalized at fixed intervals $\Delta t = 1.0$, to obtain an estimate of the largest short-time Lyapunov exponent (cf. Ref. [7]). When focusing on time-independent Hamiltonian systems, it is customary to record a running Lyapunov exponent that is a numerical approximation to the quantity

$$\chi(t) = \lim_{\delta Z(0) \to 0} \frac{1}{t} \ln\left[\frac{\|\delta Z(t)\|}{\|\delta Z(0)\|}\right], \quad (14)$$

with $\|\cdot\|$ the natural Euclidean norm, which converges towards the true Lyapunov exponent χ in a $t \to \infty$ limit. In the context of a time-dependent potential, it is more natural to record short-time Lyapunov exponents (cf. Ref. [17]) $\chi(\Delta t_i)$ for each interval Δt, which, for an integration begun at time $t = 0$, are related to $\chi(t)$ by [18]

$$\chi(\Delta t_i) = \frac{\chi(t_i + \Delta t)(t_i + \Delta t) - \chi(t_i)t_i}{\Delta t}. \tag{15}$$

Given such $\chi(\Delta t_i)$'s, the partial sums

$$\xi(t_i) = \frac{1}{\Delta t}\sum_{j=1}^{i=1} \chi(\Delta t_j) = \frac{1}{\Delta t}\ln\left[\frac{\|\delta Z(t_i + t_0)\|}{\|\delta Z(t_0)\|}\right] \tag{16}$$

capture the net growth of the initial perturbation within a time t_i.

Plots of $\chi(\Delta t_i)$ and $\xi(t_i)$ for individual orbit segments were examined visually in an effort to identify clear distinctions between regular and chaotic behavior. For those orbit segments deemed chaotic, $\xi(t_i)$ was fitted to a growth law

$$\xi = a + bt^q \tag{17}$$

to determine (1) whether such a fit was reasonable and (2) whether the best fit yielded super- or subexponential growth. Orbital data $\mathbf{r}(t)$ and $\mathbf{p}(t)$, and the associated Fourier spectra, $|\mathbf{r}(\omega)|$ and $|\mathbf{p}(\omega)|$, were also inspected visually in a search for distinguishing features. One aim was to determine whether orbit segments deemed regular also had simpler topologies and/or simpler spectra than chaotic segments that manifested a sensitive dependence on initial conditions. The other was to search for evidence for abrupt transitions between chaotic and regular behavior.

RESULTS OF THE EXPERIMENTS

For values of p that are not too negative, it is often possible to distinguish relatively clearly between regular segments, where $\chi(\Delta t_i)$ fluctuates around zero, and chaotic segments, where, if one averages over several time steps, $\chi(\Delta t_i)$ is usually larger than zero. This distinction becomes especially apparent if, for an ensemble of segments in the same potential with the same value of p, one computes $N[\xi(t_{\text{fin}})]$, the distribution of the final values of ξ. This $N[\xi(t_{\text{fin}})]$ often corresponds to a bimodal distribution and, even when one seems to see only a single population, tracking the form of the distribution as a function of p usually allows one to determine whether that population is regular or chaotic.

This is illustrated in FIGURE 1, which was generated from an ensemble of 1000 initial conditions with energy $E = 1.0$ and $z = p_z = 0$, evolved for the interval $10.0 < t < 266.0$ in the potential $V = V_0[R(t)\mathbf{r}]$ of Eq. (13) with $a = 1$. The six panels correspond to different values of p ranging from $p = -0.6$ to $p = 0.45$. It is clear that, for the time-independent case with $p = 0.0$, the distributions of ξ's is bimodal, the peak near $\xi = 0$ corresponding to regular segments, and the segments with larger values

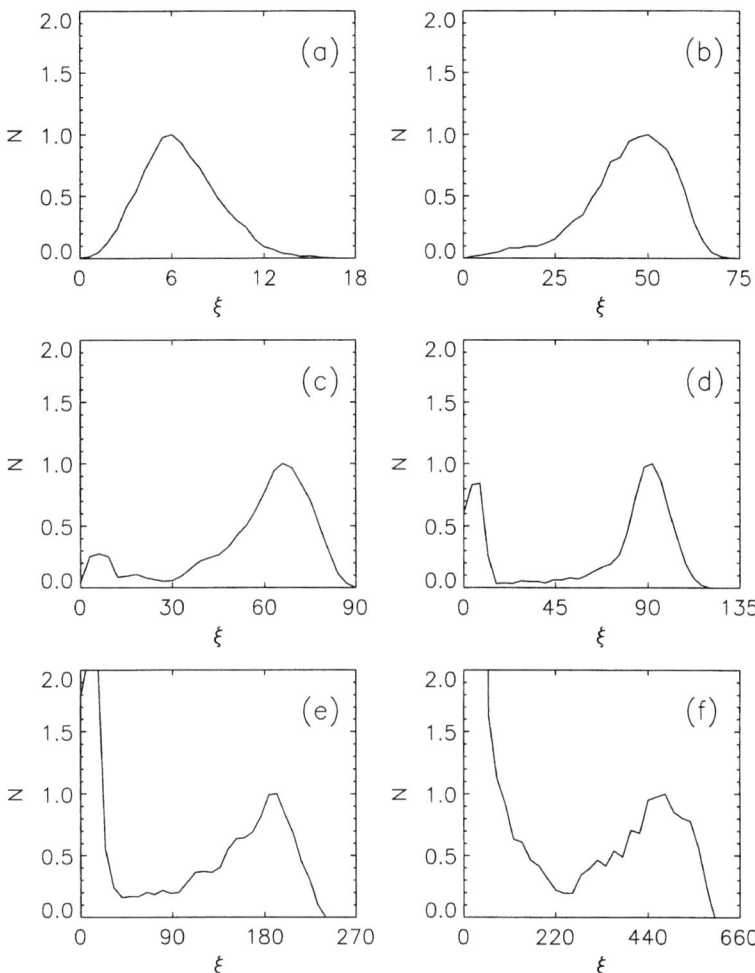

FIGURE 1. The distribution $N[\xi(t_{\text{fin}})]$, with ξ defined as in Eq. (16), generated from an ensemble of initial conditions $E = 1.0$ and $z = p_z = 0$ evolved for the interval $10 < t < 266$ in the potential (13) with $a = 1.0$ and variable p. (a) $p = -0.6$. (b) $p = -0.1$. (c) $p = -0.05$. (d) $p = 0.0$. (e) $p = 0.2$. (f) $p = 0.45$.

of ξ corresponding to chaotic orbits. (A longer time integration reveals that the segments with $20 < \xi < 60$ correspond to "sticky" orbits which, at early times, were trapped near regular islands by one or more cantori.) This bimodal behavior persists for $p > 0$, although the relative abundance of "regular" segments increases rapidly with increasing p. Alternatively, the relative abundance of regular segments decreases very rapidly when p becomes negative so that, for $p < 0.1$ or so, a sample of 1000

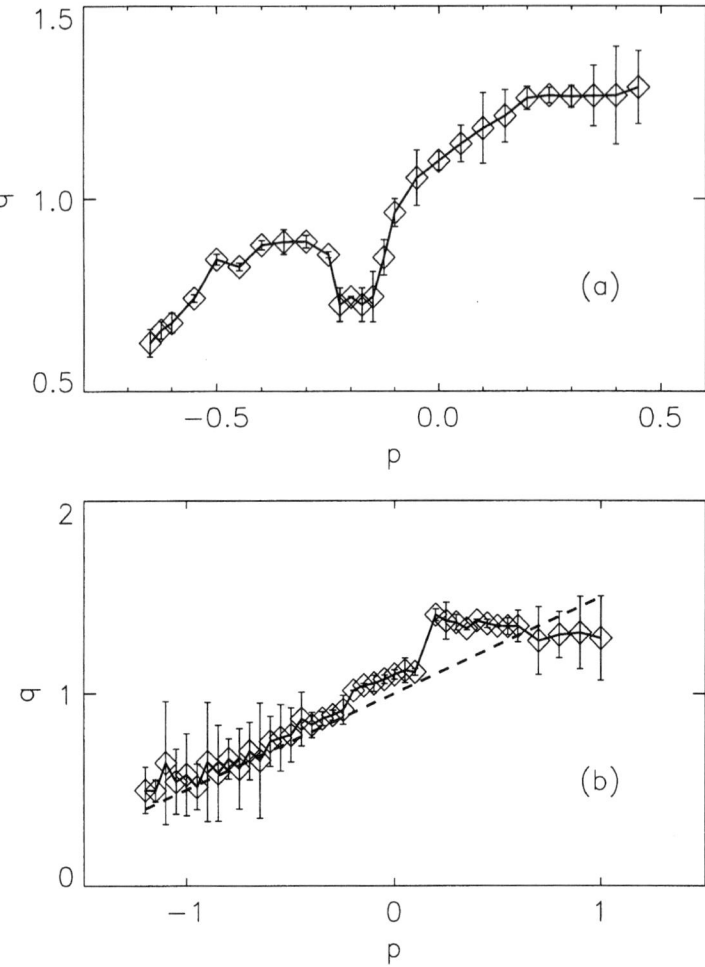

FIGURE 2. (a) The mean slope q for chaotic segments generated as in FIGURE 1, plotted as a function of p. (b) The analogue of (a) for chaotic segments generated in the potential (12), again with $a = 1.0$ and variable p.

initial conditions is too few to yield an appreciable number of regular segments. These changes in the relative abundance of regular and chaotic segments probably reflect the specific form of the time-independent potential $V_0(\mathbf{r})$. For example, $p > 0$ implies that the kinetic energy $K = \mathbf{v}^2/2$ increases, the potential energy $V = t^p V_0(x, y)$ decreases in magnitude, and the total energy $E = \mathbf{v}^2/2 + t^p V_0(x, y)$ exhibits a modest systematic increase, with the net result that the orbits tend to evolve in a regular or near-regular fashion in the "trough" of the dihedral potential $V(t)$.

Another generic feature, also apparent in FIGURE 1, is that, even though increasing p implies fewer chaotic segments, those segments that remain chaotic tend to be more unstable in the sense that the final $\xi(t_{\text{fin}})$ is larger. In part, this trend reflects the fact that, overall, the values of $\chi(\Delta t_i)$ tend to be larger for larger values of p. However, this trend also reflects the fact that, as expected, $p > 0$ yields perturbations that exhibit superexponential growth whereas $p < 0$ yields subexponential growth. For fixed sets of initial conditions, this latter assertion was confirmed for each value of p by (1) identifying a minimum value of ξ that (seemingly) represents a sufficient criterion for chaotic behavior, (2) fitting the computed $\xi(t_i)$ for each chaotic segment to the power law (17), and (3) determining a mean slope for all the chaotic segments with given p. The results are exhibited in FIGURE 2(a), where the error bars reflect the effects of reasonable variations in the value ξ_{min} used to identify chaotic behavior.

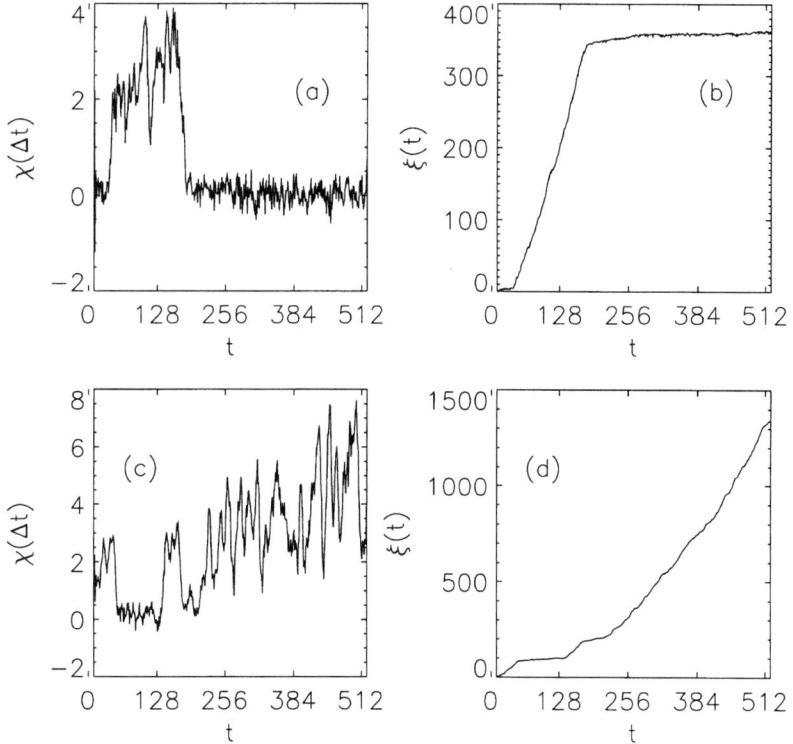

FIGURE 3. (a) and (b): The short-time Lyapunov exponent $\chi(\Delta t_i)$ and the cumulative $\xi(t_i)$ computed for one orbit with initial energy $E = 1.0$ and $z = p_z = 0$ evolved for the interval $10 < t < 266$ in the potential (13) with $p = 0.5$. (c) and (d): The same quantities for another orbit, again with $E = 1.0$ and $z = p_z = 0$, evolved for the same interval in the same potential.

FIGURE 2 agrees with predictions in the sense that $p > 0$ and $p < 0$ yield, respectively, super- and subexponential growth. However, there is one new, not completely expected feature, namely that q is not a monotonic function of p. In particular, there appears to be a range of values of p, say $-0.25 < p < -0.15$ where, as probed by the aforementioned diagnostic, the growth of a small initial perturbation is weaker than for both somewhat smaller and somewhat larger values of p. Given this behavior, it is especially important to check whether the predicted behavior for the simpler potential $V = t^p V_0(\mathbf{r})$ is confirmed by experiments. FIGURE 2(b), which presents the analogue of FIGURE 2(a) for the same set of initial conditions now evolved in the potential (12), indicates that, overall, the agreement between theory and experiment is quite good, although some systematic differences are seen for $p > 0$. Note that the larger error bars for especially large and small p reflect the fact that, for these values of p, most of the orbits seem regular or near-regular, with very small values of $\chi(\Delta t_i)$ and $\xi(t_i)$.

A third general feature, also apparent from FIGURE 1, is that, for p's somewhat larger than zero, a larger fraction of the segments have values of ξ well separated from both the low and high ξ peaks than is the case for $p = 0$. In most cases, these intermediate values appear to correspond to segments which change from chaotic to regular or, in some cases, from regular to chaotic. That this is the case is easily seen by computing either $\chi(\Delta t_i)$, which can exhibit abrupt systematic increases and decreases, or $\xi(t_i)$, which can exhibit a nearly stepwise growth. Two examples of this behavior are provided in FIGURE 3, both corresponding to segments computed with $p = 0.5$. The top two panels correspond to an orbit segment which makes an abrupt transition from chaotic to regular behavior at $t \sim 160$; the lower panels correspond to a segment which exhibits a more erratic behavior early on. It should also be evident that, during the chaotic phases, $\chi(\Delta t_i)$ is evidencing a systematic increase, so that $\xi(t_i)$ grows faster than linearly in time, this corresponding to a perturbation that evolves superexponentially.

Inspection of individual orbit segments also reveals that segments which are chaotic in the sense that they exhibit a sensitive dependence on initial conditions tend to be manifestly more irregular in visual appearance. In particular, regular segments typically have identifiable shapes and topologies which persist for relatively long periods of time, even as the orbital energy changes by an order of magnitude or more. Pieces of two representative regular orbits evolved it the potential (13) with $a = 1$ and, respectively, $p = 0.3$ and $p = 0.5$, are exhibited in the top four panels of FIGURES 4 and 5. Viewed over relatively short intervals $\Delta t < 50$ or so, the first segment closely resembles a loop orbit in a time-independent potential. If, however, the orbit is tracked over longer intervals, one sees significant changes as the "radius" of the loop slowly decreases. The second segment, corresponding to the orbit used to generate FIGURE 3(a) and (b), exhibits more distinct variability than the loop orbit, but it is evident once again that the overall shape and topology are robust.

These regularities imply that, even though regular segments are not periodic, their Fourier spectra are distinctly different, and simpler, than the spectra for chaotic segments. For example, like true loop orbits in a time-independent potential, regular segments that look loopy are characterized by spectra $|x(\omega)|$ and $|y(\omega)|$ which are very similar in amplitude and shape. Moreover, in many cases the overall form of the spectrum can be interpreted as involving one or more peak frequencies ω whose val-

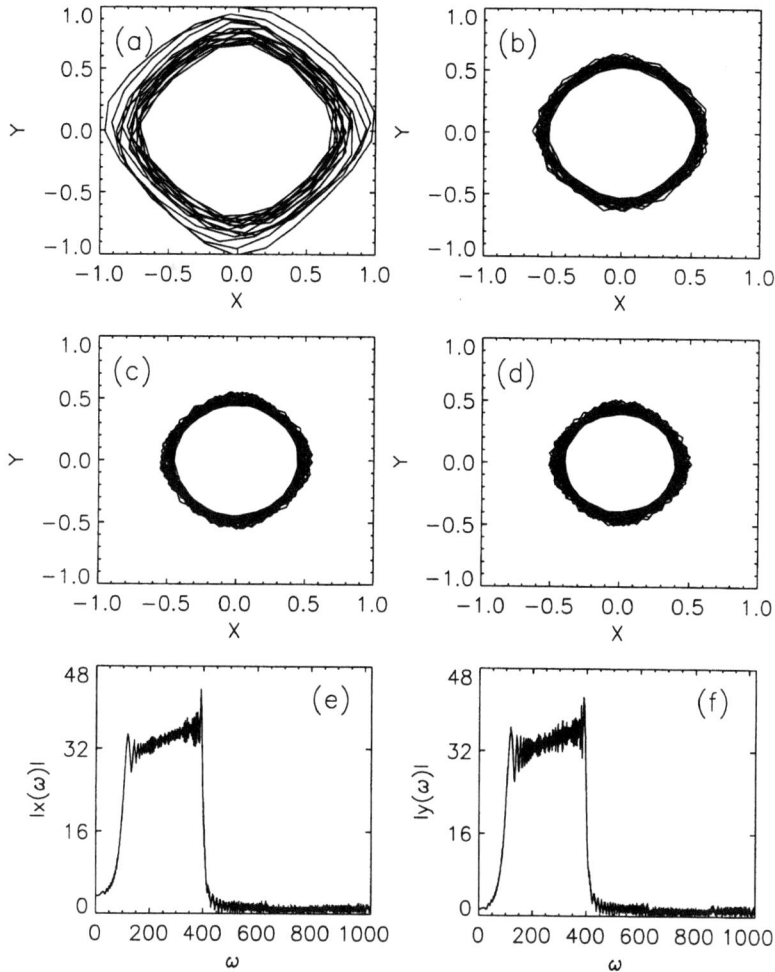

FIGURE 4. Segments of a single trajectory with $E = 1.0$ and $z = p_z = 0$ evolved with $p = 0.3$ in the potential (13) for the interval $10 < t < 266$, along with the total power spectra, $|x(\omega)|$ and $|y(\omega)|$. (a) $0 < t < 32$. (b) $64 < t < 96$. (c) $128 < t < 160$. (d) $192 < t < 224$. (e) $|x(\omega)|$. (f) $|y(\omega)|$.

ues exhibit a systematic drift over the course of time. This is particularly evident in the final two panels of FIGURE 4, which exhibit $|x(\omega)|$ and $|y(\omega)|$ for the loopy regular orbit. At early times, when the orbit rotates relatively slowly, the power for both $|x(\omega)|$ and $|y(\omega)|$ is concentrated at relatively low values of ω but, as time elapses and the orbit begins to rotate more rapidly, power slides up to high values of ω. The composite spectra in FIGURES 4(e) and (f) can be understood, at least approximately,

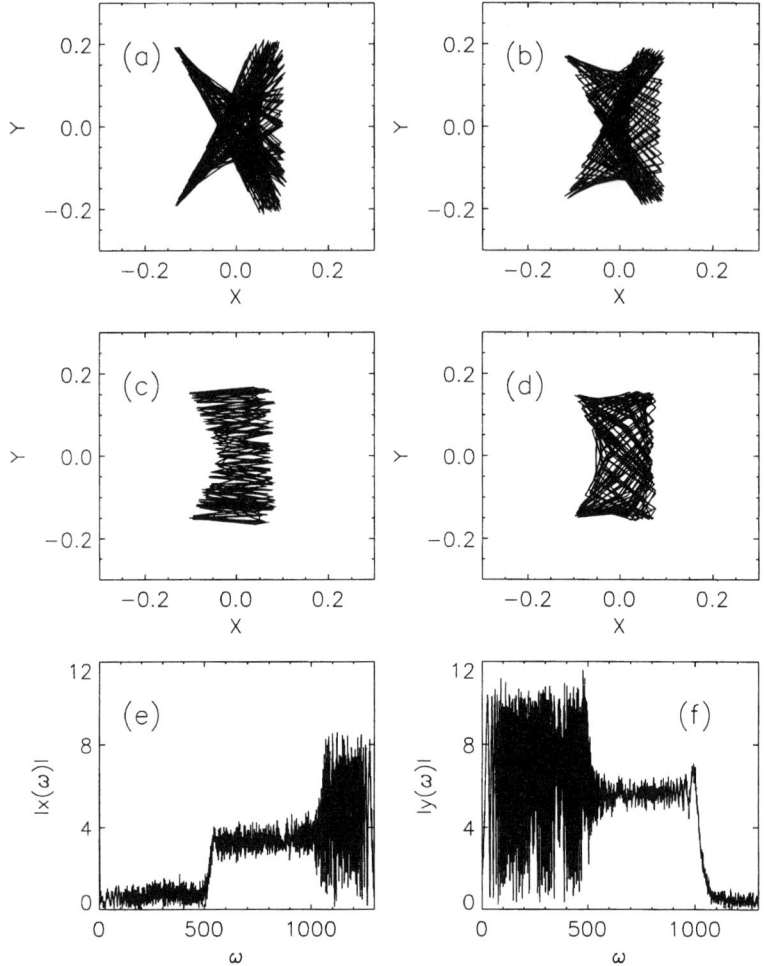

FIGURE 5. The analogue of FIGURE 4 for another orbit with $E = 1.0$ and $z = p_z = 0$, now evolved with $p = 0.5$.

as representing the time integral of a set of narrow peaks which, as time elapses, move systematically towards higher frequency.

CONCLUSIONS

This paper summarized a numerical investigation which focused on identifying meaningful definitions of regular and chaotic behavior in time-dependent Hamiltonian systems of the type that one might expect to encounter in a cosmological setting. Special attention focused on two- and three-dimensional Hamiltonians of the form $H(\mathbf{r}, \mathbf{p}, t) = \mathbf{p}^2/2 + V(\mathbf{r}, t)$, with $V = R(t)V_0(\mathbf{r})$ or $V = V_0[R(t)\mathbf{r}]$, where V_0 is a polynomial in x, y, and z and $R(t) \propto t^p$ represents a time-dependent scale factor. When p is not too negative, one can distinguish between regular and chaotic behavior by determining whether, over the time interval in question, an orbit segment exhibits a sensitive dependence on initial conditions. However, the time-dependence of H complicates the physics in several important ways.

(1) A single orbit can exhibit intermittent behavior, changing from chaotic to regular and/or visa versa.

(2) A small perturbation of a chaotic segment will not in general exhibit an average exponential growth. Rather, a computation of suitably defined short-time Lyapunov exponents shows that the phase-space perturbation δZ^i is often well fit by a growth law $\ln|\delta Z(t)| = a + bt^q$, with $q > 1$ for $p > 0$ and $q < 1$ for $p < 0$. An expanding Universe makes the effects of chaos milder; a contracting Universe makes them stronger.

(3) Regular segments are not periodic and, as such, do not have sharply peaked Fourier spectra. However, the topology of regular segments *is* robust, so that, e.g., a loop orbit continues to look loopy even as $|\mathbf{r}|$ and $R(t)|\mathbf{r}|$ grow or shrink systematically. Moreover, the spectrum of a regular segment is simpler than that for a chaotic segment since, in many cases, the regular segment can be approximated as a sum of a few contributions of the form $Z(t) \sim Z(0)\exp[i\Omega(t)t]$, where $\Omega(t)$ exhibits a simple secular variation.

All these observed phenomena can be understood in terms of a simple theoretical model based on a time-dependent generalization of the Matthieu equation.

ACKNOWLEDGMENTS

Work on this manuscript was completed while H.E.Kandrup was a visitor at the Aspen Center for Physics, the hospitality of which is acknowledged gratefully.

REFERENCES

1. BUCHLER, J.R., S.L. GOTTESMAN, J.H. HUNTER & H.E. KANDRUP. 1998. Nonlinear Dynamics and Chaos in Astrophysics. Ann. N.Y. Acad. Sci. **867**. This volume.
2. KANDRUP, H.E. & H. SMITH. 1991. Astrophys. J. **374**: 255.
3. GOODMAN, J., D. HEGGIE & P. HUT. 1993. Astrophys J. **415**: 715.
4. KANDRUP, H.E., M.E. MAHON & H. SMITH. 1994. Astrophys. J. **428**: 458.
5. MELOTT, A. 1998. Private communication.
6. KOFMAN, L., A. LINDE & A. A. STAROBINSKY. 1994. Phys. Rev. Lett. **73**: 3195.
7. LICHTENBERG, A.J. & M.A. LIEBERMAN. 1992. Regular and Chaotic Dynamics. Springer-Verlag. Berlin.
8. ARNOLD, V.I. 1989. Mathematical Methods of Classical Mechanics. Springer-Verlag. Heidelberg.
9. PEEBLES, P.J.E. 1993. Principles of Physical Cosmology. Princeton University Press. Princeton.
10. TABOR, M. 1989. Chaos and Nonintegrability in Nonlinear Dynamics. Wiley. New York.
11. TODA, M. 1967. J. Phys. Soc. Japan **22**: 431.

12. PETTINI, M. 1993. Phys. Rev. E **47:** 828.
13. CERRUTI-SOLA, M. & M. PETTINI. 1995. Phys. Rev. E **53:** 179.
14. KANDRUP, H.E. 1997. Phys. Rev. E. **56:** 2722.
15. WHITTAKER, E.T. & G.H. WATSON. 1965. A Course of Modern Analysis. Cambridge University Press. Cambridge.
16. ARMBRUSTER, D., J. GUCKENHEIMER & S. KIM. 1989. Phys. Rev. Lett. A **140:** 416.
17. GRASSBERGER, P., R. BADII & A. POLITI. 1988. J. Stat. Phys. **51:** 135.
18. KANDRUP, H.E. & M.E. MAHON. 1994. Astron. Astrophys. **290:** 762.

Phase Space Transport in Noisy Hamiltonian Systems

HENRY E. KANDRUP[a]

Department of Astronomy and Department of Physics and Institute for Fundamental Theory, University of Florida, Gainesville, Florida 32611

> ABSTRACT. This paper analyzes the effect of low amplitude friction and noise in accelerating phase space transport in time-independent Hamiltonian systems that exhibit global stochasticity. Numerical experiments reveal that even very weak non-Hamiltonian perturbations can dramatically increase the rate at which an ensemble of orbits penetrates obstructions like cantori or Arnold webs, thus accelerating the approach toward an invariant measure, i.e., a microcanonical population of the accessible phase space region. An investigation of first passage times through cantori leads to three conclusions, namely: (i) that, at least for white noise, the detailed form of the perturbation is unimportant, (ii) that the presence or absence of friction is largely irrelevant, and (iii) that, overall, the amplitude of the response to weak noise scales logarithmically in the amplitude of the noise.

WHY CONSIDER FRICTION AND NOISE?

In general, very weak non-Hamiltonian perturbations will only have very weak effects on properties of flows in time-independent Hamiltonian systems which are integrable or near-integrable and admit no global stochasticity. This also seems to be true for systems completely dominated by chaos where regular orbits are virtually nonexistent. However, low amplitude non-Hamiltonian perturbations *can* have important qualitative effects on more complex Hamiltonian systems that admit significant measures of both regular and chaotic orbits. For example, weak noise will serve as a source of extrinsic diffusion that can dramatically accelerate phase space transport through cantori (for $D = 2$) or along Arnold webs (for $D \geq 3$).

Consider, for example, flows in two-dimensional systems. Here one knows that, in the absence of friction and noise, cantori [1], [2], fractured *KAM* tori associated with the breakdown of integrability that contain a cantor set of holes, can partition a single connected chaotic phase space region into separate parts which, albeit not completely disjoint, are distinct in the sense that a chaotic orbit starting in one part of the phase space will remain stuck in that part for a long time before wending its way through one or more holes in the cantori to access another region [3], [4]. However, introducing even very weak friction and noise can dramatically increase the rate at which orbits pass through these holes, thus allowing orbits to probe the entire

---

[a] Supported in part by National Science Foundation Grant No. PHY92-03333 and by Los Alamos National Laboratory through the Institute of Geophysics and Planetary Physics. Some of the numerical calculations described here were facilitated by computer time provided by IBM through the Northeast Regional Data Center (Florida).

accessible phase space much more quickly. That noise can accelerate phase space diffusion has been long known to dynamicists studying various maps, dating back at least to the work of Lieberman and Lichtenberg [5] in the 1970s. However, the details have not received all that much attention, especially for continuous systems.

But why should one care? Why is this phenomenon important in the real world? The crucial observation here is that there is no such thing as a truly isolated system. Every system in nature is coupled to at least some degree to its surrounding environment. The important point then is that, in many cases, one should expect that the coupling of a system to an external environment can be modeled as resulting in friction and noise, related by a Fluctuation-Dissipation Theorem [6], [7] (although there *are* examples where such a picture is *not* justified [8]). Indeed, the assumption of a system coupled by a Fluctuation-Dissipation Theorem to an environment, idealized as a heat bath characterized by some temperature $\Theta = k_B T$, is one powerful starting point for modern theories of nonequilibrium statistical mechanics (see, e.g., the textbook by Kubo, Toda, and Hashitsume [9]).

Most modeling of this sort involves the assumption of a composite entity of system plus environment which is characterized by a time-independent Hamiltonian H. One might therefore worry that this picture does not extend naturally to a cosmological setting where, in the average comoving frame, the Hamiltonian H typically acquires an explicit time-dependence. Fortunately, however, the assumption of a time-independent H is not essential [10]. Following Caldeira and Legett [6], [7], it is natural to write the composite Hamiltonian H as a sum

$$H = H_{sys} + H_{bath} + H_{int}, \qquad (1)$$

where the system Hamiltonian H_{sys} is completely arbitrary, H_{bath} is idealized as the Hamiltonian for a collection of linearized excitations, i.e., "phonons," with (in general) time-dependent frequencies, and the interaction H_{int} is an arbitrary function of the system variables but linear in the bath variables. (These restrictions on H_{bath} and H_{int} assure that each mode of the environment is only weakly coupled to the system, so that the environment can be visualized as a thermal "bath.") In this setting, one can always integrate out the explicit dependence on bath variables to derive an exact nonlocal (in time) Langevin equation for the system; and, if H_{bath} is time-independent, this exact equation (and, presumably, any reasonable Markov approximation thereunto) will satisfy a Fluctuation-Dissipation Theorem, regardless of the possible time-dependence of H_{sys} and H_{int}. This implies in particular that, for the case of a conformally static Friedmann cosmology, one has a cosmological Fluctuation-Dissipation Theorem whenever the environment can be approximated as a conformally coupled field, for example, electromagnetic blackbody radiation.[1]

The natural inference is that, in a variety of different settings, including many relevant to astronomical systems, weak couplings to an external environment will result

---

[1] In this connection, it should be noted [11] that, in the dipole approximation, the Hamiltonian describing the interaction of an electron with a radiation field is, when formulated in an inertial frame, equivalent to the independent oscillator model [12], [13], which is perhaps the simplest "realistic" example of the Hamiltonian (1). Transforming to comoving coordinates makes H_{sys} and H_{int} time-dependent, but the conformal invariance of the electromagnetic field implies that H_{bath} remains time-independent.

in non-Hamiltonian perturbations which could have important physical implications. However, the proper form for the noise is not always obvious. The simplest model in terms of which to couple a system to its surroundings, the independent oscillator model [12], [13] (which can be shown to be equivalent to many other phenomenological models that have been considered in the past [11]), leads immediately to additive white noise, that is, state-independent noise that is delta-correlated in time. However, this simple form for the noise is a direct consequence of the assumptions (i) that H_{int} is linear in the system variables and (ii) that the bath phonons are characterized by an ohmic distribution, that is, a spectral distribution $\propto \omega^2 d\omega$ with appropriate cutoffs. Allowing H_{int} to involve the system variables in a more complicated fashion leads to multiplicative (i.e., state-dependent) noise; allowing for a different spectral density leads to colored noise (i.e., noise that is not delta-correlated in time).

When considering a galaxy embedded in a rich cluster, one is confronted with a system which, in many cases, is significantly impacted by its surrounding environment. Particularly close encounters between galaxies in a cluster, for example those resulting in physical collisions, probably cannot be viewed as "random" events. However, large numbers of relatively weak interactions probably *can* be viewed as a source of friction and noise, although there is no obvious reason why the noise associated with these interactions can be approximated as delta-correlated in time.

Another piece of physics which one might hope to model as friction and noise, again arising in galactic astronomy, is discreteness effects reflecting the fact that a galaxy is comprised of a collection of nearly point mass stars, rather than the smoothed-out continuum assumed in the context of a description based on the collisionless Boltzmann (i.e., gravitational Vlasov) equation. In the context of a smooth one-particle distribution function, these discreteness effects are typically described by a Fokker-Planck, or Landau, equation, which involves a velocity-dependent coefficient of dynamical friction and multiplicative noise (diffusion) related by a self-consistent Fluctuation-Dissipation Theorem [14]. Superficially this source of friction and noise might seem completely different from the aforementioned effects associated with an external environment. However, this is not really so! In this setting, one can view the full many-particle dynamics as the composite entity of system plus environment, the reduced one-particle dynamics as the system, and couplings to higher-order correlations ignored in a collisionless description as interactions that serve as a source of friction and noise [15].

In the past, a good deal of work has focused on the effects of relatively strong friction and noise in triggering barrier penetration and other phenomena which proceed on the natural relaxation time t_R associated with the system's approach toward thermal equilibrium (see, e.g., [14], [16] and numerous references cited therein.). This is *not* the problem of interest here. Rather, the objective of the work described in this paper has been to focus on much weaker perturbations, where t_R is much longer than any time scale of interest, and to determine the extent to which friction and noise have significant statistical effects on the evolution of ensembles of orbits already on time scales $\ll t_R$.

The numerical experiments described in the next two sections were performed with the aim of answering three basic questions:

1. How should one visualize the effects of accelerated phase space transport induced by friction and noise?

2. How does the size of the effect scale with the amplitude of the perturbation?

3. To what extent do the details of the perturbation matter? One knows, for example, that multiplicative noise can drive a system toward thermal equilibrium much more quickly than additive noise [17], and, as such, it would seem natural to ask whether multiplicative noise can also accelerate diffusion through cantori and Arnold webs more than additive noise.

INVARIANT AND NEAR-INVARIANT DISTRIBUTIONS

The computations described in this paper involved integrating Langevin equations of the form

$$\frac{d\mathbf{x}}{dt} = \mathbf{v} \quad \text{and} \quad \frac{d\mathbf{v}}{dt} = -\nabla \Phi - \eta \mathbf{v} + \mathbf{F}, \tag{2}$$

these corresponding to motion in a time-independent Hamiltonian $H = v^2/2 + \Phi(\mathbf{r})$ which is perturbed by friction and noise. The quantity η represents a coefficient of dynamical friction which, in general, can be a nontrivial function of both $\mathbf{x}$ and $\mathbf{v}$. The quantity $\mathbf{F}$ is a "random" force, idealized as Gaussian white noise, which is characterized completely by the statistical properties of its first two moments. Specifically,

$$\langle F_i(t) \rangle = 0 \quad \text{and} \quad \langle F_i(t_1) F_j(t_2) \rangle = 2\eta \Theta \delta_{ij} \delta_D(t_1 - t_2), \tag{3}$$

where i and j label vector components and angle brackets denote an ensemble average. The first of these conditions ensures that the average force vanishes identically. The second ensures that the autocorrelation function is delta-correlated in both direction and time. The normalization in Eq. (3) imposes a Fluctuation-Dissipation Theorem which ensures that, for $t \to \infty$, an arbitrary ensemble of orbits evolved with Eqs. (2) will approach a canonical distribution with temperature Θ.

To date, integrations have focused on three specific two-dimensional potentials, namely the sixth-order truncation of the Toda lattice potential [18],

$$\begin{aligned}\Phi(x,y) = \tfrac{1}{2}(x^2+y^2) + x^2 y - \tfrac{1}{3}y^3 &+ \tfrac{1}{2}x^4 + x^2 y^2 + \tfrac{1}{2}y^4 + x^4 y + \tfrac{2}{3}x^2 y^3 - \tfrac{1}{3}y^5 \\ &+ \tfrac{1}{5}x^6 + x^4 y^2 + \tfrac{1}{3}x^2 y^4 + \tfrac{11}{45}y^6,\end{aligned} \tag{4}$$

the so-called dihedral potential [19] for one particular set of parameter values, that is,

$$\Phi(x, y) = -(x^2 + y^2) + \tfrac{1}{4}(x^2 + y^2)^2 - \tfrac{1}{4}x^2 y^2, \tag{5}$$

and the sum of isotropic and anisotropic Plummer potentials [20] for specified core radii and anisotropy parameters, that is,

$$V(x, y) = \frac{1}{(c^2 + x^2 + y^2)^{1/2}} - \frac{m}{(c^2 + x^2 + ay^2)^{1/2}}, \tag{6}$$

with $c = 20^{2/3} \approx 0.136$, $a = 0.1$, and $m = 0.3$. In all three cases, the constants were so chosen that, in absolute units, a characteristic crossing time $t_{cr} \sim 1$. This implies that, if one visualizes these potentials as representing large galaxies like the Milky Way, the Hubble time $t_H \sim 100-200$.

As discussed more carefully elsewhere [20], these three potentials manifest very different symmetries. Indeed, the only obvious feature which they share is that, for a variety of energies, they admit significant measures of both regular and chaotic orbits, so that the chaotic phase space regions are significantly impacted by cantori. The fact that, nevertheless, similar qualitative results were obtained for orbits evolved in all three potentials can thus be interpreted as evidence that the basic conclusions are probably robust.

Because all three potentials yielded similar results, the largest number of calculations were performed for the dihedral potential, the most inexpensive computationally, which corresponds physically to a slightly "squared" Mexican hat potential. In all cases, the orbits were computed using a fourth order Runge-Kutta algorithm, noise being implemented using an algorithm developed by Griner et al. [21]. Most of the integrations were performed using a time step $\delta t = 10^{-3}$. It was verified that a shorter time step $\delta t = 10^{-4}$ does not yield significantly different results.

The first class of experiments to be performed involved tracking the evolution of ensembles of chaotic initial conditions of fixed energy E, selected from some small phase space region in the center of the stochastic sea far from any important cantori. These initial conditions were first evolved into the future in the absence of any friction or noise by integrating the deterministic Hamilton equations. They were then re-integrated allowing for friction and noise of variable amplitude. All these experiments assumed additive white noise and friction characterized by a constant η, the two quantities being related by a Fluctuation-Dissipation Theorem. The temperature was frozen at a value $\Theta \sim E$ and the amplitude of the perturbing influences was varied by systematically changing the value of η.

In the absence of friction and noise, such orbit ensembles exhibit a two-stage evolution [20], [22]. The first stage involves a rapid coarse-grained evolution, proceeding exponentially in time, toward a phase space distribution which is near-invariant in the sense that, once achieved, it only exhibits significant changes on a much longer time scale. Basically, this near-invariant distribution corresponds to a distribution characterized by a nearly constant number density in those portions of the constant energy phase space hypersurface that are not blocked by cantori and a near-zero density everywhere else. The second stage involves a much slower evolution toward what appears to be a true invariant distribution, as orbits in the ensemble diffuse through cantori to access phase space regions that were avoided systematically over shorter time scales. This final invariant distribution corresponds to a microcanonical population of the accessible chaotic regions, that is, a uniform (in canonical coordinates) population of those portions of the phase space that are accessible to orbits with the specified initial conditions. The time scale for the first stage of the evolution is set at least approximately by the largest short time Lyapunov exponents for the orbits in the ensemble, which determine how fast the ensemble will disperse. Typically this time $\sim t_{cr}$. The time scale for the second stage is set by the time scale on which orbits diffuse through cantori, typically $\gg t_{cr}$.

Suppose now that the orbits are perturbed by friction and noise with $\eta \sim 10^{-9} - 10^{-4}$, the limiting values here corresponding, respectively, to the typical amplitude associated with discreteness effects in very large and very small galaxies [23]–[25]. In this case, one finds that the time scale associated with the first stage of the evolution is essentially unchanged, i.e., ensembles still approach a near-invariant distribution on a time scale $\sim t_{cr}$, but that the time scale for the second stage decreases dramatically! In the absence of friction and noise, the time scale associated with diffusion through cantori typically satisfies $t_{diff}(\eta = 0) \sim 10^3$–$10^5 t_{cr}$, but even very weak friction and noise can decrease $t_{diff}(\eta)$ by orders of magnitude. For example, $\eta \sim 10^{-9} - 10^{-6}$ can result in a diffusion time as short as $\sim 100 t_{cr}$, an interval which, for large galaxies, corresponds to the Hubble time t_H. Indeed, for values of η as large as $\eta \sim 10^{-4}$, the diffusion time scale t_{diff} is often so short that one cannot clearly distinguish between two different stages of evolution. For values of η so large, noisy ensembles exhibit a rapid approach toward a near-invariant distribution that differs significantly from the near-invariant distribution associated with a purely Hamiltonian evolution but is comparatively similar to the true invariant distribution associated with a Hamiltonian evolution. Grey-scale plots comparing representative deterministic and noisy near-invariant distributions in the dihedral and truncated Toda potentials are exhibited, respectively, in FIGURE 8 in [25] and FIGURE 2 in [24].

In the absence of friction and noise, one anticipates that a generic ensemble of initial conditions will ultimately evolve toward a microcanonical distribution, i.e., a uniform population of the accessible portions of the constant energy hypersurface, but this will only happen on the relatively long time scale $t_{diff}(\eta = 0)$. Alternatively, if one allows for friction and noise and integrates for a time $\sim t_R$, one anticipates an evolution toward a canonical distribution with temperature Θ. Noisy integrations performed for a time $\ll t_R$ but still much longer than the time required to breech cantori will result in a near-invariant distribution that can be reasonably visualized as a slightly "thickened" version of a constant energy microcanonical distribution. Because E is not exactly conserved, this near-invariant distribution is not exactly microcanonical, that is, not proportional to a delta function in energy. However, because E is almost conserved and the orbits have succeeded in breeching cantori, this noisy near-invariant distribution is much closer to the purely Hamiltonian invariant distribution than to either a canonical distribution or the purely Hamiltonian near-invariant distribution [25]. In this sense, weak friction and noise can accelerate an approach toward a near-microcanonical equilibrium.

FIRST PASSAGE TIME EXPERIMENTS

To quantify the rate at which individual trajectories diffuse through cantori, a collection of first passage time experiments was also performed. These involved four components:

1. Select individual initial conditions corresponding in the absence of friction and noise to *sticky* or *confined chaotic orbits*, that is, chaotic orbits which, because of cantori, are trapped near regular regions for relatively long times.

2. Specify the form and amplitude of the friction and noise.

3. For each choice of form and amplitude, perform a large number ($\sim$ 2000–5000) of different noisy realizations of the same initial condition;

and, for each noisy realization, determine the time at which the orbit escapes through one or more cantori to become unconfined.

4. Analyze the data to extract $N(t)$, the fraction of the orbits that have not yet escaped within a time t.

The results quoted below involve orbits in the dihedral potential (5) with $E = 10$ where, in the absence of any friction or noise, the diffusion time $t_{\text{diff}} \sim 1000$. Other choices of potential or energy can yield results that differ quantitatively, but the principal qualitative conclusions seem unchanged. Estimating when an orbit has escaped was done by identifying a "masked" region in the configuration space and recording the first time that the orbit left this region. That this mask criterion is reasonable was tested in two ways: (1) It was verified that changing slightly the shape and location of the mask had no appreciable effects. (2) For the case of purely Hamiltonian trajectories, escape from the masked region was shown to correspond to an abrupt increase in the value of the largest short time Lyapunov exponent. This is in accord with the fact that, albeit still chaotic, confined chaotic orbits are less unstable exponentially than are unconfined chaotic orbits [20]. One interesting variant of the preceding, also considered, involved tracking a localized ensemble of initial conditions corresponding to confined chaotic orbits evolved into the future both with and without friction and noise. These experiments yielded results very similar to those obtained from multiple integrations of individual initial conditions.

Six different forms of friction/noise were considered, namely: (1) additive white noise and a constant coefficient of dynamical friction η, related by a Fluctuation-Dissipation Theorem at temperature $\Theta = E = 10$; (2) multiplicative white noise and dynamical friction with $\eta = \eta_0 v^2$, related by a Fluctuation-Dissipation Theorem with $\Theta = E = 10$; (3) multiplicative white noise and dynamical friction with $\eta = \eta_0 v^{-2}$, again related by a Fluctuation-Dissipation Theorem with $\Theta = E = 10$; and (4)–(6) the same noises as in (1) – (3) but vanishing friction. In all six cases, the individual noisy realizations were generated using the same pseudorandom seeds.

In analyzing the effects of friction and noise, attention focused on three principal issues, namely:

1. What is the functional form of $N(t)$, the fraction of the orbits that have not yet escaped?

2. How does $N(t)$ depend on the amplitude of the perturbation?

3. To what extent does the form of the friction and the noise actually matter?

Overall, in these experiments escape is a two-stage process. Early on, there are no escapes. All that one sees is that, as one might expect [25], [26], different noisy realizations of the same initial condition diverge exponentially at a rate set by the value of the largest short time Lyapunov exponent for the unperturbed deterministic trajectory. Eventually, however, once the noisy ensemble has dispersed to the extent that the root mean squared $\delta r_{\text{rms}} \sim 1.0$, individual noisy orbits begin to escape through holes in the cantori. This onset of escape is a comparatively abrupt phenomenon, the interval during which the first 5% of the orbits escape typically being only a small fraction of the time T before the first escape occurs. It is also clear that, at least early on, escapes can be well approximated as a Poisson process, with the confined orbits becoming unconfined at a nearly constant rate, i.e.,

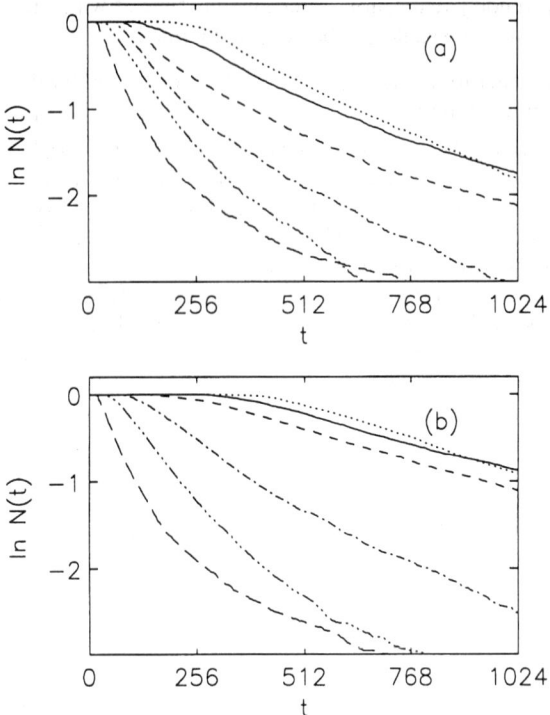

FIGURE 1. (a) $N(t)$, the fraction of confined chaotic orbits that have not yet escaped to become unconfined, for ensembles of 4000 noisy realizations of the same initial condition evolved in the dihedral potential with $E = 10.0$, $x = 0.0$, $y = 1.3$, $v_y = 1.75$, and $v_x = v_x(x, y, v_y, E) > 0$. Each orbit was subjected to additive white noise and friction with constant η, related by a Fluctuation-Dissipation Theorem with temperature $\Theta = 10$. Passing from top to bottom at small t, the six curves represent ensembles with $\eta = 10^{-9}$, 10^{-8}, 10^{-7}, 10^{-6}, 10^{-5}, and 10^{-4}. (b) The same quantities generated for a different initial condition, namely $E = 10.0$, $x = 0.0$, $y = 2.7$, and $v_y = 2.25$.

$$N(t) \approx \begin{cases} N(0), & \text{if } t \leq T; \\ N(0)\exp[-\Lambda(t-T)], & \text{if } t > T. \end{cases} \qquad (7)$$

This behavior is illustrated in FIGURES 1(a) and 1(b), which exhibit $\ln N(t)$ as a function of time t for two different initial conditions integrated in the presence of a constant η and additive white noise. Each panel summarizes multiple noisy realizations of a single initial condition evolved for a total time $t = 1024$ with $\Theta = 10$. The six different curves in each panel, each summarizing 4000 noisy realizations, represent six different values of η, namely $\log \eta = -9, -8, -7, -6, -5$, and -4.

It is clear from FIGURE 1 that, although $\ln N(t)$ originally decreases linearly in time, it eventually develops nontrivial curvature indicating that the escape rate is

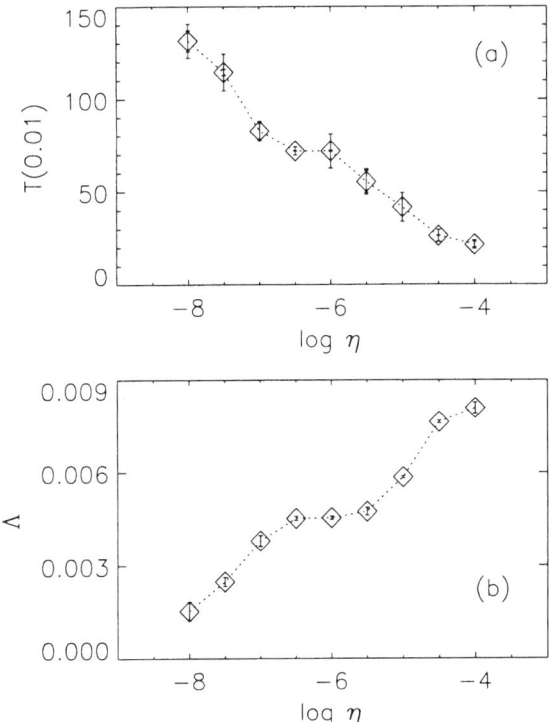

FIGURE 2. (a) $T(0.01)$, the time required for 1% of the members of an ensemble of 4000 noisy realizations of the same unconfined chaotic orbit with $E = 10.0$, $x = 0.0$, $y = 1.3$, $v_y = 1.75$, and $v_x = v_x(x, y, v_y, E) > 0$ to become unconfined. Each orbit was subjected to additive white noise and friction with constant η, related by a Fluctuation-Dissipation Theorem with $\Theta = 10$. (b) Λ, the rate at which orbits in the ensemble escape, fit to the interval $T(0.01) < t < 256$.

slowly *decreasing*. Exactly why this is so is not completely clear. However, two important points should be noted. (1) In every case where it is observed, this curvature arises at a time sufficiently late that changes in energy δE have become appreciable, ~10% or more. This suggests strongly that, at least in part, this change in escape rate reflects changes in the "effective" Hamiltonian phase space in which the noisy orbits evolve. (2) In at least some cases, the curvature reflects the fact that, because of the perturbations, some originally chaotic orbits have become trapped by *KAM* tori. (If, for the nonescapers, the friction and noise are turned off at some time τ and the trajectories integrated for a significantly longer time, it becomes apparent that many of the orbits have become regular!).

So how, overall, does the form of $N(t)$ scale with η, the amplitude of the perturbation? When probing the time T required before escapes begin or the rate Λ at which escapes initially proceed after they begin, one finds a roughly logarithmic dependence on η. In other words, when plotting $T(\eta)$ or $\Lambda(\eta)$ the natural independent variable, i.e., the abscissa, is $\ln \eta$, *not* η. An example thereof is provided in FIGURES

2(a) and 2(b), which summarize data generated from a single initial condition with friction and additive white noise related by a Fluctuation-Dissipation Theorem. The top panel exhibits $T(0.01)$, the time required for 1% of the orbits in a 4000 orbit ensemble to become unconfined. The lower panel exhibits the best fit value of the escape rate Λ of Eq. (7), as fit to the interval $T(0.01) < t < 256$. The curvature observed in both panels is statistically significant, so one cannot assert that T or Λ are linear functions of $\ln \eta$. However, it *is* clear that, overall, T and Λ should be visualized as functions of $\ln \eta$ rather than η.

Perhaps the most important conclusion derived from these experiments is that the computed $N(t)$ is nearly independent of the presence or absence of friction, and that $N(t)$ is also largely independent of whether the noise is additive or multiplicative!

First perform 4000 noisy realizations of the same initial condition, all with the same Θ and the same $\eta(\mathbf{v})$, and analyze the resulting data to extract $N(t)$. Then repeat these experiments with exactly the same noise (generated from the same pseudorandom seeds!) but without friction, and once again compute $N(t)$. A comparison of the two $N(t)$'s then shows virtually no appreciable differences. At early times, there are absolutely no statistically significant differences. Later on, one *can* see some tiny differences. However, these can be attributed entirely to the fact that the energies of the orbits with and without friction will be slightly different, and that slightly different energies can give rise to slightly different escape statistics.

Comparing additive and multiplicative noise is a bit more subtle since one must worry about normalizations. Suppose, however, that, when introducing multiplicative noise, one selects η_0 so that the "average" $\eta \equiv \eta_0 \langle v^2 \rangle$ or $\eta \equiv \eta_0 \langle v^{-2} \rangle$ coincides with the white noise constant η. In this case, one finds that the form of the noise matters very little. Plots of $N(t)$ for additive white noise, multiplicative noise $\propto v^2$, and multiplicative noise $\propto v^{-2}$ yield no statistically significant differences.

Two examples of this behavior are exhibited in FIGURES 3(a) and 3(b) which compare the effects of additive and multiplicative noise for two different initial conditions at two different perturbation levels. Each panel contains four curves, representing (1) additive white noise and friction with a constant η, (2) the same additive white noise with vanishing friction, (3) multiplicative noise and friction with $\eta \propto v^2$, and (4) multiplicative noise and friction with $\eta \propto v^{-2}$. It is evident that, at late times, the curves do not completely overlap. However, it is also clear that none of the curves is extremely different from the others.

These numerical experiments suggest two obvious inferences which, however, remain to be checked more carefully for larger orbit ensembles, different frictions and noise, and other potentials: (1) Smooth non-Hamiltonian perturbations like friction play only a minimal role in accelerating phase space transport through cantori. (2) At least assuming that the noise is white, its details seem comparatively unimportant. In particular, additive noise and multiplicative noise depending on the orbital velocity $\mathbf{v}$ exhibit only minimal differences. Overall, when perturbing the Hamiltonian trajectories what seems important is that the orbits be subjected to highly "irregular" perturbations that violate Liouville's Theorem at some given amplitude. In this context, it should perhaps be noted explicitly that a rapidly varying time-dependent Hamiltonian need not be as efficient in triggering accelerated phase space transport as random noise. Specifically, when considering a Hamiltonian of the form $H = H_0 + \epsilon H_1(t)$, with $H_1(t)$ periodic in time and ϵ an adjustable parameter,

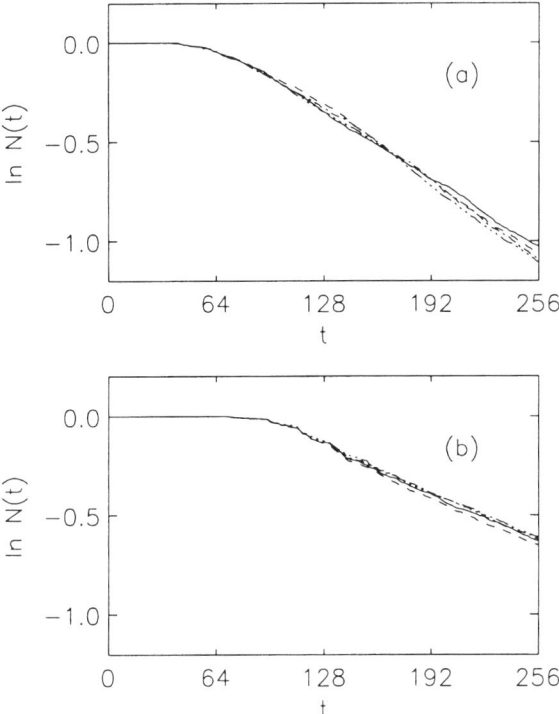

FIGURE 3. (a) $N(t)$, the fraction of confined chaotic orbits that have not yet escaped to become unconfined, for ensembles of 4000 noisy realizations of the same initial condition evolved in the dihedral potential with $E = 10.0$, $x = 0.0$, $y = 1.1$, $v_y = 3.35$, and $v_x = v_x(x, y, v_y, E) > 0$. Each orbit was evolved with $\Theta = 10$ and $\eta_0 = 10^{-5}$. The four curves represent additive white noise and a constant η (*solid line*), additive noise but no friction (*dashed*), multiplicative noise and friction with $\eta \propto v^2$ (*dot-dashed*), and multiplicative noise and friction with $\eta \propto v^2$ (*triple-dot-dashed*). (b) The same quantities generated for a different initial condition, namely $E = 10.0$, $x = 0.0$, $y = 1.3$, $v_y = 1.75$, and $v_x = v_x(x, y, v_y, E) > 0$, now allowing for $\Theta = 10$ and $\eta_0 = 10^{-7}$.

one finds in at least some cases [27] that, for periods $\ll t_{cr}$, one must allow for relatively large values of ϵ to dramatically accelerate diffusion through cantori.

WORK IN PROGRESS AND POTENTIAL IMPLICATIONS

The experiments described above still need to be generalized in two important ways.

One obvious tack involves extending the computations to three-dimensional systems. Arnold webs can serve as partial phase space obstructions in the same sense as can cantori; and one might anticipate that friction and noise could accelerate phase space transport through such barriers in three-dimensional systems in the same ways as they do through cantori in two dimensions. This remains, however, to be checked.

Indeed, for orbit ensembles in three-dimensional Hamiltonian systems, even the purely deterministic evolution toward an invariant or near-invariant distribution is not completely understood. Preliminary investigations performed by Merritt and Valluri [28] suggest that the approach toward equilibrium can closely resemble what is observed in two-dimensional systems [20], [22]. However, one would anticipate that, for a generic three-dimensional system with two positive Lyapunov exponents, the situation could be more complicated — and interesting — than in two dimensions since unequal Lyapunov exponents could induce "mixing" that proceeds in different directions at different rates [29]!

Another equally important objective is to allow for the effects of colored noise, where the autocorrelation function $\langle F_i(t_1)F_j(t_2)\rangle$ is not delta-correlated in time. The assumption of singular, delta-correlated noise is an idealization never exactly realized in nature, even when modeling high frequency phenomena like discreteness effects in systems interacting via short-range forces; and the assumption seems especially unreasonable when trying to model objects like galaxies embedded in a dense cluster environment where individual "random" interactions would seem characterized dimensionally by a time scale $\sim t_{\rm cr}$ or even larger! Allowing for colored noise could also be important by providing some insights into the question of exactly how and why non-Hamiltonian perturbations result in accelerated phase space transport. Diffusion through cantori, either in the absence or presence of noise, must be related to some "natural" time scale(s), but the precise nature of these time scales has not yet been established. Systematically increasing the autocorrelation time from zero (white noise) to values substantially larger will allow one to determine the point at which a finite correlation time actually begins to matter.

In summary, the numerical experiments described in this paper lead to at least three tentative conclusions:

1. At least for systems that admit a coexistence of regular and chaotic behavior, even weak couplings to an external environment, modeled as friction and noise, can dramatically accelerate evolution toward a (near-)microcanonical equilibrium. This suggest in particular that idealizing a complex Hamiltonian system as a completely isolated entity may be a very bad idea.

2. There is reason to think that the detailed form of the friction and noise is comparatively unimportant. When assessing the effects of friction and noise in accelerating phase space transport, all that really matters may be the amplitude of the perturbation. If true, this would suggest that it may not be all that hard to satisfactorily model the coupling of one's system to its surrounding environment. The details which are hard to determine may not be very important!

3. "Collisionality," that is, discreteness effects, may be significantly more important in galactic dynamics than generally recognized. For example, such graininess could serve to destabilize quasi-equilibria which use confined chaotic orbits to support interesting structures such as bars [30] or triaxial cusps [31].

ACKNOWLEDGMENTS

It is a pleasure to acknowledge useful interactions with my collaborators, Katja Lindenberg, Elaine Mahon, Ilya Pogorelov, and, especially, Salman Habib. I am also grateful to James Meiss and Donald Lynden-Bell for useful comments and critiques. The final draft of this manuscript was written at the Aspen Center for Physics, the hospitality of which I acknowledge gratefully.

REFERENCES

1. AUBRY, S. & G. ANDRE. 1978. *In* Solitons and Condensed Matter Physics. A.R. Bishop and T. Schneider, Eds.: 264. Springer, Berlin.
2. MATHER, J.N. 1982. Topology **21**: 457.
3. MACKAY, R.S., J.D. MEISS & I.C. PERCIVAL. 1984. Phys. Rev. Lett. **52**: 697.
4. MACKAY, R.S., J.D. MEISS & I.C. PERCIVAL. 1984. Physica D **13**: 55.
5. LIEBERMAN, M.A. & A.J. LICHTENBERG. 1972. Phys. Rev. A **5**: 1852.
6. CALDEIRA, A.O. & A.J. LEGETT. 1983. Physica A **121**: 587.
7. CALDEIRA, A.O. & A.J. LEGETT. 1983. Ann. Phys. (NY) **149**: 374.
8. BARONE, P.M.V.B. & A.O. CALDEIRA. 1991. Phys. Rev. A **43**: 57.
9. KUBO, R., M. TODA & N. HASHITSUME. 1991. Statistical Physics II: Nonequilibrium Statistical Mechanics, 2nd Edit. Springer. Berlin.
10. HABIB, S. & H.E. KANDRUP. 1992. Phys. Rev. **D 46**: 5303.
11. FORD, G.W., J.T. LEWIS & R.F. O'CONNELL. 1988. Phys. Rev. **A 37**: 4419.
12. FORD, G.W., M. KAC & P. MAZUR. 1965. J. Math. Phys. **6**: 504.
13. ZWANZIG, R. 1973. J. Stat. Phys. **9**: 215.
14. CHANDRASEHHAR, S. 1943. Rev. Mod. Phys. **15**: 1.
15. KANDRUP, H.E. 1989. Comments on Astrophys. **13**: 325.
16. HONERKAMP, J. 1994. Stochastic Dynamics Systems. VCH Publishers. New York.
17. LINDENBERG, K. & V. SESHADRI. 1981. Physica A **109**: 481.
18. TODA, M. 1967. J. Phys. Soc. Japan **22**: 431.
19. ARMBRUSTER, D., J. GUCKENHEIMER & S. KIM. 1989. Phys. Lett. **A 140**: 416.
20. MAHON, M.E., R.A. ABERNATHY, B.O. BRADLEY & H. E. KANDRUP. 1995. Mon. Not. R. Astr. Soc. **275**: 443.
21. GRINER, A., W. STRITTMATTER & J. HONERKAMP. 1988. J. Stat. Phys. **51**: 95.
22. KANDRUP, H.E. & M.E. MAHON. 1994. Phys. Rev. **E 49**: 3735.
23. KANDRUP, H.E. & M.E. MAHON. 1995. Ann. N. Y. Acad. Sci. **751**: 93.
24. HABIB, S., H.E. KANDRUP & M.E. MAHON. 1996. Phys. Rev. **E 53**: 5473.
25. HABIB, S., H.E. KANDRUP & M.E. MAHON. 1997. Astrophys. J. **480**: 155.
26. KANDRUP, H.E. & D.E. WILLMES. 1994. Astron. Astrophys. **283**: 59.
27. KANDRUP, H.E., R.A. ABERNATHY & B.O. BRADLEY. 1995. Phys. Rev. **E 51**: 5287.
28. MERRITT, D. & M. VALLURI. 1996. Astrophys. J. **471**: 82.
29. KANDRUP, H.E. 1998. Mon. Not. R. Astr. Soc. **301**: 960.
30. WOZNIAK, H. 1993. *In* Ergodic Concepts in Stellar Dynamics. V.G. Gurzadyan & D. Pfenniger, Eds. Springer. Berlin.
31. MERRITT, D. & T. FRIDMAN. 1996. Astrophys. J. **460**: 136.

Papers by George Contopoulos

NOTATION: (D), Stellar Dynamics-Galactic Dynamics; (R), Relativity-Cosmology; (C), Celestial Mechanics; (Q), Classical and Quantum Mechanics

1954–1960

1. CONTOPOULOS, G. 1954. Beitrag zur Dynamik der Kugelsternhaufen. Z. Astrophys. **35**: 67. (D)
2. CONTOPOULOS, G. 1956. On the isophotes of ellipsoidal nebulae. Z. Astrophys. **39**: 126. (D)
3. CONTOPOULOS, G. 1956. On the motions of stars in an ellipsoidal stellar system. Astrophys. J. **124**: 643. (D)
4. CONTOPOULOS, G. 1956. Study of the potential in the plane of symmetry of a stellar system. Ann. Acad. Athens **31**: 21. (D)
5. CONTOPOULOS, G. 1957. Der Einfluss des Strahlungsdruckes auf die Dynamik der interstellaren Koerner. Z. Astrophys. **42**: 75. (D)
6. CONTOPOULOS, G. 1957. On the relative motions of stars in a galaxy. Stockholm Obs. Ann. **19**, No. 10. (D)
7. CONTOPOULOS, G. 1957. Astronomien in Grekland. Pop. Astron. Tidskrift 1.
8. CONTOPOULOS, G. 1958. On the vertical motions of stars in a galaxy. Stockholm Obs. Ann. **20**, No. 5. (D)
9. CONTOPOULOS, G. 1958. Space and time in general relativity. Ann. Fac. Sci. Univ. Thessaloniki **8**: 45. (R)
10. CONTOPOULOS, G. 1958. A solution of the clock paradox. Ann. Fac. Sci. Univ. Thessaloniki **8**: 23. (R)
11. CONTOPOULOS, G. 1958. Abstracts of Talks of the Kiel Meeting of the Astronomische Gesellschaft. (D)
12. CONTOPOULOS, G. 1960. A third integral of motion in a galaxy. Z. Astrophys. **49**: 273. (D)

1962–1970

13. CONTOPOULOS, G. & G. BOZIS. 1962. Perturbations of stars in a galaxy. Ann. Fac. Sci. Univ. Thessaloniki **11**: 11. (D)
14. CONTOPOULOS, G. & B. BARBANIS. 1962. An application of the third integral of motion. Observatory **82**: 80. (D)
15. CONTOPOULOS, G. 1962. Notes added in proof. *In* The Third Integral of Motion for Low-Velocity Stars, H.C. van de Hulst, Bull. Astr. Netherlands **16**: 235. (D)
16. CONTOPOULOS, G. 1963. On the existence of a third integral of motion. Astron. J. **68**: 1. (D)

17. CONTOPOULOS, G. 1963. A classification of the integrals of motion. Astrophys. J. **138:** 1297. (D)
18. CONTOPOULOS, G. 1963. Resonance cases and small divisors in a third integral of motion. Astron. J. **68:** 763. (D)
19. CONTOPOULOS, G. 1963. Some applications of the third integral of motion [Abstract]. Astron. J. **68:** 70. (D)
20. CHANDRASEKHAR, S. & G. CONTOPOULOS. 1963. The virial theorem in general relativity in post-Newtonian approximation. Proc. Nat. Acad.Sci. **49:** 608. (R)
21. CONTOPOULOS, G. & G. BOZIS. 1964. Escape of stars during the collision of two galaxies. Astrophys. J. **139:** 1239. (D)
22. CONTOPOULOS, G. & L. WOLTJER. 1964. The "third" integral in non-smooth potentials. Astrophys. J. **140:** 1106. (D)
23. CONTOPOULOS, G. & B. STROEMGREN. 1965. Tables of plane galactic orbits. NASA Institute for Space Studies. (D)
24. CONTOPOULOS, G. 1965. The "third" integral in the restricted three-body problem. Astrophys. J. **142:** 802. (C)
25. CONTOPOULOS, G. 1965. Periodic and tube orbits. Astron. J. **70:** 526 (D)
26. CONTOPOULOS, G. & M. MOUTSOULAS. 1965. Resonance cases and small divisors in a third integral of motion, II. Astron. J. **70:** 817. (D)
27. CONTOPOULOS, G. 1966. Recent developments in stellar dynamics. IAU Symposium **25:** 3. (D)
28. CONTOPOULOS, G. 1966. Adiabatic invariants and the third integral. J. Math. Phys. **7:** 788. (D)
29. CONTOPOULOS, G. 1966. Problems of stellar dynamics. *In* Space Mathematics, Part I. J.B. Rosser, Ed.: 169. American Mathematical Society. Providence, RI (D)
30. CONTOPOULOS, G. 1966. Tables of the third integral. Astrophys. J. Suppl. **13:** 503. (D)
31. CONTOPOULOS, G. & M. MOUTSOULAS. 1966. Resonance cases and small divisors in a third integral of motion, III. Astron. J. **71:** 687. (D)
32. CONTOPOULOS, G. 1966. Spiral structure and the third integral. *In* Outline of Talks Presented at the Columbia Nov. 4 Meeting, F. Shu, Ed.: 30. (D)
33. CONTOPOULOS, G. 1966. Rules for Scientific Meetings. Astronomer's Handbook, IAU : 284.
34. CONTOPOULOS, G. 1966. Style Book. Astronomer's Handbook, IAU: 254.
35. CHANDRASEKHAR, S. & G. CONTOPOULOS. 1967. On a post-Galilean transformation appropriate to the post-Newtonian theory of Einstein, Infeld and Hoffman. Proc. Roy. Soc. A. **298:** 23. (R)
36. CONTOPOULOS, G. 1967. Stellar orbits and the stability of spiral structure. *In* Gravitational Instability and the Formation of Stars and Galactic Structure. Proc. 14th Liège Symposium: 213. (D)

37. CONTOPOULOS, G. 1967. Applications of the third integral in the galaxy. *In* Relativity Theory and Astrophysics 2, Galactic Structure, J. Ehlers, Ed.: 98. American Mathematical Society. (D)

38. CONTOPOULOS, G. 1967. Integrals of motion in the three-dimensional restricted three-body problem. Astron. J. **72:** 191. (C)

39. CONTOPOULOS, G. 1967. Resonance phenomena and the non-applicability of the "third" integral. Besancon Symposium 1966. Bull. Astron. 3e Ser. 2, Fasc. **1:** 223. (D)

40. CONTOPOULOS, G. 1967. Integrals of motion in the elliptic restricted three-body problem. Astron. J. **72:** 669. (C)

41. BOK, B. & G. CONTOPOULOS. 1967. Report of Commission 33. Transactions IAU **13A.** Reidel Publishing Co.

42. CONTOPOULOS, G. 1968. New integrals of motion and the orbital history of the Moon. Nature **220:** 1018 (Letter to the Editor). (C)

43. CONTOPOULOS, G. & J. HADJIDEMETRIOU. 1968. Characteristics of invariant curves of plane orbits. Astron. J. **73:** 86. (D)

44. CONTOPOULOS, G. 1968. Resonant periodic orbits. Astrophys. J. **153:** 83. (D)

45. CONTOPOULOS, G. & B. BARBANIS. 1968. Is the third integral a function of the Hamiltonian? Astrophys. Space Sci. **2:** 134. (D)

46. CONTOPOULOS, G. 1968. Families of periodic orbits [Abstract]. Astron. J. **73:** 172. (D)

47. CONTOPOULOS, G. 1968. Report on the work by K. Prendergast and R. Miller, "Numerical experiments with a very large number of bodies." Bull. Astron. **3:** 309. (D)

48. CONTOPOULOS, G. 1970. Orbits in highly perturbed dynamical systems 1. Periodic orbits. Astron. J. **75:** 96. (D)

49. CONTOPOULOS, G. 1970. Orbits in highly perturbed dynamical systems 11. Stability of periodic orbits. Astron. J. **75:** 108. (D)

50. CONTOPOULOS, G. 1970. Resonance effects in spiral galaxies. Astrophys. J. **160:** 113. (D)

51. CONTOPOULOS, G. 1970. Gravitational theories of spiral structure. IAU Symposium. **38:** 303. (D)

52. CONTOPOULOS, G. 1970. Resonance phenomena in spiral galaxies. *In* Periodic Orbits, Stability and Resonances. G.E.O. Giacaglia, Ed.: 332. Reidel Publishing Co. (D)

53. CONTOPOULOS, G. & S. W. MCCUSKEY. 1970. Structure and dynamics of the galactic system. Contr. Astron. Dept. Univ. Thessaloniki No. 53.

1971–1980

54. CONTOPOULOS, G. 1971. Recent developments in galactic dynamics. *In* Structure and Evolution of the Galaxy, L.N. Mavridis, Ed.: 198. Reidel Publishing Co. (D)

55. CONTOPOULOS, G. 1971. Preference of trailing waves. Astrophys. J. **163:** 181. (D)
56. CONTOPOULOS, G. 1971. Gravitational N-body problem. Earth Extraterr. Sci. **1:** 185. (D)
57. CONTOPOULOS, G. 1971. Orbits in highly perturbed dynamical systems, III. Non-periodic orbits. Astron. J. **76:** 147. (D)
58. CONTOPOULOS, G. 1971. Collisionless stellar dynamics. Astrophys. Space Sci. **13:** 337. (D)
59. CONTOPOULOS, G. 1972. The dynamics of spiral structure. Lecture Notes. University of Maryland. (D)
60. CONTOPOULOS, G. 1973. The particle resonance in spiral galaxies. Non-linear effects. Astrophys. J. **181:** 657. (D)
61. CONTOPOULOS, G. 1973. Problems of stellar dynamics. *In* Recent Advances in Dynamical Astronomy, B.D. Tapley & V. Szebehely, Eds.: 177. Reidel Publishing Co. (D)
62. CONTOPOULOS, G. 1973. Theory of spiral structure. Resonances. *In* Proc. First European Astron. Meeting, L.N. Mavridis, Ed. **3:** 104. Springer Verlag. (D)
63. CONTOPOULOS, G. & M. ZIKIDES. 1973. Periodic orbits of the restricted problem for various values of the mass-ratio. *In* Proc. First European Astron. Meeting, B. Baxbanis & J.D. Hadjidemetriou, Eds. **2:** 279. Springer Verlag. (C)
64. CONTOPOULOS, G. 1973. The density wave theory of spiral structure. *In* Dynamical Structure and Evolution of Stellax Systems, L. Martinet & M. Mayor, Eds. Geneva Observatory. (D)
65. CONTOPOULOS, G. 1973. Topological methods in stellax dynamics. *In* Dynamical Structure and Evolution of Stellax Systems, L. Martinet & M. Mayor, Eds. Geneva Observatory. (D)
66. CONTOPOULOS, G. 1974. Some recent developments in the theory of spiral structure. IAU Symposium **58:** 413. (D)
67. CONTOPOULOS, G. 1974. Formal integrals of Hamiltonian systems in resonance and near resonance cases. *In* Volume in Honor of S. Placidis. D. Kotsakis, Ed. Athens. (D)
68. CONTOPOULOS, G. 1974. Stellar dynamics. *In* Summer Institute on Galactic Structure. E. A. Kreiken & Z. Tufekcioglu, Eds. Ankara. (D)
69. CONTOPOULOS, G. 1974. IAU Information Bulletin No **31:** 1–37.
70. CONTOPOULOS, G. 1974. IAU Information Bulletin No **32:** 1–13.
71. CONTOPOULOS, G. 1975. IAU Information Bulletin No **33:** 1–43.
72. CONTOPOULOS, G. 1975. IAU Information Bulletin No **34:** 1–15.
73. CONTOPOULOS, G. & L. VLAHOS. 1975. Integrals of motion and resonances in the magnetic dipole field. J. Math. Phys. **16:** 1469.
74. CONTOPOULOS, G. 1975. Integrals of motion. IAU Symposium **69:** 209. (D)
75. CONTOPOULOS, G. 1975. The theory of resonances. *In* La Dynamique des Galaxies Spirales. L. Weliachew, Ed. : 17. CNRS. Paris. (D)

76. CONTOPOULOS, G. 1975. Inner Lindblad resonance in galaxies. Non-linear theory I. Astrophys. J. **201**: 566. (D)
77. CONTOPOULOS, G. 1976. IAU Information Bulletin No **35**: 1–35.
78. CONTOPOULOS, G. 1976. IAU Information Bulletin No **36**: 1–43.
79. CONTOPOULOS, G. 1976. The relativistic restricted three-body problem. *In* In Memoriam D. Eginitis. D. Kotsakis, Ed. Athens. (R)
80. CONTOPOULOS, G. & P. GROSBOL. 1976. The past positions of the spiral arms of our Galaxy. Proc. 3rd European Astron. Meeting. Tbilissi. (D)
81. CONTOPOULOS, G. 1976. Strongly perturbed dynamical systems. *In* Long-Time Predictions in Dynamics, V. Szebehely & B.D. Tapley, Eds: 43. Reidel Publishing Co. (D)
82. CONTOPOULOS, G. & N. SPYROU. 1976. The center of mass in the post-Newtonian approximation of general relativity. Astrophys. J. **205**: 592. (R)
83. CONTOPOULOS, G. & C. BANOS. 1976. The new Greek 48-inch Cassegrain-Coude telescope. Sky and Telescope. : 154.
84. CONTOPOULOS, G. 1977. Closing speech. Transactions of the IAU. **16B**: 44. Grenoble, 1976.
85. CONTOPOULOS, G. & C. MERTZANIDES. 1977. Inner Lindblad resonance in spiral galaxies. Non-linear theory II. Bars. Astron. Astrophys. **61**: 477. (D)
86. CONTOPOULOS, G. The non-linear theory of spiral galaxies. IAU Colloquium **45**: 229. (D)
87. CONTOPOULOS, G. 1977. Adiabatic invariants, stochasticity, and the third integral. Book of Abstracts. N.C. Christofilos International Summer School and Conference in Plasma Physics: 34a. U.S. Energy R&D Administration. (D)
88. CONTOPOULOS, G. 1978. Stellar dynamics. *In* Theoretical Principles in Astrophysics and Relativity, N.P. Levobitz, W. Reid & P.O. Vandervoort, Eds.: 93. Univ. of Chicago Press. (D)
89. CONTOPOULOS, G. 1978. The disappearance of integrals in systems of more than two degrees of freedom. Celest. Mech. **17**: 167. (D)
90. CONTOPOULOS, G. 1978. Periodic orbits near the particle resonance in galaxies. Astron. Astrophys. **64**: 323. (D)
91. CONTOPOULOS, G. 1978. Integrable and stochastic behaviour in dynamical astronomy. *In* Stochastic Behaviour in Classical and Quantum Hamiltonian Systems. G. Casati & J. Ford Eds.: 1. Springer-Verlag. (D)
92. CONTOPOULOS, G., L. GALGANI & A. GIORGILLI. 1978. On the number of isolating integrals in Hamiltonian systems. Phys. Rev. **A18**: 1183. (D)
93. CONTOPOULOS, G. 1978. The dynamics of the spiral structure in galaxies. *In* Astronomical Papers dedicated to B. Strömgren. A. Reiz & T. Andersen, Eds.: 387. Copenhagen Univ. Observatory. (D)
94. CONTOPOULOS, G. 1979. The four-armed response near the Lindblad resonances in galaxies. IAU Symposium **84**: 194. (D)

95. CONTOPOULOS, G. 1979. Higher-order resonances in dynamical systems. Celest. Mech. **18:** 195. (D)
96. CONTOPOULOS, G. 1979. Instabilities in systems of 3 degrees of freedom. *In* Instabilities in Dynamical Systems. Applications to Celestial Mechanics. V.G. Szebehely, Ed.: 25. Reidel Publishing Co. (D)
97. CONTOPOULOS, G. 1979. Inner Lindblad resonance in galaxies. Non-linear theory III. The response density. Astron. Astrophys. **71:** 229. (D)
98. CONTOPOULOS, G. 1979. Stellar dynamics of barred spirals. *In* Photometry, Kinematics and Dynamics of Galaxies. D.S. Evans, Ed.: 425. Univ. of Texas, Austin. (D)
99. CONTOPOULOS, G. 1980. How far do bars extend? Astron. Astrophys. **81:** 198. (D)
100. CONTOPOULOS, G. & P. MICHAELIDIS. 1980. Bifurcations of triple periodic orbits. Celest. Mech. **22:** 403. (D)
101. CONTOPOULOS, G. & T. PAPAYANNOPOULOS. 1980. Orbits in weak and strong bars. Astron. Astrophys. **92:** 33. (D)
102. CONTOPOULOS, G. & M. ZIKIDES. 1980. Periodic orbits and ergodic components of a resonant dynamical system. Astron. Astrophys. **90:** 198. (D)
103. CONTOPOULOS, G. 1980. A dispersion relation for open spiral galaxies. J. Astrophys. Astron. **1:** 79. (D)

1981–1990

104. CONTOPOULOS, G. 1981. The 4:1 resonance. Celest. Mech. **24:** 355. (D)
105. CONTOPOULOS, G. 1981. Do successive bifurcations in Hamiltonian systems have the same universal ratio? Lett. Nuovo Cim. **30:** 498. (D)
106. CONTOPOULOS, G. 1981. Inner Lindblad resonance. Non-linear theory. IV. Self-consistent bars. Astron. Astrophys. **104:** 116. (D)
107. CONTOPOULOS, G. 1981. The effects of resonances near corotation in barred galaxies. Astron. Astrophys. **102:** 265. (D)
108. CONTOPOULOS, G. 1981. The date of easter. Synodica **5:** 53.
109. CONTOPOULOS, G. 1982. Trapping of orbits in barred galaxies. *In* Compendium in Astronomy. E.G. Mariolopoulos, P.S. Theochaxis & L.N. Mavridis, Eds.: 55. Reidel Publishing Co. (D)
110. CONTOPOULOS, G., P. MAGNENAT & L. MARTINET. 1982. Invariant surfaces and orbital behaviour in dynamical systems of 3 degrees of freedom II. Physica **D6:** 126. (D)
111. CONTOPOULOS, G. 1982. Non-linear dynamical astronomy, applications and extrapolations. Bull. Amer. Astron. Soc. **14:** 647. (D)
112. CONTOPOULOS, G. 1983. Ordered and ergodic motions of stars in galaxies. Astron. Astrophys. **117:** 89. (D)

113. CONTOPOULOS, G. 1983. Relativistic stellar dynamics. IAU Symposium **104**: 417. (R)
114. CONTOPOULOS, G. 1983. Inverse Feigenbaum sequences in Hamiltonian systems. Lett. Nuovo Cim. **37**: 149. (D)
115. CONTOPOULOS, G. 1983. Infinite bifurcations, gaps and bubbles in Hamiltonian systems. Physica **D8**: 142. (D)
116. CONTOPOULOS, G. 1983. The genealogy of periodic orbits in a model galaxy. Celest. Mech. **31**: 193. (D)
117. CONTOPOULOS, G. 1983. Bifurcations, gaps and stochasticity in barred galaxies. Astrophys. J. **275**: 511. (D)
118. CONTOPOULOS, G. 1983. Termination of sequences of bifurcations in 3-dimensional Hamiltonian systems. Lett. Nuovo Cim. **38**: 257. (D)
119. CONTOPOULOS, G. 1983. The dynamics of galaxies. Mitt. Astron. Gesellschaft **60**: 31. (D)
120. CONTOPOULOS, G. 1984. Orbits though the ergosphere of a Kerr black hole. Gen. Rel. Gravitation **16**: 33. (R)
121. CONTOPOULOS, G. Theoretical periodic orbits in 3-dimensional Hamiltonians. Physica **D11**: 179. (D)
122. CONTOPOULOS, G. & A. PINOTSIS. 1984. Infinite bifurcations in the restricted three-body problem. Astron. Astrophys. **113**: 49. (C)
123. CONTOPOULOS, G. 1985. Bifurcations in Hamiltonian dynamical systems. *In* Singularities and Dynamical Systems, S. Pnevmaticos, Ed.: 375. North Holland. (D)
124. CONTOPOULOS, G. 1985. The transition to chaos in galactic models of two and three degrees of freedom. *In* Chaos in Astrophysics, J.R. Buchler, J. Perdang & E.A. Spiegel, Eds.: 259. Reidel Publishing Co. (D)
125. CONTOPOULOS, G. 1985. Non-linear problems in stellar dynamics. *In* Nonlinear Phenomena in Physics, F. Claro, Ed.: 238. Springer-Verlag. (D)
126. CONTOPOULOS, G. 1985. Bifurcations and stability in three-dimensional systems. *In* Stability of the Solar System and its Minor Natural and Artificial Bodies, V. Szebehely, Ed.: 97. Reidel Publishing Co. (D)
127. CONTOPOULOS, G. 1985. Spiral galaxies end at the 4/1 Resonance. Comments on Astrophysics **11**: 1. (D)
128. CONTOPOULOS, G. & B. BARBANIS. 1985. Resonant systems of three degrees of freedom. Astron. Astrophys. **153**: 44. (D)
129. CONTOPOULOS, G. & P. MAGNENAT. 1985. Simple three-dimensional periodic orbits in a galactic-type potential. Celest. Mech. **37**: 387. (D)
130. CONTOPOULOS, G. 1986. Instabilities in Hamiltonian systems of 2 and 3 degrees of freedom. Particle Accelerators **19**: 107. (D)
131. CONTOPOULOS, G. & P. GROSBOL. 1986. Stellar dynamics of spiral galaxies: Non-linear effects at the 4/1 resonance. Astron. Astrophys. **155**: 11. (D)

132. CONTOPOULOS, G. 1986. Bifurcations in Systems of 3 Degrees of Freedom. Celest. Mech. **38:** 1. (D)
133. CONTOPOULOS, G. 1986. Qualitative changes in 3-D dynamical systems. Astron. Astrophys. **161:** 244. (D)
134. CONTOPOULOS, G., H. VARVOGLIS & B. BARBANIS. 1987. Large degree stochasticity in a galactic model. Astron. Astrophys. **172:** 55. (D)
135. CONTOPOULOS, G. & C. POLYMILIS. 1987. Approximations of the 3-particle Toda lattice. Physica **D24:** 328. (D)
136. CONTOPOULOS, G. 1987. Nonlinear phenomena in galaxies. *In* The Galaxy. G. Gilmore & B. Carswell, Eds.: 199. Reidel Publishing Co. (D)
137. CONTOPOULOS, G. 1987. Stochasticity in galactic models. *In* Chaotic Phenomena in Astrophysics. J.R. Buchler & H. Eichhorn, Eds. **475:** 1. Ann. N.Y. Acad. Sci. (D)
138. CONTOPOULOS, G. 1987. Astron. Astrophys. **183:** Editorial.
139. CONTOPOULOS, G. 1987. Demetrios Kotsakis. Quart. J. Roy. Astron. Soc. **28:** 397.
140. CONTOPOULOS, G. 1987. In memoriam Demetrios Kotsakis. Mitt. Astr. Gesellschaft. **70:** 11.
141. CONTOPOULOS, G. 1988. The formation of gaps along families of periodic solutions. Volume in Memory of D. Kotsakis: 85. Athens. (D)
142. CONTOPOULOS, G. 1988. Escapes of stars from stellar systems. IAU Colloquium **96:** 265. (D)
143. CONTOPOULOS, G. 1988. Qualitative characteristics of dynamical systems. *In* Long Term Behaviour of Natural and Artificial N-Body Systems, A. Roy, Ed.: 301. Kluwer Academic Publishers. (D)
144. CONTOPOULOS, G. 1988. Critical cases in 3-dimensional systems. Celest. Mech. **42:** 239. (D)
145. CONTOPOULOS, G. & P. GROSBOL. 1988. Stellar dynamics of spiral galaxies: Self-consistent models. Astron. Astrophys. **197:** 83. (D)
146. CONTOPOULOS, G. & A. GIORGILLI. 1988. Bifurcations and complex instability in a 4-dimensional symplectic mapping. Meccanica **23:** 19. (D)
147. CONTOPOULOS, G. 1988. The 4/1 resonance in barred galaxies. Astron. Astrophys. **201:** 44. (D)
148. CONTOPOULOS, G. 1988. Short and long period orbits. Celest. Mech. **43:** 147. (D)
149. CONTOPOULOS, G. 1988. Resonant integrable galactic models. *In* Integrability in Dynamical Systems. J.R. Buchler, J.R. Ipser & C.A. Williams, Eds. **536:** 1. Ann. N.Y. Acad. Sci. (D)
150. CONTOPOULOS, G. 1988. Nonuniqueness of families of periodic solutions in a four dimensional mapping. Celest. Mech. **44:** 393. (D)
151. CONTOPOULOS, G. 1988. Astron. Astrophys. **200.** Editorial.
152. CONTOPOULOS, G. 1989. Astron. Astrophys. **211.** Editorial.

153. CONTOPOULOS, G., S.T. GOTTESMANN, J.H. HUNTER, JR. & M.N. ENGLAND. 1989. Comparison of stellar and gas dynamics of a barred galaxy. Astrophys. J. **343**: 608. (D)
154. FOUNARIOTAKIS, M., S.C. FARANTOS, G. CONTOPOULOS & G. POLYMILIS. 1989. Periodic orbits, bifurcations and quantum mechanical eigenfunctions and spectra. J. Chem. Phys. **91**: 1389. (Q)
155. CONTOPOULOS, G. & L. ZACHILAS. 1989. Complex instability in 3-D systems. *In* Singular Behaviour and Nonlinear Dynamics. T. Bountis, S. Pnevmaticos & St. Pnevmaticos, Eds.: 130. World Scientific. (D)
156. CONTOPOULOS, G. & B. BARBANIS. 1989. Lyapunov characteristic numbers and the structure of phase space. Astron. Astrophys. **222**: 329. (D)
157. CONTOPOULOS, G. & P. GROSBOL. 1989. Orbits in barred galaxies. Astron. Astrophys. Review **1**: 261. (D)
158. CONTOPOULOS, G. 1990. Self-Consistent models of spiral galaxies. *In* Galactic Models. J.R. Buchler, S. Gottesmann & J.H. Hunter, Jr. Eds. **596**: 101. Ann. N. Y. Acad. Sci. (D)
159. CONTOPOULOS, G. & J. SEIMENIS. 1990. Application of the Prendergast method to a logarithmic potential. Astron. Astrophys. **227**: 49. (D)
160. CONTOPOULOS, G. 1990. Asymptotic curves and escapes in Hamiltonian systems. Astron. Astrophys. **231**: 41.
161. CONTOPOULOS, G., N. VOGLIS & N. HIOTELIS. 1990. Galaxy formation: Gas dynamics versus stellar dynamics. *In* Nonlinear Astrophysical Fluid Dynamics. J.R. Buchler & S.T. Gottesman, Eds. **617**: 178. Ann. N. Y. Acad. Sci. (D)
162. CONTOPOULOS, G. 1990. Periodic orbits and chaos around two fixed black holes. Proc. Roy. Soc. **A431**: 183. (R)

1991–1998

163. CONTOPOULOS, G. 1991. The generation of spiral characteristics. Celest. Mech. Dyn. Astron. **50**: 251. (D)
164. PATSIS, P., G. CONTOPOULOS & P. GROSBOL. 1991. Self-consistent spiral galactic models. Astron. Astrophys. **243**: 373. (D)
165. HIOTELIS, N., N. VOGLIS & G. CONTOPOULOS. 1991. Hydrodynamics in a collapsing gaseous protogalaxy. Astron. Astrophys. **242**: 69. (D)
166. CONTOPOULOS, G. 1991. Transition to instability and chaos in 3-D Hamiltonians. *In* Non-Linear Problems in Future Particle Accelerators. W. Scandale & G. Turchetti, Eds.: 163. World Scientific, Singapore. (D)
167. CONTOPOULOS, G. 1991. Gas and star dynamics in galaxies. IAU Symposium. **146**: 335. (D)
168. CONTOPOULOS, G. 1991. A new route to chaos generation of spiral chaxacteristics. *In* Predictability, Stability and Chaos in *N*-Body Dynamical Systems. A. Roy, Ed.: 35. Plenum Press. (D)

169. CONTOPOULOS, G. 1991. Chaos around two fixed black holes. *In* Nonlinear Problems in Relativity and Cosmology. J.R. Buchler, S.L. Detweiler & J.R. Ipser, Eds. **631**: 143. Ann. N. Y. Acad. Sci. (R)

170. CONTOPOULOS, G. 1991. Periodic orbits and chaos around two fixed black holes II. Proc. Roy. Soc. **A435**: 551. (R)

171. CONTOPOULOS, G. 1991. Stavros Plakidis. Quart. J. Roy. Astron. Soc. **32**: 483.

172. CONTOPOULOS, G. & P.O. VANDERVOORT. 1992. A rotating Stäckel potential. Astrophys. J. **389**: 118. (D)

173. CONTOPOULOS, G. & D. KAUFMANN. 1992. Types of escapes in a simple Hamiltonian system. Astron. Astrophys. **253**: 389. (D)

174. CONTOPOULOS, G. & D. KAUFMANN. 1992. Nonlinear self-consistent models of barred galaxies. *In* Astrophysical Discs, S.F. Dermott, J.H. Hunter, Jr. & R.E. Wilson, Eds. **675**: 126. Ann. N. Y. Acad. Sci. (D)

175. RESVANIS, L.K., G. CONTOPOULOS *et al.* 1992. NESTOR. A neutrino particle astrophysics underwater laboratory for the mediterranean. *In* Proceedings of the High Energy Neutrino Astrophysics Workshop: 325. Univ. of Hawaii.

176. CONTOPOULOS, G. & H. PAPADAKI. 1993. Newtonian and relativistic periodic orbits around two fixed black holes. Celest. Mech. Dyn. Astron. **55**: 47. (R)

177. CONTOPOULOS, G. 1993. Classical periodic orbits and quantum mechanical eigenvalues and eigenfunctions. Celest. Mech. Dyn. Astron. **56**: 325. (Q)

178. CONTOPOULOS, G. & C. POLYMILIS. 1993. Geometrical and dynamical properties of homoclinic tangles in a simple Hamiltonian system. Phys. Rev. **E47**: 1546. (D)

179. CONTOPOULOS, G., H. KANDRUP & D. KAUFMANN. 1993. Fractal properties of escape from a two-dimensional potential. Physica **D64**: 310. (D)

180. CONTOPOULOS, G. 1993. Noise and chaos in galactic dynamics. *In* Stochastic Processes in Astrophysics. R. Buchler & H. Kandrup, Eds. **706**: 207. Ann. N. Y. Acad. Sci. (D)

181. CONTOPOULOS, G., B. GRAMMATICOS & A. RAMANI. 1993. Painleve analysis for the Mixmaster Universe model. J. Phys. **A26**: 5795. (R)

182. CONTOPOULOS, G. 1994. The structure of chaos. *In* Hamiltonian Mechanics, Integrability and Chaotic Behaviour. J. Seimenis, Ed.: 51. Plenum Press. (D)

183. CONTOPOULOS, G. 1994. Order and chaos in galaxies. *In* Galactic Dynamics and *N*-body Simulations, G. Contopoulos, N. Spyrou & L. Vlahos, Eds.: 33. Springer-Verlag. (D)

184. PATSIS, P.A., N. HIOTELIS, G. CONTOPOULOS & P. GROSBOL. 1994. Hydrodynamic simulations of open normal spiral galaxies. Astron. Astrophys. **286**: 46. (D)

185. CONTOPOULOS, G., B. GRAMMATICOS & A. RAMANI. 1994. Integrability of the Mixmaster Universe. *In* Chaos in Relativity. D. Hobill, Ed.: 423. Plenum Press. (R)

186. CONTOPOULOS, G. 1994. Chaos in the case of two fixed black holes. *In* Chaos in Relativity. D. Hobill, Ed.: 129. Plenum Press. (R)

187. CONTOPOULOS, G., H. PAPADAKI & C. POLYMILIS. 1994. The structure of chaos in a potential without escapes. Celest. Mech. Dyn. Astron. **60:** 249. (D)

188. CONTOPOULOS, G. & B. BARBANIS. 1994. Periodic orbits and their bifurcations in a 3-D system. Celest. Mech. Dyn. Astron. **59:** 279. (D)

189. CONTOPOULOS, G., S. FARANTOS, H. PAPADAKI & C. POLYMILIS. 1994. Complex unstable periodic orbits and their manifestation in classical and quantum dynamics. Phys. Rev. **E50:** 4399. (Q)

190. VOGLIS, N. & G. CONTOPOULOS. 1994. Invariant spectra of orbits in dynamical systems. J. Phys. **A27:** 4899. (D)

191. CONTOPOULOS. G., B. GRAMMATICOS & A. RAMANI. 1994. The Mixmaster Universe model revisited. J. Phys. **A27:** 5357. (R)

192. CONTOPOULOS, G. 1995. Order in chaos. *In* From Newton to Chaos: Modern Techniques for Understanding and Coping with Chaos in *N*-Body Dynamical Systems, A. Roy, Ed.: 425. Plenum Press. (D)

193. PAPADAKI, H., G. CONTOPOULOS & C. POLYMILIS. 1995. Complex instability. *In* From Newton to Chaos: Modern Techniques for Understanding and Coping with Chaos in *N*-Body Dynamical Systems, A. Roy, Ed.: 485. Plenum Press. (D)

194. CONTOPOULOS, G. 1995. Order and chaos in three-dimensional systems. *In* Three-Dimensional Systems, S. Gottesman, J. Ipser Jr. & H. Kandrup, Eds. **751:** 112. Ann. N. Y. Acad. Sci. (D)

195. SIOPIS, C., G. CONTOPOULOS & H.E. KANDRUP. 1995. Escape probabilities in a Hamiltonian with two channels of escape. *In* Three-Dimensional Systems, S. Gottesman, J. Ipser & H. Kandrup, Eds. **751:** 205. Ann. N. Y. Acad. Sci. (D)

196. BARBANIS, B. & G. CONTOPOULOS. 1995. Order in the distribution of 3-D periodic orbits. Astron. Astrophys. **294:** 33. (D)

197. CONTOPOULOS, G., N. VOGLIS, C. EFTHYMIOPOULOS & E. GROUSOUSAKOU. 1995. Invariant spectra of dynamical systems. *In* Waves in Astrophysics. J.H. Hunter, Jr. & R.Wilson, Eds. **773:** 145. Ann. N. Y. Acad. Sci. (D)

198. SMITH, H. & G. CONTOPOULOS. 1995. Spectra and Lyapunov numbers in pulsating systems. *In* Waves in Astrophysics. J.H. Hunter, Jr. & R. Wilson, Eds. **773:** 189. Ann. N. Y. Acad. Sci. (D)

199. SIOPIS, C., H. KANDRUP, G. CONTOPOULOS & R. DVORAK. 1995. Universal properties of escapes. *In* Waves in Astrophysics. J.H. Hunter, Jr. & R. Wilson, Eds. **773:** 221. Ann. N. Y. Acad. Sci. (D)

200. CONTOPOULOS, G., E. GROUSOUSAKOU & N. VOGLIS. 1995. Invariant spectra in Hamiltonian systems. Astron. Astrophys. **304:** 374. (D)

201. CONTOPOULOS, G. & N. VOGLIS. 1996. The role of chaos in barred galaxies. IAU Colloquium 157. Astron. Soc. Pacific Conf. Series **91:** 321. (D)

202. CONTOPOULOS, G., B. GRAMMATICOS & A. RAMANI. 1995. The last remake of the Mixmaster Universe model. J. Phys. **A28:** 5313. (R)

203. CONTOPOULOS, G. 1995. Message from the President. Hipparchos. The Hellenic Astronomical Society Newsletter **1:** No. 1.

204. KAUFMANN, D. & G. CONTOPOULOS. 1996. Self-consistent models of barred spiral galaxies. Astron. Astrophys. **309**: 381. (D)
205. SKOKOS, C., G. CONTOPOULOS & A. GIORGILLI. 1996. Study of the effective stability of the restricted three-body problem. Proc. 2nd Hellenic Astron. Meeting: 526. (D)
206. SKOKOS, C. G. CONTOPOULOS & C. POLYMILIS. 1996. Non-periodic orbits in a four-dimensional symplectic map. Proc. 2nd Hellenic Astron. Meeting: 578. (D)
207. GROUSOUSAKOU, E., G. CONTOPOULOS & C. POLYMILIS. 1996. Distribution of periodic orbits in a 2-D Hamiltonian system. Proc. 2nd Hellenic Astron. Meeting: 515 (D)
208. CONTOPOULOS, G. & N. VOGLIS. 1996. Spectra of stretching numbers and helicity angles in dynamical systems. Celest. Mech. Dyn. Astron. **64**: 1. (D)
209. SMITH, H. & G. CONTOPOULOS. 1996. Spectra of stretching numbers in oscillating galaxies. Astron. Astrophys. **314**: 795. (D)
210. CONTOPOULOS, G. & C. POLYMILIS. 1996. Recurrence time in the homoclinic tangle. Celest. Mech. Dyn. Astron. **63**: 189. (D)
211. CONTOPOULOS, G., N. VOGLIS & C. EFTHYMIOPOULOS. 1996. Orbits in barred galaxies. *In* Barred Galaxies and Circumnucleax Activity, A. Sandqvist & P.O. Lindblad, Eds. Nobel Symposium **98**: 19. Springer-Verlag. (D)
212. CONTOPOULOS, G., E. GROUSOUSAKOU & C. POLYMILIS. 1996. Distribution of periodic orbits and the homoclinic tangle. Celest. Mech. Dyn. Astron. **64**: 363. (D)
213. CONTOPOULOS, G. 1996. Message from the President, Hippaxchos, The Hellenic Astronomical Society Newsletter **1**: No. 2.
214. CONTOPOULOS, G. 1996. Contribution to the General Discussion of the IAU Colloquium 157. *In* Astr. Soc. Pacific Conf. Ser., R. Buta, D.A. Crocker and B.G. Elmegreen, Eds. **91**: 454–455. (D)
215. CONTOPOULOS, G. & N. VOGLIS. 1997. A fast method for distinguishing between ordered and chaotic orbits. Astron. Astrophys. **317**: 73. (D)
216. CONTOPOULOS, G. & E. GROUSOUSAKOU. 1997. Regular and irregular periodic orbits. Celest. Mech. Dyn. Astron. **65**: 33. (D)
217. CONTOPOULOS, G. & N. VOGLIS. 1997. Spectra of stretching numbers and helicity angles. *In* Analysis and Modelling of Discrete Dynamical Systems. D. Benest & C. Froeschlé, Eds.: 55. Gordon and Breach. (D)
218. EFTHYMIOPOULOS, C., N. VOGLIS & G. CONTOPOULOS. 1997. Diffusion and transient spectra in a 4-D symplectic mapping. *In* Analysis and Modelling of Discrete Dynamical Systems, D. Benest & C. Froeschlé, Eds.: 91. Gordon and Breach. (D)
219. GROUSOUSAKOU, E. & G. CONTOPOULOS. 1997. Distribution of periodic orbits in 2-D dynamical systems. In Analysis and Modelling of Discrete Dynamical Systems. D. Benest & C. Froeschlé, Eds.: 107. Gordon and Breach. (D)

220. SIOPIS, C., H.E. KAUNDRUP, G. CONTOPOULOS & R. DVORAK. 1997. Universal properties of escape in dynamical systems. Celest. Mech. Dyn. Astron. **65**: 57. (D)

221. SKOKOS, C., G. CONTOPOULOS & C. POLYMILIS. 1997. Structures in the phase space of a four-dimensional symplectic map. Celest. Mech. Dyn. Astron. **65**: 223. (D)

222. CONTOPOULOS, G. 1997. Relativity, Cosmology and Celestial Mechanics. Commission 7. Reports of Astronomy. IAU Transactions **23A**. (D)

223. PATSIS, P., C. EFTHYMIOPOULOS, G. CONTOPOULOS & N. VOGLIS. 1997. Dynamical spectra of barred galaxies. Astron. Astrophys. **326**: 493. (D)

224. CONTOPOULOS G. 1997. Message from the President. Hipparchos, The Hellenic Astronomical Society Newsletter **1**: No. 3.

225. CONTOPOULOS, G., N. VOGLIS, C. EFTHYMIOPOULOS, C. FROESCHLÉ, R. GONCZI, E. LEGA, R. DVORAK & E. LOHINGER. 1997. Transition spectra from order to chaos. Celest. Mech. Dyn. Astron. **67**: 293. (D)

226. EFTHYMIOPOULOS, C., G. CONTOPOULOS, N. VOGLIS & R. DVORAK. 1997. Stickiness and cantori. J. Phys. **A30**: 8167. (D)

227. FOUNARIOTAKIS, M., S.C. FARANTOS, C. SKOKOS & G. CONTOPOULOS. 1997. Bifurcation diagrams of periodic orbits for unbound molecular systems: FH_2. Chem. Phys. Lett. **277**: 456. (Q)

228. EFTHYMIOPOULOS, C., G. CONTOPOULOS, N. VOGLIS & R. DVORAK. 1997. Stickiness, cantori and lobe dynamics. JENAM **97**: 6 (Abstracts).(D)

229. GROUSOUSAKOU, E., G. CONTOPOULOS & C. POLYMILIS. 1997. Distribution of periodic orbits in 2-D Hamiltonian systems. JENAM **97**: 9 (Abstracts). (D)

230. SKOKOS, C. & G. CONTOPOULOS. 1997. Formal integrals of one-dimensional time dependent Hamiltonians. JENAM **97**: 24 (Abstracts). (D)

231. DVORAK, R., G. CONTOPOULOS, G. EFTHYMIOPOULOS & N. VOGLIS. 1998. "Stickiness" in mappings and dynamical systems. Planet. Space Sci. **46**. In press. (D)

232. CONTOPOULOS G., N. VOGLIS & C. EFTHYMIOPOULOS. 1998. Method for distiguishing between ordered and chaotic orbits in four-dimensional maps. Phys. Rev. **E57**: 372. (D)

233. CONTOPOULOS, G. 1998. Message from the President. Hipparchos, The Hellenic Astronomical Society Newsletter, **1**, No. 4.

234. CONTOPOULOS, G., N. VOGLIS & C. EFTHYMIOPOULOS. 1998. Order and chaos in 3-D systems. *In* Hamiltonian Systems with Three or more Degrees of Freedom, C. Simo, Ed.: 24. Plenum Press. (D)

235. VOGLIS N., C. EFTHYMIOPOULOS & G. CONTOPOULOS. 1998. Invariant spectra of orbits in multidimensional symplectic maps. *In* Hamiltonian Systems with Three or more Degrees of Freedom. C. Simo, Ed.: 356. Plenum Press. (D)

236. SKOKOS C., G. CONTOPOULOS & C. POLYMILIS. 1998. Numerical study of the phase space of a 4-D symplectic mapping. *In* Hamiltonian Systems with Three or More Degrees of Freedom, C. Simo, Ed.: 599. Plenum Press. (D)

237. CONTOPOULOS, G. 1998. A Message from the President. Hipparchos, The Hellenic Astronomical Society Newsletter, **1:** No 5.
238. CONTOPOULOS, G. 1998. From one GS of the IAU to the next. Remembering Edith Alice Miiller. Kluwer Academic Publishers.
239. CONTOPOULOS, G. & N. VOGLIS. 1998. Dynamical spectra. *In* The Dynamics of Small Bodies in the Solar System: A Major Key to Solar System Studies, A. Roy, Ed. Kluwer Academic Publishers. In press. (D)
240. EFTHYMIOPOULOS C., N. VOGLIS & G. CONTOPOULOS. 1998. Angular dynamical spectra and their applications. *In* The Dynamics of Small Bodies in the Solar System: A Major Key to Solar System Studies, A. Roy, Ed. Kluwer Academic Publishers. In press. (D)
241. GROUSOUSAKOU, E. & G. CONTOPOULOS. 1998. Periodic orbits in 2-D Hamiltonian systems. *In* The Dynamics of Small Bodies in the Solar System: A Major Key to Solar System Studies, A. Roy, Ed. Kluwer Academic Publishers. In press. (D)
242. CONTOPOULOS, G. & R. DVORAK. 1998. Studies of dynamical systems. Introduction. *In* The Dynamics of Small Bodies in the Solar System: A Major Key to Solar Studies, A. Roy, Ed. Kluwer Academic Publishers. In press. (D)
243. CONTOPOULOS, G. 1998. Dynamical spectra and the onset of chaos. *In* Nonlinear Dynamics and Chaos in Astrophysics, R. Buchler, S. Gottesman & H. Kandrup, Eds. Ann. N.Y. Acad. Sci. **867:** 14–40. (D)
244. CONTOPOULOS, G., C. EFTHYMIOPOULOS & N. VOGLIS. 1998. The form and significance of dynamical spectra. *In* Modern Astrometry and Astrodynamics, R. Dvorak, Ed. Austrian Acad. Sci. In press. (D)
245. CONTOPOULOS, G. 1998. Laudatio for Prof. H. Eichhorn. *In* Modern Astrometry and Astrodynamics, R. Dvorak, Ed. Austrian Acad. Sci. In press.
246. GROUSOUSAKOU, E. & G. CONTOPOULOS. 1998. Distribution of regular and irregular orbits. Open Systems and Information Dynamics **6**. In press. (D)
247. CONTOPOULOS, G. & N. VOGLIS. 1998. Spectra of dynamical systems. Open Systems and Information Dynamics **6**. In press. (D)
248. CONTOPOULOS, G., N. VOGLIS & C. EFTHYMIOPOULOS. 1998. Chaos in relativity and cosmology. *In* The Impact of Modern Dynamics in Astronomy. J. Henrard, Ed. IAU Colloquium 172. In press. (R)
249. VOGLIS, N., G. CONTOPOULOS & C. EFTHYMIOPOULOS. 1998. Detection of ordered and chaotic orbits using the dynamical spectra. *In* The Impact of Modern Dynamics in Astronomy, J. Henraxd, Ed. IAU Colloquium 172. In press. (D)
250. EFTHYMIOPOULOS, C., N. VOGLIS & G. CONTOPOULOS. 1998. Cantori and asymptotic curves in the stickiness region. *In* The Impact of Modern Dynamics in Astronomy. J. Henrard, Ed. IAU Colloquium 172. In press. (D)

Plus about 30 annual reviews and 70 publications in Greek.

BOOKS

1. CONTOPOULOS G., Ed. 1966. The Theory of Orbits in the Solar System and in Stellar Systems. Proc. IAU Sumposium 25. Academic Press.
2. CONTOPOULOS, G. & W. BECKER, Eds. 1970. The Spiral Structure of Our Galaxy. Proc. IAU Symposium 38. Reidel Publishing Co.
3. CONTOPOULOS, G., Ed. 1974. Highlights of Astronomy 1973. Reidel Publishing Co.
4. CONTOPOULOS, G. & A. JAPPEL, Eds. 1974. Transactions of the International Astronomical Union. Reidel Publishing Co.
5. CONTOPOULOS, G., Ed. 1976. Reports on Astronomy (3 Volumes). Reidel Publishing Co.
6. CONTOPOULOS, G. 1965. Lectures in Celestial Mechanics (in Greek). University of Thessaloniki.
7. CONTOPOULOS, G. 1976. Introduction to Astrophysics. Cosmology (in Greek). University of Thessaloniki.
8. CONTOPOULOS, G. 1976. Introduction to Astrophysics. Stellar Systems (in Greek). University of Thessaloniki.
9. CONTOPOULOS, G. & D. KOTSAKIS. 1987. Cosmology. Springer-Verlag. (Translation from the 3rd Edition in Greek, Athens, 1986).
10. CONTOPOULOS, G., B. BARBANIS & P. LASKARIDES, Eds. 1988. In Memoriam D. Kotsakis. Athens.
11. CONTOPOULOS, G., N. SPYROU & L. VLAHOS, Eds. 1994. Galactic Dynamics and N-Body Simulations. Springer-Verlag.

Index of Contributors

Athanassoula, E., 141–155

Berentzen, I., 200–216
Buchler, J.R., vii–viii
Burns, A.M., 61–84

Casti, A.R.R., 93–108
Chavanis, P.-H., 120–140
Contopoulos, G., 14–40, 321–335

Drury, J., 306–320
Dupke, A., 253–257

Eckstein, B.L., 41–60
Eichhorn, H., 1–2
Erickson, L.K., 173–199

Fridman, A.M., 156–172

Gottesman, S.T., vii–viii, 173–199

Heller, C., 200–216
Hua, X.-M., 283–297
Hunter, C., 61–84
Hunter, J.H., Jr., 173–199

Kandrup, H.E., vii–viii, 41–60, 306–320, 320-1–320-13
Kazanas, D., 283–297

Khoruzhiv, O.V., 156–172

Lin, C.C., 229–252
Lovelace, R.V.E., 217–228
Lynden-Bell, D., 3–13

Miller, B.N., 268–282
Morrison, P.J., 93–108, 109–119
Murante, G., 258–267

Porchia, D., 61–84
Provenzale, A., 258–267

Shlosman, I., 200–216
Siopis, C., 41–60
Spiegel, E.A., 93–108, 258–267, 298–305

Tao, L., 298–305
Terzić, B., 61–84, 85–93
Thieberger, R., 258–267
Thiffeault, J.-L., 109–119

Umurhan, O.M., 298–305

Yawn, K., 268–282
Youngkins, P., 268–282

Zink, C., 61–84